FISHERIES ECOLOGY

FISHERIES ECOLOGY

TONY J. PITCHER
University of British Columbia, Canada
AND PAUL J.B. HART
University of Leicester, UK

CHAPMAN & HALL

London · Glasgow · Weinheim · New York · Tokyo · Melbourne · Madras

Published by Chapman & Hall, 2-6 Boundary Row, London SE1 8HN

Chapman & Hall, 2-6 Boundary Row, London SE1 8HN, UK

Blackie Academic & Professional, Wester Cleddens Road, Bishopbriggs, Glasgow G64 2NZ, UK

Chapman & Hall GmbH, Pappelallee 3, 69469 Weinheim, Germany

Chapman & Hall USA., One Penn Plaza, 41st Floor, New York, NY10119, USA

Chapman & Hall Japan, ITP - Japan, Kyowa Building, 3F, 2-2-1 Hirakawacho, Chiyoda-ku, Tokyo 102, Japan

Chapman & Hall Australia, Thomas Nelson Australia, 102 Dodds Street, South Melbourne, Victoria 3205, Australia

Chapman & Hall India, R. Seshadri, 32 Second Main Road, CIT East, Madras 600 035, India

First edition 1982
Reprinted 1983, 1985, 1987, 1990, 1992, 1993, 1994

Printed in Great Britain by the Ipswich Book Co., Ipswich

ISBN 0 412 38260 1

A Catalogue record for this book is available from the British Library
Library of Congress Cataloging-in-Publication Data available

CONTENTS

PREFACE

This book is about the ecology of fish which are caught by man. Its aim is to set out the processes which must be identified, described, measured, analysed and ultimately predicted in order to provide optimal management of exploited fisheries. The text is designed for undergraduates and masters level courses in biology, ecology and zoology and attempts to provide a self-contained account of one of the major areas of applied ecology.

A major stumbling block for potential students of fisheries science is the mathematics; many students are dismayed at first meeting a yield equation. In order to ease this problem, we have attempted to reduce the mathematics to a necessary minimum, have avoided calculus where possible, and have explained most equations in words as well as symbols. We have written this book in the belief that existing fishery treatments are usually too specialised and restricted in their outlook. Consequently, we have endeavoured to consider fish as components of delicately balanced ecosystems, and have included sections on nutrition, evolution, fish farming, energy use, and fishery economics as well as covering the main types of fishery models and the methods of fishery management. One major omission enforced through lack of space is the impact of fish behaviour on fisheries, notably the reactions of fish to gear and the effect of fish migrations on the location, duration, and yields to man.

We hope that the broader outlook of our approach will make it easier for students, still at a relatively undifferentiated stage of their careers, to fit fisheries science more easily into its biological context. However, we hope too that we have shown that fisheries ecology is one area where ecology is directly connected with the daily activities of man.

We would like to thank the following colleagues for criticising drafts, although of course the views expressed in the book and any mistakes remain our own responsibility: Dr H. Greenwood (Chapter 1), Dr P. Reay (Chapter 10), Dr A. Grimm (Chapter 10), Dr A. Magurran (Chapters 6, 7 and 11), Dr R. Reddy (Chapter 4) and Mr P. Lawton (Chapter 6). We would also like to thank typists Miss L. Owen and Miss E. Williams, Dr. A. Magurran who checked all the references, and especially Miss Julia Polonski who prepared all the figures. We are particularly grateful to Mrs S. Hart who assisted with the index.

Finally, we would like to adopt a poem by Peter Larkin as a theme for the book:

Here lies the concept, MSY,
It advocated yields too high,
And didn't spell out how to slice the pie,
We bury it with the best of wishes,
Especially on behalf of fishes.
We don't know yet what will take its place,
But we hope it's as good for the human race.

(P.A. Larkin, 1977, 'An Epitaph for the Concept of Maximum Sustained Yield', *Transactions of the American Fisheries Society*, 106:1-11.)

ACKNOWLEDGEMENTS

We would like to thank the undermentioned people and organisations for permission to use the following material: Academic Press Inc. (who hold the copyright) and Dr J.R. Brett for Figures 4.5, p. 126, 4.8, p.131, 4.9, p. 132 and 4.10, p. 133; the American Association for the Advancement of Science and Dr J.R. Beddington for Figure 7.8, p. 240, from *Science 197*: 463-5 (1977), copyright 1977 by AAAS, and the AAAS and Prof. C.W. Clark for Figure 9.4, p. 302, from *Science 181*: 630-4 (1973), copyright 1973 by AAAS; Blackwell Scientific Publications Ltd for Figure 6.1(d), p. 176, from E.D. LeCren and M.W. Holdgate (eds), *The Exploitation of Natural Animal Populations* (1962) and for Figures 6.1(f), p. 176 and 6.9, p. 195, from S.D. Gerking (ed.), *The Ecology of Freshwater Fish Production* (1978); the California Dept of Fish and Game for Figure 6.1(a), p. 176, from *California Fish and Game 48*: 123-40 (1962); Cambridge University Press for Figure 5.4, p. 163, from D.H. Cushing, *J. Mar. Biol. Assoc. UK 47*: 193-208 (1967); Elsevier Scientific Publishing Co. for Figure 10.4, p. 324, from *Aquaculture 5*: 19-29 (1975); the Government of Canada, Scientific and Publications Branch for Figures 4.7(a and b), p. 130, 5.5, p. 164, 6.10, p. 199, 7.6, p. 231, 7.9, p. 245 and 9.5, p. 305, from the *Journal of the Fisheries Research Board of Canada* or the *Canadian Journal of Fisheries and Aquatic Sciences* (see Figures for references), and Figure 6.7, p. 190, from the *Bulletin of the Fisheries Research Board of Canada no. 191* by W.E. Ricker (1975); Sir Alister Hardy FRS for Figure 1.7, p. 37; Harvard University Press and E.O. Wilson for Figure 5.1, p. 149, from *Sociobiology: The Modern Synthesis* (1975); Her Majesty's Stationary Office, London for Figure 8.10, p. 275, from *Fishery Investigations, Ser. II, 27*(1) (1972) and Figure 8.14, p. 282, from *Fishing Prospects 1979/1980*; Holt, Rinehart and Winston for Figure 4.4, p. 121, from *The Vertebrate Body*, Third Edition, by Alfred Sherwood Romer, copyright © 1962 by W.B. Saunders Company, copyright 1955 and 1949 by W.B. Saunders Company, reprinted by permission of Holt, Rinehart and Winston; the Institute for Marine Environmental Research for Figure 5.3, p. 162, from J.M. Colebrook and G.A. Robinson, *Bull. Mar. Ecol. 6*: 123-9 (1965); the Institute of Mathematics and its Applications for Figure 6.12, p. 207, from M.S. Bartlett and R.W. Hiorns (eds), *The*

Mathematical Theory of the Dynamics of Biological Populations (1973), Academic Press; the International Council for the Exploration of the Sea for Figures 6.6, p. 187, 8.11, p. 276 and 8.9(b), p. 272, from D.J. Garrod and B.W. Jones, *J. Cons. Int. Explor. Mer. 36*: 35–41 (1974), and Figure 3.4, p. 85, from D. Sahrhage and G. Wagner, *Rapp. Proc-Verb. Reun. Int. Comm. Explor. Mer. 172*: 72–85; the International Thomson Educational Publishing Inc., for Figures 2.4(a), p. 65, 2.5, p. 66 and 2.6, p. 67, from *The Encyclopedia of Marine Resources*, edited by Frank E. Firth, © 1969 by Litton Educational Publishing Inc., reprinted by permission of van Nostrand Reinhold Company; the Linnean Society of London for Figures 1.11, p. 47 and 1.12, p. 48 from R.H. Lowe-McConnell, *J. Linn. Soc. (Zool.) 44*: 669–700 (1962); the Ministry of Agriculture (Fisheries Division), Israel for Figure 10.6, p. 339, from Barmidgeh, *Bull. Fish. Cult. Israel 27*: 85–99 (1975); the Sport Fishing Institute, USA for Figures 4.2, p. 119 and 4.3, p. 120, from R.H. Stroud and H. Clepper (eds), *Predator Prey Systems in Fishery Management*, Sport Fishing Institute (1978); the US Dept of Commerce, National Marine Fishery Service for Figures 6.2, p. 179 and 6.11, p. 206, from *Fishery Bulletin 74*: 517–30 (1976) and *75*: 529–46 (1977); to J. Wiley & Sons Ltd for Figures 6.1(b,c,e,i), p. 176 and 6.8, p. 194, from J.A. Gulland, *Fish Population Dynamics* (1977); the Zoological Society of London for Figures 6.13, p. 210, 6.15, p. 214 and 6.16, p. 216, from P.J. Miller (ed.), *Symposium of the Zoological Society of London 44* (1979).

1 FISH DESIGN PLANS AND FISH COMMUNITIES

The rational and scientific management of fisheries must depend on a fundamental understanding of fish biology and ecology; that is, what sort of animals fish are, and where and how they live. This chapter briefly describes the basic design plan of fishes, introduces the diversity of different forms, and gives examples of fish as members of the aquatic communities in which they live. Since most fisheries are for bony fish (mainly teleosts), we concentrate on them although cartilaginous fish (elasmobranchs) are mentioned. Important 'fisheries' for mammals, notably the whales, also exist and later in the book we mention their salient features where appropriate.

The Fish Body

Predators are often considered objects of displeasure, but to see a pike stalk and capture its prey is to witness all the skills which have made bony fish as agile in their aquatic environment as birds are in the air. For a moment let us watch as the attack develops. The black and gold eyes of the pike have sighted a school of minnows; it swings round to face them and waits, perhaps assessing the coming task. Then with a brisk tail beat, the slim pike glides forward a few centimetres only to stop suddenly dead in its tracks, neatly braking by throwing forward its pectoral and pelvic fins slung underneath the body. Fixing the prey in stereoscopic vision by looking down the two sighting grooves on its snout, it hangs motionless, camouflaged in the water. Soon, the trailing edges of the dorsal, anal and tail fins begin to flutter almost imperceptibly so that the pike inches forward so slowly that it is hard to see it moving forward at all. It brakes and then continues its patient stalk. The minnows, sensing the impending attack, group uneasily, individuals darting out from the school and returning at unpredictable intervals. The slowly advancing pike is now within two of his body lengths from the prey and the action is about to start. Gradually, the pike gathers his lithe body into a tight spring-like S-shape, and then in a flash the spring is released and the pike accelerates with lightning speed, opening his lethally-toothed mouth at the last moment. Equally fast, the minnow school explodes in all directions in a confusing mass of black and silver. The pike has missed, but soon takes up the hunt again, perhaps this time to be successful.

An attack like this illustrates several points about teleosts. They are streamlined for moving efficiently through water and are able to position themselves with great precision. Many can use their fins to manoeuvre in the most delicate fashion, or can accelerate rapidly either to escape predators or to capture prey. The eyes are particularly well developed as is the visual-analysis area of the brain; teleosts can resolve impressive detail, feature and colour. Olfaction and a spatial lateral line sense, are also exceptionally acute. Many species have developed schooling, a form of social behaviour that acts as a defence against predators and serves to improve the efficiency of food gathering. These capabilities of teleost fish are brought about by a unique design plan and by specialised internal organs. It is the purpose of this short section to outline the equipment by which these beautifully designed machines carry out their lives' tasks. For more detail, the reader is referred to Alexander (1974, 1975) and to Marshall (1971).

The structure of a bony fish is shown in Figures 1.1 and 1.2. Because water is 800 times denser than air, the body has become streamlined to a shape which best reduces turbulence and drag. Bony fish swim by 'posturing' into a series of S-bends which pass back along the body, increasing in amplitude as they go. The tail fin increases the area on which the thrust acts. The smoothness of swimming is brought about by precisely-timed contractions in the 40–70 paired muscle blocks along the length of the body. Each muscle pulls on the strong but flexible vertebral column (see Figure 1.2a). The primitive sinusoidal motion has been greatly modified in fast swimmers such as mackerel, to a system where only the tail moves significantly. Acting analogously to a ship's propeller, the tail is oscillated rapidly back and forth by the main swimming muscles in the mid-body, transmitting their power through tendons which run along a much reduced caudal peduncle.

The swimming muscles make up about half the fish's weight and are divided into two sorts: the cruising motors and the emergency acceleration machinery. A transverse section through a teleost reveals a wedge of darker muscle lying along the lateral edges on each side (Figure 1.2b). It can be noticed as dark grey flesh in the cooked fillet of cod or herring. These are the red swimming muscles, whose colour derives from their high myoglobin content. Red muscle is employed for continuous cruising, functioning aerobically using fat for fuel. The muscle cells are packed with mitochondria for this purpose. Using this red muscle, fish such as cod can swim virtually indefinitely at about

Figure 1.1: A. The Cod Showing the External Features Typical of a Teleost. B. The Mackerel Shark Showing the External Features Typical of an Elasmobranch.

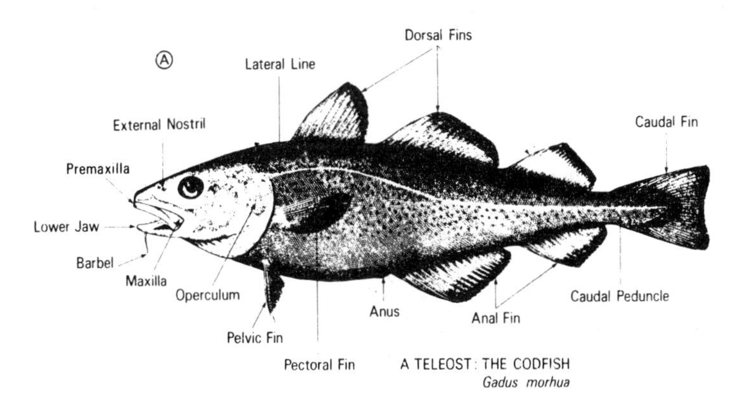

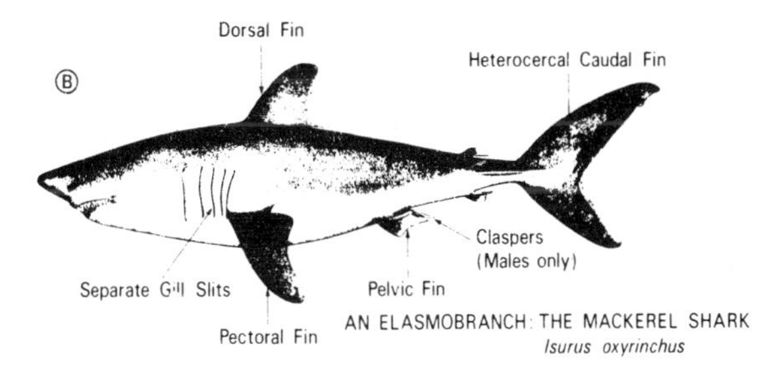

two lengths per second. The amount of red muscle is governed physiologically by the oxygen supply and therefore by the size of the gills; fish could not be 'all red muscle' without excessively large gills, which would destroy the streamlining. Fish like mackerel, tuna and swordfish (*Scombroidei*) which are adapted to cruise continuously at high speed (more than three lengths per second) have so much red muscle that the oxygen supply is critical. They have to cruise without stopping in order to get the oxygen their red muscles require; mackerel die of anoxia if they stop swimming. So the evolution of efficient fast cruising brought its own constraints on the roles these fish could perform. Many of the scombrid fish are of great commercial importance; this adaptation to fast cruising illustrates why they occupy far-ranging feeding niches.

The ceaseless activity of the swimming muscles has led to even more

Figure 1.2: A. Diagrammatic Drawing of a Perch that has been Dissected to Show the Major Internal Organs of a Teleost. B. Sections Cut through the Tail Region of the Mackerel and the Haddock to Show the Difference in the Proportion of Red and White Muscle in the Two Species.

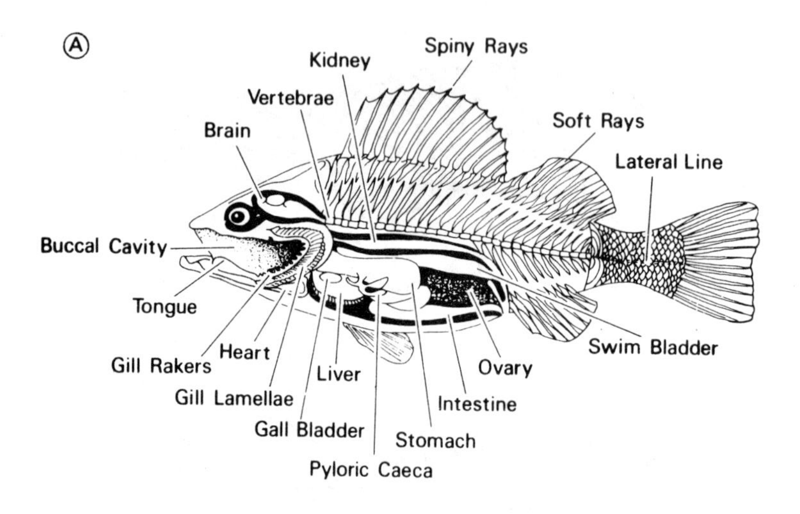

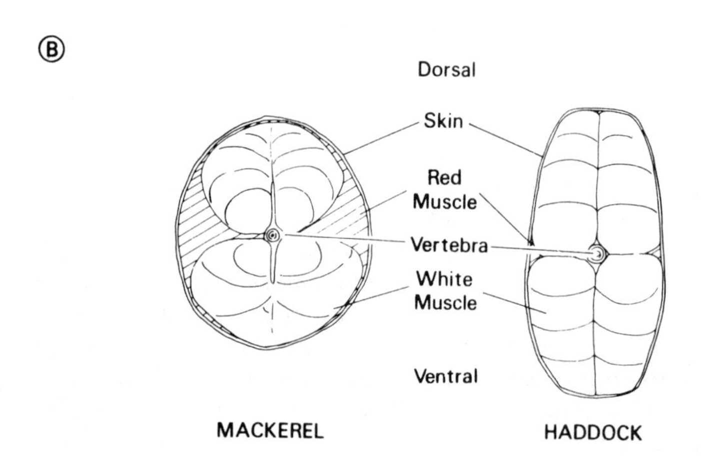

Source: Drawn from Photographs in Love (1970).

effective fast cruising by generating heat. Muscular contraction is more efficient at producing power at 30°C than at lower temperatures. In some scombroids metabolic heat is conserved within the body by a countercurrent heat exchanger in which venous and arterial blood vessels run alongside each other. Using this mechanism, tuna muscle can be 10°C higher than the temperature of the water. Interestingly, a few large sharks use the same countercurrent principle to increase muscle efficiency, probably because their design plan and niche puts a similar premium on swimming ability. Separate red muscles, not arranged in myotomes, operate the fins in many fishes such as cyprinids and sticklebacks. The tench even has a red muscle along its gut, presumably to provide churning to aid digestion of its vegetable diet!

The white muscles make up the bulk of fish like cod and nearly all of those of pike, which rarely cruise for long periods. White muscles have no myoglobin, poor blood supply but good lymph supply, and few mitochondria. They are adapted to rapid powerful contraction, obtaining energy by reducing glycogen to lactic acid anaerobically. As a result they build up an oxygen debt and tire quickly. The zig-zag pattern of overlapping muscle blocks allow each muscle fibre to contract by the optimal amount to produce efficient power. The white muscles produce large amplitude, powerful contractions for a limited period of time; during a 'kick-start' a cod's body is flexed into almost 90° bends and the tail is usually spread to maximise the thrust as it accelerates at over ten lengths per second. Such anaerobic activity soon catches up with the fish though and within a few seconds it is forced to come to a halt, exhausted. It may take several hours to recover as the lactate is pumped out of the white muscle through the lymph system. Less drastic use of the white muscle can be kept up for longer periods as the fish then goes into oxygen debt more slowly; it is possible to swim for some hours at speed just above that provided by the cruising motors by using the white muscles just a little, but eventually the fish will have to halt and recuperate. It is this exhaustion of the white muscle emergency system that determines the time an angler has to 'play' a fish before he can land it. It is also a crucial factor to consider in the design of commercial fishing gear.

Minnows, and most other bony fish, are able to react very quickly to danger signals. This is because they have special large-diameter neurons, the 'Mauthner system', running the length of the spinal cord while giving off branches to each white muscle block. Like the famous squid giant axons, the nerves of the Mauthner system have a faster rate of impulse conduction than normal nerve axons, by virtue of their

Table 1.1: A List of the Orders of Fish Having Living Representatives. Also shown is the number of species and the number of species found in freshwater. An F after the order means that the order has species that are of commercial importance.

Order	Common Name[a] (where known)	Families	Genera	Species	Freshwater Species
Petromyzoniformes	lamprey	1	9	31	24
Myxiniformes	hagfishes	1	5	32	0
Heterodontiformes	sharks	1	1	6	0
Hexanchiformes	sharks	2	4	6	0
Lamniformes F	sharks	7	56	199	0
Squaliformes F	dogfish sharks	3	19	76	0
Rajiformes F	rays, skates	8	49	315	10
Chimaeriformes	chimaeras	3	6	25	0
Ceratodiformes	Australian lungfish	1	1	1	1
Lepidosireniformes	lungfish	2	2	5	5
Coelacanthiformes	coelacanth	1	1	1	0
Polypteriformes	bichir	1	2	11	11
Acipenseriformes F	sturgeon	2	6	25	15
Semionotiformes	gars	1	1	7	7
Amiiformes	bowfins	1	1	1	1
Osteoglossiformes	boneytongues	4	9	15	15
Mormyriformes F	elephant fish	2	11	101	101
Clupeiformes F	herrings	4	72	292	25
Elopiformes	tenpounders	3	5	11	0
Anguilliformes F	eels	22	133	603	0
Notacanthiformes	deep-sea eels	3	6	24	0
Salmoniformes F	salmons	24	145	508	80
Gonorynchiformes F	milkfish	4	7	16	14

Cypriniformes F	carps	26	634	3,000	3,000
Siluriformes F	catfishes	31	470	2,000	1,950
Myctophiformes F	lanternfishes	16	73	390	0
Polymixiiformes	beardfish	1	1	3	0
Percopsiformes	trout-perch, cave fish	3	5	8	8
Gadiformes F	cods	10	168	684	5
Batrachoidoformes	toadfish	1	18	55	2
Lophiiformes F	anglerfish	15	57	215	0
Indostomiformes		1	1	1	1
Atheriniformes	flying fish, sauries	16	167	827	500
Lampridiformes	opah	10	18	35	0
Beryciformes	lanterneye fish	15	39	143	0
Zeiformes	dories	6	25	50	0
Syngnathiformes	pipefishes	6	44	200	2
Gasterosteiformes	sticklebacks	2	7	10	3
Synbranchiformes	swamp eels	3	7	13	8
Scorpaeniformes F	scorpion fish	21	260	1,000	100
Dactylopteriformes	flying gurnard	1	4	4	0
Pegasiformes	seamoth	1	2	5	0
Perciformes F	perches	147	1,257	6,880	950
Gobiesociformes F	dragonets	3	42	144	2
Pleuronectiformes F	flatfishes	6	117	520	3
Tetraodontiformes	triggerfishes	8	65	320	8
Totals		450	4,032	18,818	6,851

Note. a. Or name of a representative species.
Source: After Nelson (1976).

large diameter. This, coupled with the immediacy of the link between the Mauthner cell bodies atop the medulla and the muscles, enables a teleost to 'kick-start' in less than one-fiftieth of a second.

Teleosts can be distinguished from elasmobranchs by reproductive strategy and their greater manoeuvrability. The rigid fins and tail of a shark are committed to keeping the fish from sinking (see Figure 1.1b). Most sharks are heavier than water so that they would sink without the lift generated by the sculling action of the tail and the hydroplane-like pectoral fins. Whereas teleosts have perfected good vision, swimming and suction-action jaws, the elasmobranch machine has specialised in electric and olfactory detection of prey but has retained relatively primitive jaws and methods of swimming. The paramount (and ancient) adaptation of the elasmobranchs is internal fertilisation and the specialised reproductive apparatus which equips them to produce small numbers of advanced embryos well protected with large energy reserves. Teleosts on the other hand have relatively primitive reproduction, typically shedding large numbers of eggs into the water to be fertilised and to develop with little protection or energy store. Some sharks and rays are truly viviparous, giving birth to fully-formed live young; most produce small numbers of well-protected shelled eggs containing advanced embryos provided with an extensive and rich yolk supply – the familiar 'mermaids' purses' of the sea shore.

Teleost fins are thin sheets of tissue supported by flexible hard rays, each of which is free to move. The fins can be employed as paddles or brakes, or used for forward propulsion by means of waves passing along their length: the gymnarchid electric fishes (Mormyriformes, Table 1.1) of African swamps swim entirely by such undulations of the dorsal fin and the South American gymnotids do the same using an elongated anal fin. In advanced teleosts greater manoeuvrability has also come about through the evolutionary shift of the pectoral fins to a position higher up the sides of the body just behind the operculum (Figure 1.1a). The pectorals then act in concert with the pelvic fins which have come to lie forward almost directly underneath the pelvics, and the arrangement allows precise control over braking, and rapid short-radius turns.

The ancestors of the teleost fish evolved air-breathing lungs in warm shallow freshwater pools where the oxygen supply was depleted. These lungs were secondarily useful as buoyancy organs and this function was perfected in the early marine teleosts after air-breathing had been lost. The thin elastic-walled swimbladder means that the fish does not have to expend energy in swimming to maintain its depth in the water.

Lower teleosts have a connecting tube from the swimbladder to the rear of the pharynx through which they can take or lose air from the swimbladder and adjust buoyancy to be neutral at any depth. For example, carp can adjust it to swim deeper in the water by swimming up to the surface and taking in more air: when they return to the desired depth the air will be compressed by the greater water pressure to the volume giving neutral buoyancy. Conversely, air is merely expelled via the mouth to adjust for shallower depths. The need to go to the surface is dispensed with in advanced teleosts who have evolved swimbladders with specialised gas secretion and gas absorption glands. Here the tube to the pharynx is lost and the gas glands work on economical countercurrent principles. These higher teleosts have exact control over their buoyancy. Some teleosts have lost the swimbladder, e.g. lurking, bottom-living fish like angler fish (lophiids) and fast-moving swimmers which change depth often (tuna).

Precision control over fins and buoyancy means that advanced teleosts, such as many of the Perciforms on coral reefs, can hover and dart with impressive agility. The evolution of the pipette-action mouth, which has to be exactly positioned to work properly, was probably the trigger for this fascinating set of linked adaptations. Manoeuvrability *per se* has impiications for fishing gear design and performances, and of course the particular design plan of the fish determines its overall mode of life, and therefore the type of fishery which can be operated.

Much of the adaptive radiation in fish has been connected with feeding habits. In the Great Lakes of Africa, for instance, the cichlids have radiated to form numerous species that have between them exploited every conceivable way of gaining food (see Table 1.2). Specialisations range from detritus or algal feeding through cichlids which eat molluscs, crustaceans, or smaller fish, to such esoteric habits as feeding on the scales and perhaps even the eyes of other fish. Such fine differentiation of feeding niches in this and other species-assemblages of teleosts (e.g. coral reef butterfly fishes) has been made possible by the evolutionary potential of the flexible swinging jaw design in teleosts. By contrast, typical elasmobranchs have awkward one-purpose jaws suited only to tearing or crushing; hence most are piscivores or molluscivores. With the exception of changes in the jaws for filter feeding with gillbars in whale-sharks, basking sharks and manta rays, elasmobranch jaw structure remains crude and conservative throughout the group, perhaps because any great changes might compromise the swimming ability which is so essential to their buoyancy.

Among bony fish, the jaws of teleosts became more adaptable by the

Table 1.2: Trophic Groups of *Haplochromis* from Lake Victoria.

Group	Trophic Status	Species
1	Insectivores	H. bloyeti
		H. saxicola
		H. chilotes
		H. chromogynous
		H. empodisma
2	Phytophagous species	H. nigricans
		H. nuchisquamulatus
		H. lividus
		H. obliquidens
		H. cinctus
		H. paropius
		H. erythrocephalus
		H. phytophagus
		H. acidens
3	Scale eaters	H. welcommei
4	Benthic crustacean feeders	H. dolichorynchus
		H. tyrianthinus
		H. chlorochrous
		H. cryptogramma
		H. tridens
		H. melichrous
5	Mollusc eaters	H. sauvagei
		H. prodromus
		H. granti
		H. xenognathus
		H. ishmaeli
		Macropleurodes bicolor
		H. pharyngomylus
		H. obtusidens
		H. humilor
		H. pallidus
		H. riponianus
		H. theliodon
		H. aelocephalus
6	Predators on eggs and larvae	H. cronus
		H. barbarae
		H. parvidens
		H. microdon
		H. cryptodon
		H. melanopterus
		H. maxillaris
		H. obesus
7	Piscivores	H. serranus complex (11 spp)
		H. prognathus complex (20 spp)
		H. squamulatus
		H. michaeli
		H. martini
		H. guiarti

Table 1.2: Continued.

Group	Trophic Status	Species
8	Mixed (Species mostly found at depths greater than 20 m. See Greenwood and Gee, 1969, for ecological details)	H. megalops
		H. piceatus
		H. paropius
		H. cinctus
		H. erythrocephalus (2)[a]
		H. melichrous (4)
		H. laparogramma
		H. fusiformis
		H. dolichorhynchus (4)
		H. tyrianthinus
		H. chlorochrous (4)
		H. cryptogramma (4)
		H. arcanus
		H. decticostoma
		H. gilberti
		H. paraplagiostoma

Note. a. Also listed under another group the number of which is shown in the bracket.
Sources: Greenwood (1974).

loss of the rigid connection between the maxilla and the skull, at the same time as a reduction in cheek bones which allows muscles operating the jaw mechanism maximum leverage and size. The essential feature of the teleost jaw is the swinging maxilla pivoted at front and rear. Unlike the bony pivot at the front, the rear 'hinge' is formed of pleated connective tissue (Figure 1.1). When the early teleosts radiated in the Cretaceous, at a grade represented by the salmonid fishes today, the swinging jaw and muscles of the mouth cavity increased the buccal volume by 5–10 per cent when the mouth was opened, so aiding prey capture at a vital moment by a back-draught of water. Presumably, small invertebrates are prevented from escaping in this fashion, so that a fish with the swinging-jaw design is able to capture successfully a wider range of prey sizes with less effort than when the suction effect is not present.

The basic swinging-jaw design had tremendous evolutionary potential in the perfection of the pipette-like suction mechanism; advanced teleosts with fully protrusible jaws do not have to grab prey with the mouth at all but can creep up and capture it by pure suction! The diagram in Figure 1.3 shows how one type of protrusible jaw works in the perches; the principle is that the maxilla no longer forms part of the jaw margin, but operates as a swinging lever being pulled down as the

Figure 1.3: Three Types of Jaw Mechanism in Teleost Fish – Main Bones of Skull and Jaws Only. Note that (3) is a rather specialised wrasse – most wrasse do not exhibit such an extreme adaptation to nibbling.

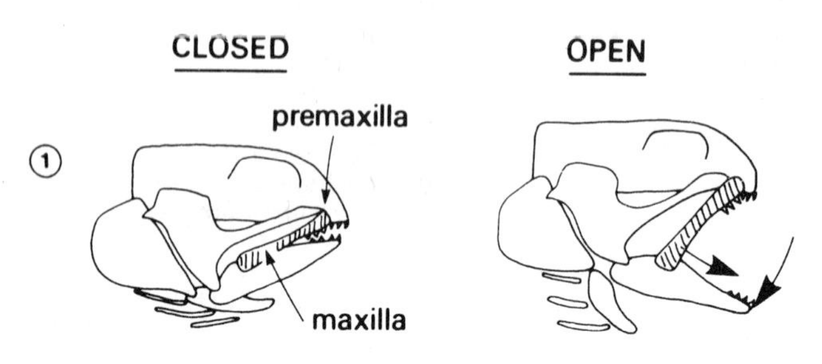

Salmon type – swinging maxilla

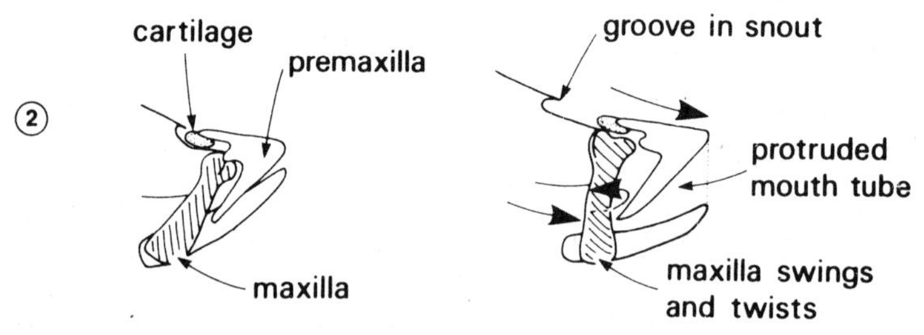

Perch type – one form of protrusion (note lack of teeth on jaw)

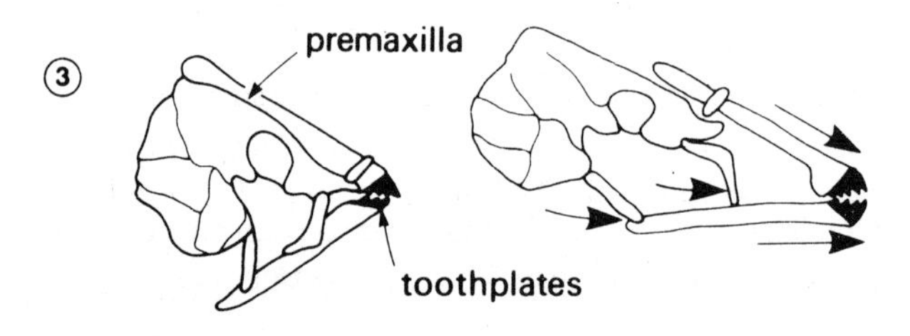

Wrasse – protrusion derivative for nibbling coral

lower jaw pulls the mouth open, and by twisting as it swings, pushing the pre-maxilla forwards on a special cartilage suspension. The pre-maxilla is now the only bone on the margin of the upper jaw. As the upper jaw is pushed forwards (hence the term 'protrusion'), the pleated connective between the bones is unfolded forwards to make a tube.

The opercular cavity is expanded by muscles at the same time. All this happens so rapidly that the buccal volume increases and pipettes a jet of water into the mouth. This suction mechanism is fascinating to watch in action; we can watch as in a large tank in a marine research laboratory a metre-long cod swims lazily up to a smaller pollack who veers off and sculls rapidly away. The cod languidly opens his cavernous mouth and one can visibly see the poor pollack dragged inexorably into the cod's mouth, clamped, and then swallowed whole within a few moments. Some advanced teleosts became lurking, static predators with vast suction-acting mouths, the angler fishes (lophiids) for example. Others perfected the small aperture pipette action (e.g. the pipefishes, syngnathids), while yet others modified the protrusible jaws into 'remote control' nibbling structures with great leverage to exploit coral and rock-living invertebrates (e.g. parrot fishes and the wrasses, see Figure 1.3). Fully protrusible jaws were so advantageous that they evolved independently at least six times from the basic teleost design – in the perches, the cods, the herrings, the whitefish, the carp and the guppy families. Each of these groups uses a slightly different method of protrusion.

One of the direct consequences of protrusion was that teeth tended to be much reduced on the jaw margins since the fish no longer captured prey by biting, although a few teeth were often retained on the pre-maxilla and inside the upper palate to help prevent escape of larger prey clamped before swallowing. New chewing and grinding areas therefore evolved by elaborating structures elsewhere in the alimentary tract; for example the carp family developed beautifully designed tooth-plates (pharyngeal teeth) at the rear of the gill region just in front of the stomach. Each species of cyprinid has subtly different pharyngeal teeth to exploit a slightly different range of foods (and the teeth can be used to identify the species taxonomically). In a few advanced teleosts with protrusion modified to nibbling, scraping or cutting structures, and with heavy grinding mechanisms elsewhere in the digestive tract, we find the evolution of macrophytes or algae. This is an art which the elasmobranchs have never mastered. Herbivorous teleosts such as grass carp and anchovy, and omnivores like mullet and the freshwater catfish found in the USA, have cellulase-producing bacteria in their stomachs. The consequences of the evolution of the teleost jaw mechanism were enormous!

Fish sense organs are equally well-adapted to the constraints of an aquatic environment. The lens in a fish eye is spherical in order to provide sufficient refraction to produce a sharp image. On land, light is refracted significantly as it passes into the cornea, so that a relatively weak lens will be sufficient to focus the light on the retina. In fish, however, the lens does most of the refraction and has to be more curved. In addition, the refractive index decreases out from the centre of the lens to counteract the effect of spherical aberration. Fish from shallow environments where there is plenty of light have colour vision from retinal cones as well as rods for feature vision. Fish from the middle depths have much larger eyes and have only rods in the retina to form accurate images in the dim light. Some fish that live at great depths and in caves where there is no light have lost eyes altogether, but in most teleosts vision is used to hunt prey, recognise other fish and predators and learn the characters of the home range. Vision, as in most vertebrates, is the single most important sense.

The other senses widely used are hearing, the lateral line system sense, and olfaction. In a few teleosts prey is detected and messages passed through an electric field generated by modified muscle blocks. Nearly all elasmobranchs seem to have exceptionally sensitive electroreceptors. In dogfish, experiments have shown these capable of picking up the action potentials in nerves operating the gill muscles in buried prey!

Hearing, using the typical vertebrate inner ear tympanic organ, is well developed, probably because sounds, especially lower frequency sounds, travel far in water without much attenuation. A cod can hear a trawler's engines several miles away. Hearing is particularly well-developed in the carp family. Minnows can distinguish pitch as well as the best human musicians! They have a chain of bony connectors called 'Weber's ossicles', evolved from parts of vertebrae. The ossicles connect the swimbladder and the inner ear rather like the bones of our own middle ear. The elastic-walled swimbladder acts as an amplifier over a wide range of frequencies. Amplification of sound by the swimbladder is also exploited by other groups including cod, and the drums or croakers (*Sciaenidae*) which live in tropical seas. They use the swimbladder as a resonator to increase the volume of sound produced as well as of that received. It has recently been discovered that many other fish, such as blennies, use sounds in their social behaviour.

The lateral line system (Figure 1.1a) is an organ for detection of vibrations in the water. Although its function is not fully understood, since it is rarely covered in elementary biology and is almost exclusively found in fishes, we will describe it here in greater detail. It usually lies in a tube along the length of each side of the fish just underneath the skin and scales, with small holes leading to the outside. A similar series of perforated canals often forms a network over the head, although it is not certain if their function is similar to the main canals. The lateral line organ contains many small 'neuromast' organs, containing cells which are modified cilia and having delicate projections called cupulae, connected to sensory hairs. Water vibrations move the cupulae, bending the hairs, and this changes the frequency of pulses passing back along the lateral line nerve to the hind brain, where there is a special centre to deal with the analysis. The lateral line assists some fish in detecting prey and is used to locate objects and in schooling behaviour to monitor the position of nearby fish, supplementing vision.

Finally, smell is well-developed in most teleosts and nearly all elasmobranchs, probably because food-finding and spawning both depend upon detection of minute amounts of important chemicals in the water. Experiments have proved eels capable of detecting a single molecule of some substances in the water, and minnows can distinguish water of their home aquarium from that of any other. Some species with good vision, like the pike, have a weak sense of smell, but for most it is too important to dispense with. Mature salmon home to the river of their birth using smell, and many fish associate (and mate) with members of their own species using olfactory cues and pheromones. Indeed, an understanding of the way that fish perceive the world is relevant to man's fishing gear and fisheries.

The different salinities found in various aquatic environments have not prevented the spread of fish. Modern elasmobranchs are virtually confined to salt water, but the teleosts can be found in fresh, marine, and fluctuatingly brackish habitats. One species, the cichlid *Sarotherodon grahami*, lives in a concentrated soda solution of pH 11 at about 40°C in Lake Magadi, Kenya. Salt and water balance are mediated by the gills and the kidneys (Figure 1.2a). The usual vertebrate glomerular kidney is an elongated structure in teleosts; the anterior end is modified for endocrine function. The kidney tubule system filters the blood at one end and returns necessary salts and water to it at the other. In freshwater fish, in order to counteract the rapid uptake of water by osmosis, dilute urine is produced. The loss of salts is made up by the gills. The reverse happens in the hypertonic sea; to conserve water the urine is concentrated, the fish drinks water, and sodium and chlorine which would otherwise accumulate are excreted by the gills to maintain the salt balance. The kidneys also excrete magnesium and sulphate. The main nitrogenous waste is ammonia in teleosts and in addition to the kidneys this too is excreted by the gills, which contain very specialised cells, well-supplied with mitochondria for pumping sodium and other salts against concentration and osmotic gradients. The superb efficiency of the teleost salt and water balance machinery is exemplified by the many species which migrate between the sea and freshwater (salmon, trouts, whitefish, eels, flounders) and nevertheless function equally well in the totally different water and salt regimes. One can actually take a stickleback or an eel from a freshwater aquarium and place it directly in sea water, and, after a few seconds of evident distress, it survives easily. Even pike, normally considered an exclusively freshwater species, have been found swimming along a mile offshore in the sea. Energy expenditure on salt and water balance is different in the two regimes of course, and this can be an important consideration in marine and freshwater fish farming.

Reproduction is of vital importance to the continuation of natural fisheries and to fish farming (see Chapters 3, 6 and 10). In an aquatic environment a fish needs fewer elaborate mechanisms to ensure that sperm and egg meet in appropriate conditions. Although apparently wasteful, sperm and eggs can be shed freely into the water and provided the fish has invested enough energy in the developing gonads, vast numbers of eggs will be fertilised and develop to the yolk sac and fry stage. It is the general rule for teleosts, unlike the elasmobranchs, to produce large numbers of eggs; for example, a metre-long female cod will release well over a million eggs. When fertilised the eggs live in the plankton, as do the hatched larvae, both being passively carried by the currents. This is one of the reasons for fish undertaking long spawning migrations, so that adults and young of the species can keep within tolerable environmental limits (e.g. cod, plaice, eels). After using up the energy in the small yolk sac, a critical time for the young fry is in taking their first meal; mortality at this stage is often density-dependent and can determine the number of recruits which enter a fishery. We will return to this topic in Chapters 5 and 6.

The gonads are illustrated in Figure 1.2a. The ovaries can be completely enclosed within the body cavity with a special duct to the outside, or, as in trout, the eggs can be released into the body cavity before passing to the water. The testes are less bulky than the ovaries at spawning time, and sperm passes to the outside via a special duct in teleosts. The gonads mature each year, or seasonally in the tropics, and the process is under endocrine control, mature fish spawning at the appropriate time of year according to environmental triggers such as changes in temperature and day length (see Chapter 10). The time of spawning has been 'tuned' in each species by natural selection to the best needs of the young fry; therefore many temperate species spawn in winter or early spring. Releasing factors in the hypothalamus region of the midbrain trigger the pituitary into orchestrating the whole set of metabolic changes which funnel surplus energy from food digestion into gonadal growth rather than somatic growth. This it does by triggering production of secondary hormones in the endocrine glands (gonads, thyroid etc.) Because of the two alternative pathways for the surplus energy from food, an understanding of the endocrine system is directly revelant to the somatic growth required in fish farming.

Like the elasmobranchs, a few teleosts have evolved viviparity while some have behavioural care for the eggs and young. Examples of true viviparity are found in the eelpout *Zoarces viviparus* and in the familiar aquarium guppies. Many cichlid fishes of commercial importance are 'mouthbrooders', mothers or fathers retaining the fertilised eggs in the mouth until they are quite advanced fry. For some period the young fry may use the parental mouth as a refuge when danger threatens, such intensive care providing another parallel between the advanced teleosts and the more familiar mammals and birds.

The Diversity of Fish

Fishes have adapted to a wide range of habitats. The 20,000 or so species are found from small ponds to depths of 4 km in the sea, from the Antarctic at less than 0°C to hot springs at 40°C. The exploited species make up a relatively small proportion of the total and before its collapse in 1972 the Peruvian anchovy made up about 15 per cent of the total world catch of 65 million tons (see Chapter 7). In the next chapter we will present more fully the groups that are fished; here we just give evidence of the full range of forms found. This is done in Table 1.1 which shows all the orders of fish, their diversity and how many freshwater species each has. A letter 'F' after the order name indicates the major groups that are fished. The majority of species are teleosts (which also have the greatest biomass), and within the infra-class a few orders contain most species, such as the Cypriniformes (carps), Siluriformes (catfishes), Scorpaeniformes (sculpins, red fish) and Perciformes (perch-like fishes). The next section describes some tropical and temperate aquatic communities to illustrate the roles fish play in aquatic ecosystems.

Where and How Fish Live

Much of the theory of exploited fish populations was created on the assumption that each species could be treated in isolation. A move away from such simplicity is now taking place. This section provides the backdrop against which many of the processes to be discussed later in the book should be evaluated. Fish are integral parts of the ecosystems in which they live so that reducing the population of one species can have profound consequences on others. We have no space to describe all the variants of aquatic systems, so we have chosen Lake Victoria and areas of the mid-Atlantic continental shelf to represent the tropics, and the North Sea and British lakes and rivers to represent temperate climates.

In each section we will be discussing assemblages of fish species and so it will help if we make clear what we imply by the groupings given. The term 'community' is most often used to describe a collection of different species living together in a particular area. The term has been defined in a number of different ways (Emlen, 1973; Ricklefs, 1979), each emphasising some relationship between the assembled species. In the following community descriptions we intend to emphasise the trophic relationships between the species, so that this is the type of interaction implied when we speak of, for example, the Lake Victoria fish community. In large areas such as the North Sea there is considerable spatial separation between some species which will preclude any close association, but this should be clear from the discussion.

Lake Victoria and its Fish Community

Lake Victoria straddles the equator lying between 0°21′N and 3°0′S. It has a surface area of 68,635 km^2 but is shallow with a mean depth of only 93 m (see Figure 1.4). Unlike the very deep lakes (Tanganyika 1600 m and Malawi 722 m), Victoria is not part of the rift valley system. The shallowness of the lake means that during the windy season from September to January the whole depth is mixed and there is stratification from January to May. Overturn of the water column is almost entirely wind-controlled as the lake temperature does not change appreciably during the course of the year.

Below 50 m in the central region of the lake most of the bottom is mud, although some channels between islands have gravel or sand. The sheltered gulfs and bays have bottoms covered by a two-metre-thick liquid mass of algae which is an important food of the cichlid, *Sarotherodon esculenta*. The shorelines are sandy in some places and

Figure 1.4: A Diagrammatic Cross-section across Lake Victoria to Show Climatic, Limnological, Bottom Deposit and Fish Community Characteristics. (See text for further details.)

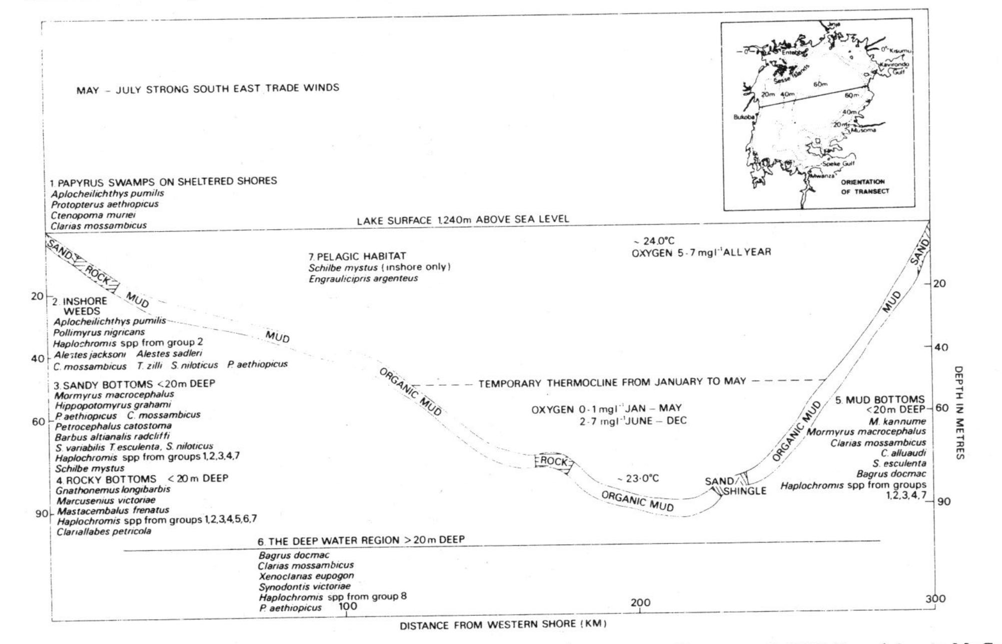

Source: Extracted from Greenwood and Gee (1969), Beadle (1974), Greenwood (1974) and Lowe-McConnell (1975).

rocky in others, but large stretches are covered by beds of papyrus (*Cyperus papyrus*) which in some areas block the mouths of incoming streams, so forming swamps containing very low levels of dissolved oxygen.

The productivity of tropical lakes has not been studied sufficiently to allow much generalisation (Beadle, 1974). The primary production of Lake Victoria and that of Windermere have been compared by Talling (1965). Gross daily production in Victoria is about five times that in Windermere and a similar difference applies to annual gross production. The latter in Victoria was estimated to be 950 g C m^{-2} yr^{-1} while in Windermere the only available data is for net production which was 20.4 g C m^{-2} yr^{-1}. Adding to this the production used up in respiration would still mean that gross production is higher in the tropical lake. Studies in Lake George, Uganda (Beadle, 1974), have shown that algal respiration is also very high in the tropics. This means that the amount of production available for other organisms is much less than the gross amounts would lead one to believe. The rate of primary production throughout the year is quite constant in tropical habitats, although in Lake Victoria the windy season in July and September is marked by increased mixing of the water column which redistributes nutrients, thus boosting production. The rate of production also varies spatially, with the bays and river mouths generating higher rates than the central parts of the lake. This is caused partly by nutrients brought down in the rivers and partly by the greater frequency of mixing in the shallow bays.

The level of Lake Victoria fluctuates in the long and short term (Beadle, 1974). The subsequent flooding of shallow marginal areas leads to significant input of detritus and nutrients and probably contributes to food for fish. The extensive macrophyte growth in the shallow bays must also contribute significantly to the total productivity of the lake.

There are 208 known species of fish in the lake, most of the fauna deriving from east-west flowing rivers which drained the Kenya highlands in the Pliopleistocene (Greenwood, 1974). Some may have come from the lake's recent association with the River Nile (Lowe-McConnell, 1975). The fish community (see Figure 1.4 and Table 1.2) is dominated by the Cichlid family, there being about 170 species of which most belong to the genus *Haplochromis* (Greenwood, 1974). There are 38 non-cichlids, 16 of which are endemic, split between five orders. Some of the most abundant species are shown in Figure 1.4 which gives general information on their habitats and depth distribution.

The pelagic zone does not contain a rich fauna although *Schilbe mystus*, a catfish, and the small cyprinid, *Engraulicypris argenteus*, inhabit the region. *Engraulicypris* is truly pelagic with floating eggs and larvae and the silvery camouflage typical of marine pelagic fish such as the herring. Lake Tanganyika has two true clupeid species in fact. There are a number of zooplankton-feeding haplochromis species. The non-cichlid groups that dominate the benthic region are the mormyrids and the siluriforms, both contributing to the catches of commercial trawls. Experiments in tanks have shown that the major species of mormyrid seem to prefer particular types of bottom as indicated on Figure 1.4. The diets of the several species are wide and largely similar; perhaps the restriction to a specific bottom type minimises competition between the species. Alternatively, the species could each be concentrating on a different size range of food organisms.

The genus *Haplochromis* has undergone extensive adaptive radiation to produce the 170 species now found (Greenwood, 1974). Speciation is largely reflected in the feeding habits, with adaptations for feeding on all possible foods (Table 1.2). The most common food is insect with about 30 per cent of the species being insectivorous or detritus feeders. Some of the insect food such as chironomid larvae is periodically very abundant (Beadle, 1974). Many insects emerge at certain phases of the moon and at these times fish can gorge themselves. Some fish have evolved specialisations for feeding on molluscs; one group prises the soft parts of the mollusc out of the shell whilst a second group has developed massive pharyngeal bones and teeth with which the whole mollusc is crushed. A few species are herbivorous, some grazing on epilithic algae (aüfwuchs) and on epiphytic algae, and one species feeding on the leaves of macrophytes. About 30 per cent of the species are carnivores, showing many strange adaptations to assist them in their calling. One group specialises in feeding on the eggs and larvae of other species of *Haplochromis*. Many of the piscivores have become elongated and slim, allowing them to accelerate rapidly to catch their fast-moving prey. The cichlids are found in all major habitats except in the papyrus swamps and at the surface of the open water.

This brief review of a tropical lake ecosystem has illustrated two features common to such habitats – stability and diversity. Production continues the whole year round and is high, but there is a smaller surplus available to the rest of the food web than is implied by the gross figures. It is clear that detritus and macrophytes also contribute significantly to production although just now it is not possible for us to give a quantitative analysis of how much production is contributed by each sector. The diversity of the cichlids, many of which appear to have very similar feeding habits, provides an interesting source of material for the biologist interested in speciation. The close similarity in diet between many species is perhaps an artifact of the lack of detailed information; although the superabundance of food in the algal feeding species may mean that food competition does not arise.

Freshwater Fish Communities in the British Isles

It is important to make a distinction between lakes and rivers. Like the sea, most large lakes have abundant phytoplankton populations which contribute most of the input of new organic matter to the system (autochthonous input). This is so for Loch Leven in Scotland for example (Morgan and McLusky, 1974). In typical rivers, plankton is relatively unimportant and higher water plants do not always contribute much production. As a result the major input of organic matter comes from outside the system (allochthonous input), very often from the land in the form of dead leaves and branches. Consider for example Bear Brook in New Hampshire, USA (Fisher and Likens, 1973). The annual input of energy to a 1700 m segment of the stream is 25.2 MJ m^{-2} of which 99 per cent is allochthonous, coming from the surrounding forest or from the upstream areas. Leaf litter and dead branches account for 44 per cent of the allochthonous input while 56 per cent comes from inflowing surface and subsurface waters. The only primary producers present are mosses which contribute 1 per cent to the energy input. The stream has a store of energy in the form of detritus which amounts to 19.7 MJ m^{-2}. A similar emphasis on allochthonous organic matter is found in the River Thames in its slow-flowing reaches (Berrie, 1972). Some fish species are heavily dependent on detritus for food; for example the diet of larger (greater than 5 cm) roach, dace and gudgeon is 55 per cent, 35 per cent and 74 per cent detritus for each species respectively. In this region of the Thames there is some significant input of autochthonous organic matter through phytoplankton (18.4 MJ m^{-2} yr^{-1}), benthic algae (17.8 MJ m^{-2} yr^{-1}) and macrophytes (0.14 MJ m^{-2} yr^{-1}) (Berrie, 1972).

British rivers can be divided into two major regions (Hynes, 1970): the rithron, which is the faster-flowing part of the river, and the slow-flowing, wide potamon of the lowland rivers. The rithron contains the rocky, stony stretches in highland regions where the water is mostly fast-moving, cool and well-oxygenated. As the terrain becomes less precipitous the flow slows and the river is characterised by a riffle and pool structure. The potamon is more like a lake environment with slow flow, particularly in the summer, lower oxygen levels and murkey deep waters. The bottom in this type of river is usually of clay or soft mud.

Lakes can be classified into oligotrophic and eutrophic. These terms are usually at the two ends of a continuous scale but are useful as general labels. Oligotrophic lakes are often deep and the water at all depths is well supplied with oxygen, even in the summer. The concentration of plant nutrients is low, as is the rate of primary production. Eutrophic lakes are often shallow and the deeper water is often without oxygen during the summer when the water column is stratified (Wetzel, 1975). Summer temperatures are warmer than those found in oligotrophic lakes.

The potamons of rivers and eutrophic lakes in the British Isles are dominated by the cyprinids; common species with some indication of habitat are shown in Figure 1.5. The cyprinid species are catholic in their diets and eat a wide range of invertebrates and plants. Accordingly it may appear that there is considerable overlap in diet, but this is not necessarily so as there is spatial separation of the species during feeding. In faster-flowing reaches dace become common together with roach, both feeding on invertebrates and plants. Observations of feeding behaviour of the two species show that dace cannot easily pick up items from the bottom whilst roach are more manoeuvrable; they can stop quickly and can easily pivot forward to pick up benthic invertebrates. This suggests that in a river dace feed in the main stream, capturing drifting items, whilst roach are happier on the bottom. Rudd is another species closely similar to roach but is found mostly near the surface and has an upturned slanting mouth allowing it to capture surface insects. Bream on the other hand are even more specialised benthic feeders. Apart from dace, faster-flowing sections of rivers will have greater numbers of chub and minnow and will lack rudd, bleak, tench, the breams and carp. Dace are most often found in riffle areas as are young chub, the older individuals of which tend to be solitary, living in deeper holes under tree roots near the bank. Faster-flowing sections may also have the bottom-living stone loach, gudgeon and bullhead, all adapted for life on the bottom. The flat-bodied stone loach and bullhead have no swimbladder and keep out of the main current by sheltering under stones. Bleak live in the surface water of pools between riffles and in larger rivers. In North America there are species similarly adapted to bottom and semi-pelagic habitats (Hynes, 1970).

The chief piscivores in these waters are the perch and pike. Perch are more common in lakes and the young fish feed mainly on

Figure 1.5A: A Diagrammatic Longitudinal Section Along the Length of a British River to Show the Type of Flow, Vegetation Type, Bottom Deposits and Gradient and the Fish Community. For futher discussion see text. (Principal Species Underlined.)

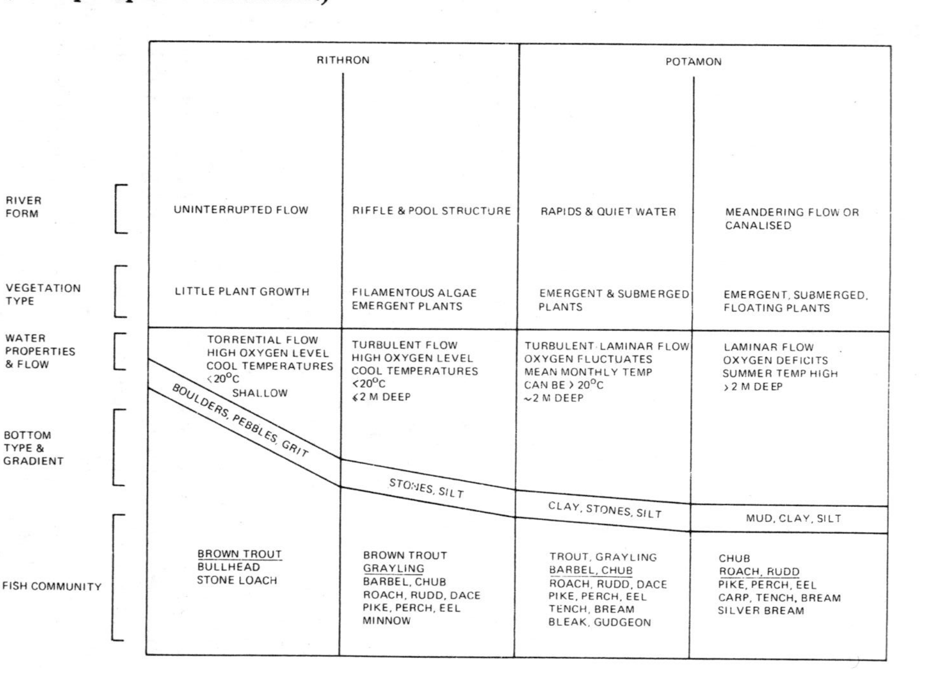

Figure 1.5B: Diagrammatic Cross-section through a Temperate Lake to Show the Differences between Oligotrophic and Eutrophic Lakes in Terms of Emergent Vegetation, Limnology, Bottom Deposits and Fish Communities. Further details are given in the text.

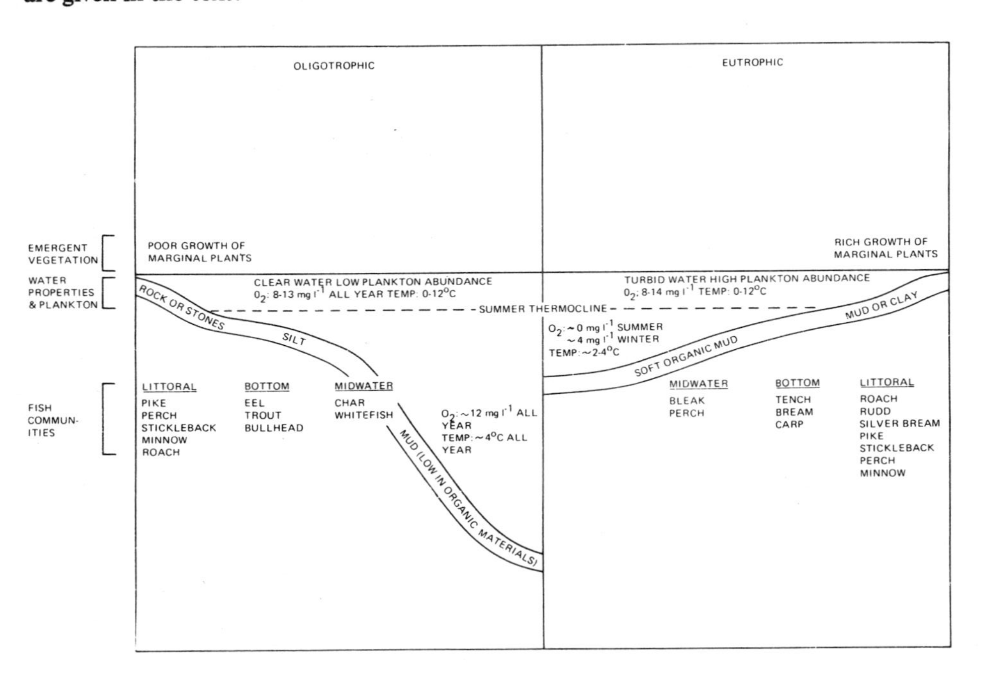

invertebrates. Pike start to feed on fish at an early age (Frost and Kipling, 1967) and remain strongly piscivorous all their lives unless fish are in short supply (Larsen, 1966). The eel is also a predator, taking some small fish and invertebrates. Other species such as the ruffe and bullhead are also piscivorous although not exclusively so.

In faster-flowing rivers such as are found in upland areas and in oligotrophic lakes the cyprinids give way to the Salmonidae. Some of the common species found in these habitats are shown in Figure 1.5a and b. The salmon spawns in the head waters of rivers and the young spend the first one to two years of their lives there. Trout living in lakes move into streams to spawn whilst the other species such as the char and whitefish spawn in the lake itself. Most salmonids prefer colder temperatures and require a high concentration of oxygen. During the summer the char in Windermere withdraw to deeper waters to avoid the warmer surface temperatures, a migration only possible in oligotrophic lakes with a well-oxygenated hypolimnion. The larger volume of a lake provides niches for pelagic fish and in deeper lakes most benthic species are restricted to the shallower margins. In Windermere this is particularly true of perch and pike, the two main predators. Both species have the same type of diet as they do in eutrophic waters and in rivers. In similar habitats in the rest of Europe and North America the dominance of salmonids prevails although different species may take the place of those found in Britain. The genus *Coregonus* is found throughout the Holarctic region presenting the taxonomist with considerable problems because of its plastic phenotype (Behnke, 1972). Local *Coregonus* populations in Europe underwent some differentiation before the last ice age, but not always sufficient to ensure that the newly-evolved groups remained genetically separate after the glaciation. Since then introgression has obscured relationships creating a set of variously related phenotypes. A particularly interesting relationship is that between the pollan in Lough Neagh, Northern Ireland (see Chapter 9) which is most closely related to the Alaskan cisco (Ferguson *et al.*, 1978), whereas the other three species in the British Isles are true whitefish and are closely related to each other. The two types probably represent different phases of glacial retreat.

The North Sea Fish Community

Physically the North Sea is a shallow extension of the North Atlantic and it shares many of its characteristics. The continental influence is greater than in the Atlantic proper. This section considers the sea areas between the Straits of Dover and about 62°N. The sea bed slopes from south to north. A large proportion of the bottom is covered with sand but some of the deeper hollows have a clay surface and there are scattered stony patches. The basin receives Atlantic water from the north around the Shetland Islands (about 90 per cent of the inflow) and through the English Channel (about 8 per cent of the water). Water from the north flows down the east coast of Scotland and England, eventually turning east in the region of the Dogger Bank. The water coming in through the English Channel joins the eastern-flowing water in the Central North Sea and the main outflow is north along the Norwegian coast (Hill and Dickson, 1978). There is also an input of low-salinity water from the Baltic. The Atlantic water is generally saltier than the water in the North Sea, which makes it possible to trace the degree of penetration by Atlantic water which covers the greatest area in winter. Dickson (1971) studied the variation in salinity in different places in the North Sea and found that the degree of penetration of Atlantic water varied significantly over the thirty years between 1945 and 1975. This variation probably has significant effects on the planktonic community in which fish eggs and larvae must spend their early months (see Chapter 6).

Temperature is also an important parameter. In summer, the shallow southern area is much warmer than the water in the north and there is rarely a thermocline. This region cools quickly in winter and is then generally cooler than the water in the north where the Atlantic water is a stabilising influence keeping the water cooler in the summer and warmer in winter. In the north the water is deeper and a thermocline is able to form. Temperature in the North Sea has changed over the past 50 years with an increase up until the mid-sixties followed by a decline since then. For a recent review of North Sea hydrography see Hill and Dickson (1978).

The North Sea has an area of approximately 500,000 km^2 so that the ecosystem within it is only crudely characterised by the diagram in Figure 1.6 (Steele, 1974). The primary producers are almost completely planktonic, shore algae providing a negligible contribution on this scale. The herbivores are also largely planktonic, the largest population being copepod crustaceans such as *Calanus finmarchicus* and *Temora longicornis*. The invertebrate carnivores are animals such as *Pleurobrachia* (sea gooseberry or comb jellies), and the Chaetognatha, commonly called arrow worms and represented around the British Isles by *Sagitta elegans* and *S. setosa*. The benthos includes all the crustaceans (crabs, shrimps, lobsters), gastropods and bivalve molluscs that live on or in

Figure 1.6: A Simplified Food Chain for the North Sea Ecosystem. The squares representing each level in the chain are proportional in area to the annual production at that level. This is not true of the square representing the human level. The figure in each square gives the annual production for a year as kilocalories per square metre (1 kilocalorie = 4,186 J). The figure by each arrowed line gives the amount of production from a particular level that is passed on to the various consumers. The catch of fish taken by man is 0.042 per cent of the amount of primary production.

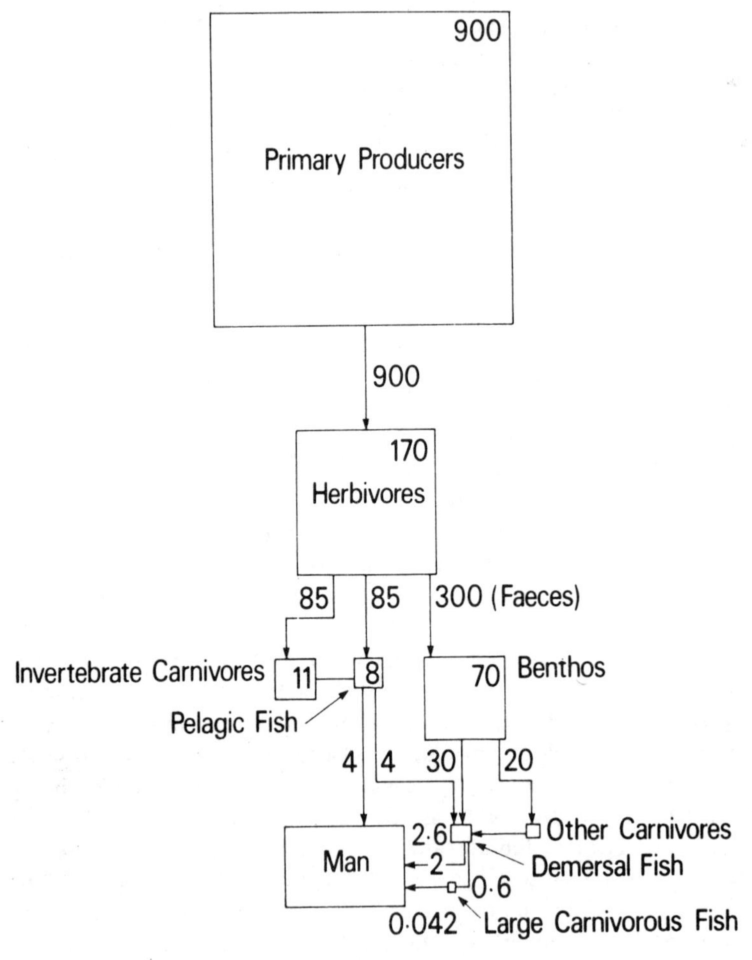

Source: Drawn from data in Steele (1974).

Figure 1.7: The Feeding Relationships for the Different Ages of Herring and Members of the Plankton. The broken lines are for those relationships defined by the work of people other than Hardy.

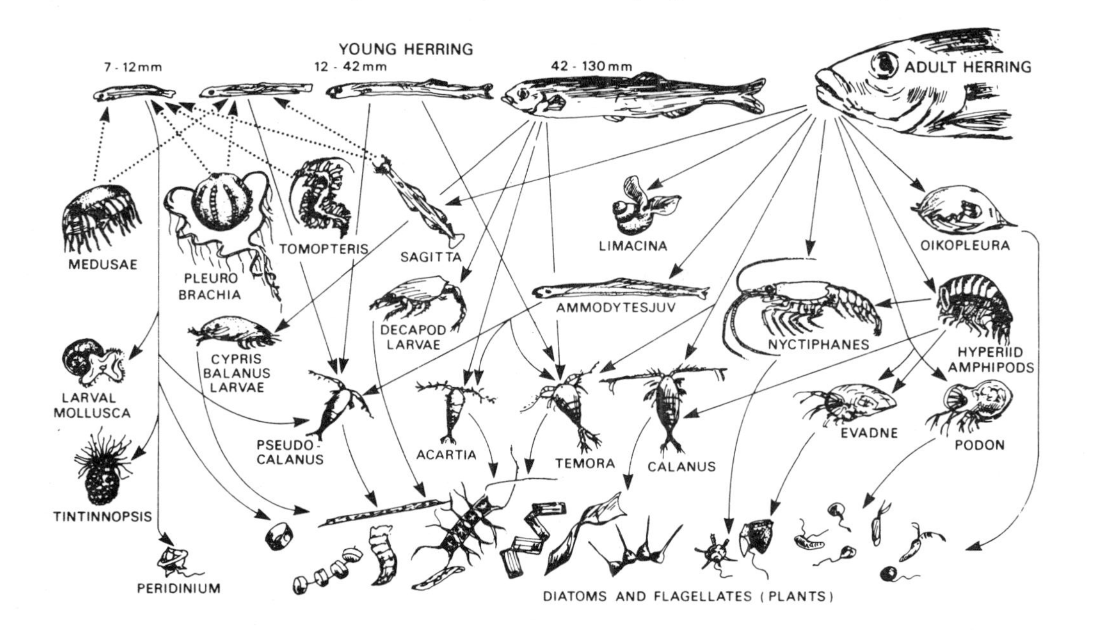

Source: Redrawn from Hardy (1958) with permission.

the bottom. These organisms are significant to many fish species as food. Transfer efficiencies of up to 20 per cent occur in the system depicted in Figure 1.6. It is also concluded (Steele, 1974) that fish production in the North Sea is high when considered in relation to the amount of production at lower levels in the food chain. The relations in Figure 1.6 are vastly simplified and many of the estimates of production are only very approximate. Many species will have complicated relationships with others as was shown by Hardy (1924) for the North Sea herring (see Figure 1.7).

Production:biomass ratios vary with the level in the food chain and the size of the organism. Values for herbivorous zooplankton vary between 0.09 and 0.3 per day with turnover times of 3 to 12 days. For carnivorous species values are higher, varying between 0.11 and 0.3 with turnover times being in the same range as for the herbivores (Tranter, 1976). Assuming that growth can be represented by the von Bertalanffy curve (Chapter 4) and mortality by an exponential decline, P:B ratios for fish are equal to the total instantaneous mortality rate (Allen, 1971). Using this approximation herring in the North Sea showed an annual P:B ratio of between 0.3 and 1.6 between 1915 and 1975 (Burd. 1978). Mean age and lifespan varied between 0.63 and 3.3. This indicates why the herring population has declined so drastically over the past 20 years. Using data in Daan (1978) on the production and biomass of North Sea cod, the P:B ratio can be calculated and averages 0.76 for the years 1920–73. The mean lifespan of a cod over this period was 1.3 years.

There are about 100 species of fish in the North Sea and the more common members of the community with their habitats are shown in Figure 1.8. It should be realised that most species will live in different places and eat different items during the separate stages of their life history. For example, cod eat copepods and then euphausiaceans (shrimp-like crustaceans) when they are in the plankton. When they first reach the bottom they eat worms and small crustaceans such as prawns, progressing as they grow to larger crustaceans such as crabs, molluscs and eventually fish. In the Arcto-Norwegian stock the larger cod can exploit the large capelin stock and this enables them to continue growing well beyond the maximum size achieved by North Sea cod. This type of food switch is quite common in fish and has an important influence on growth. In the adult stages there is also a separation in space of different species as illustrated by the gadoids cod, haddock, whiting and saithe. The haddock is the most completely benthic of the four and they have a sub-terminal mouth (i.e. ventral to

Figure 1.8: Diagrammatic Cross-section across the North Sea to Show Details of Current Types, Water Types, Oceanographic Conditions, Bottom Deposits and the Fish Community.

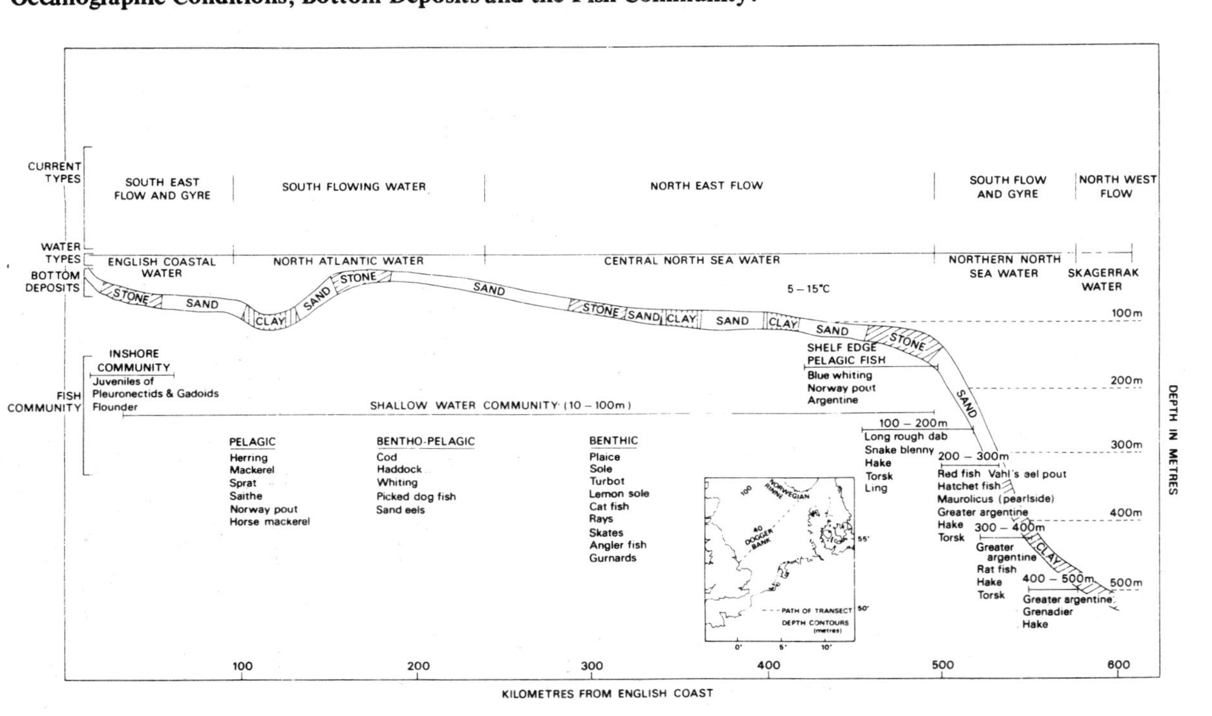

Table 1.3: The Catch of Selected Species from the North Sea.

Species or Group	Catch in Tons	Per Cent of Total	Percentage Distribution North[c]	Central[c]	South[c]
Sprat	651,591	20	5	94	1
Norway pout	642,028	19	99	1	<1
Sandeels	428,286	13	34	58	8
Herring[b]	295,307	9	37	61	3
Saithe	249,798	8	88	12	1
Mackerel	252,068	8	92	7	2
Cod	186,453	6	31	58	11
Haddock	174,163	5	64	36	<1
Whiting	140,166	4	54	30	16
Plaice	108,562	3	6	80	14
Picked dogfish	22,656	1	69	29	3
Sole	18,263	1	<1	59	41
Horse mackerel	9,933	<1	24	40	35
Angler fish	5,106	<1	57	42	1
Lemon sole	5,029	<1	16	77	7
Turbot	4,588	<1	4	73	23
Rays and skates	4,176	<1	52	27	22
Hake	3,151	<1	40	60	<1
Flounder	2,939	<1	2	31	67
Catfishes	2,299	<1	24	76	<1
Pollack	1,194	<1	69	31	<1
Redfish	876	<1	99	1	<1
Total	3,311,310	97[a]	51	44	4

Notes: a. Only those species contributing more than 1 per of the catch – twelve in all. b. Now completely banned. c. These are areas IVa, b, c of ICES.
Source: Nikolaev (1978).

the anterior end of the body) making it easier to pick up items of food from the bottom. The cod also eats benthic invertebrates but can eat fish, especially when larger; the mouth is still subterminal but larger than that of the haddock. The whiting is bentho-pelagic and feeds mostly on fish, its terminal mouth with long sharp teeth being well-adapted to this habit. The saithe is pelagic and feeds on planktonic crustacea and fish; it has a smaller superior mouth (i.e. the lower jaw provides the most anterior point of the body), making prey capture easier from below. Pollock are similar but tend to live close inshore all the year, feeding on benthic and pelagic prey.

In the North Sea about 12 species provide 95 per cent of the weight caught. The catches of the most important species for 1975 are shown in Table 1.3. This illustrates how the community is dominated by a few groups – the gadoids, the clupeids and some flatfish. The table also shows that the North Sea is too large an area to assume a homogenous distribution of species. Individuals of all the species listed in Table 1.3 could be caught throughout the sea, but most will have their greatest abundance in certain areas. It must also be assumed that fishing effort for a species is equal in all areas and this is unlikely. The division of the catch into proportions taken from the three sub-areas is then only a rough guide. A further point to realise is that the balance of species is not static. The changes in the North Sea fish stocks between 1909 and 1973 are discussed by Holden (1978), where it is shown that several species increased in abundance towards the end of the sixties, probably as a result of natural events.

The Fish Community in the Coastal Areas of the Tropical Atlantic

To complete our survey of selected fish communities we will briefly review the fish fauna found in the coastal waters of the tropical Atlantic, drawing particularly on the work of Longhurst (1969) and Lowe-McConnell (1962). The two areas considered, shown in Figure 1.9, are the Gulf of Guinea extending from the Cablerouns (about 3°N) to Cape Verde (at about 13°N) on the West African coast and the whole of the continental shelf off Guyana between about 6°N and 9°N. Off the African coast there are no living coral reefs although there is a fossil reef existing along almost the whole coast being found at the edge of the shallow parts of the continental shelf. These reefs project above a mostly soft bottom which is mainly mud to the east of Lagos, because of the many rivers, but to the west around Sierra Leone sand predominates. Off Guyana the bottom is also a mixture of mud and sand, although deposits become progressively harder towards the edge of the shelf. The soft muddy deposits are mainly found in shore regions and the bottom becomes sandier the further it is from the coast. Although this coast is like its African counterpart it is by no means typical of the whole western Atlantic region; the Carribbean zone is characterised by coral reef communities not found off Guyana.

The hydrographic character of the two regions exhibits important differences. Off the whole of the West African coast the cool eastern boundary currents converge on the tropical zone pinching it in towards the equator. This is illustrated in Figure 1.9 which also shows that on the western side of the Atlantic the major currents flow north and south from the equator expanding the width of the tropical zone. Off Africa there is a permanent shallow thermocline at a depth which varies from 10 to 50 metres. As a result, part of the continental shelf is still well-lit but has a temperature more typical of sub-tropical

Figure 1.9: A Map of the Tropical Atlantic Showing Major Surface Currents and Terrestrial Features Mentioned in the Text.

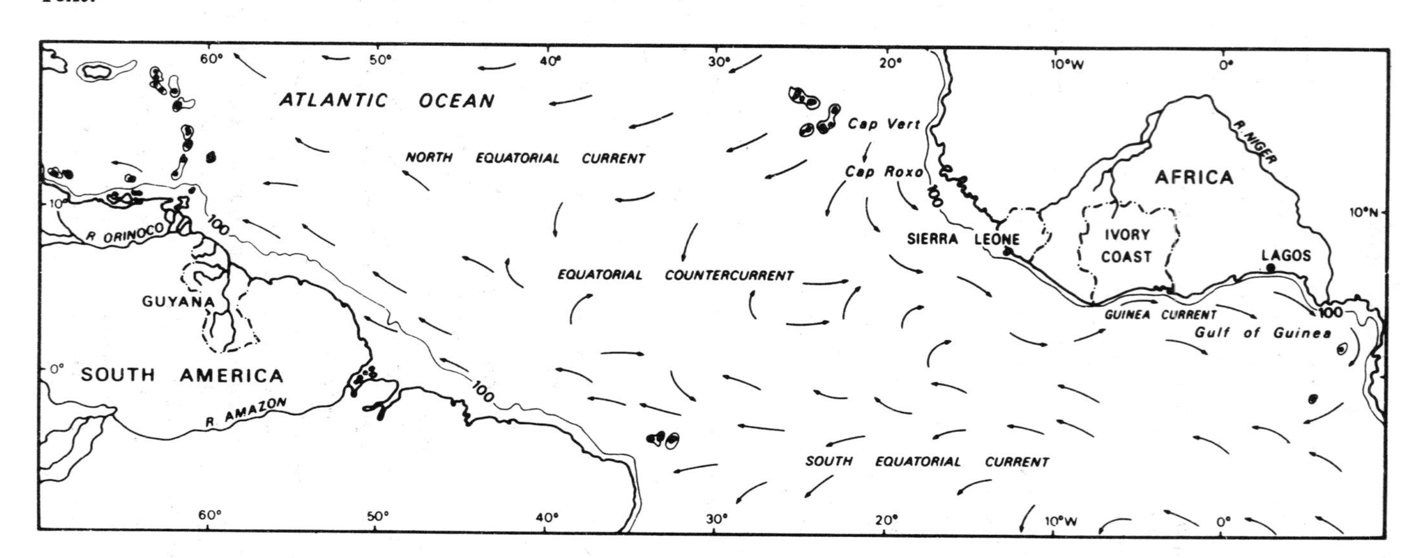

zones (water less than 20°C). This has considerable influence on the fish communities as will be discussed. The general water movements off the African coast are easterly in the form of the Guinea Current which is met at the southern limit of the area by the north and north-west flowing Benguela Current. Off Guyana the principle offshore current, the South Equatorial Current, moves in a north-westerly direction at from 1 to 4 knots; inshore water movements are the combined results of the tides and river flow. This area is also well stirred when the north-east trades are blowing between October and March. The thermocline is usually at a depth below the edge of the shelf so that sub-tropical fauna are not found in the area. In this region the mean sea temperature at 100 m can be 24°C.

As a consequence of low nutrient levels the open ocean in the tropics has one of the lowest rates of gross primary production on earth; less than 100 mg of carbon per square meter per day. Both the areas described here are inshore and as a result have higher productivity which is in the region of 100–500 mg C m^{-2} d^{-1}. In the Gulf of Guinea the lowest production rate is off Cameroun, but further west productivity increases and is highest in the region spanning the Ivory Coast to Senegal. This is caused by a period of upwelling which occurs in July and August with great reliability (Cushing, 1971; Bainbridge, 1972). Upwelling is the name given to a process whereby deep, nutrient-rich and cool water rises to the surface. This water provides an ideal medium for the growth of phytoplankton which increases greatly in abundance as a result. It is this process which usually occurs predictably off the coast of Peru and provides the productive base for the anchovy fishery. The productivity of the upwelling period in the Guinea current is contrasted with the Peru system in Table 1.4. The rate of production per unit area is the same order of magnitude in both systems, but because the Guinea upwelling covers only a small area and lasts for a shorter time it produces less in total. Transfer efficiencies between primary and secondary producers seem lower in the Peru system.

Both Longhurst (1969) and Lowe-McConnell (1962) used a bottom trawl to make their surveys, so this review will be mainly limited to species caught by this method. Off West Africa several workers before Longhurst had noticed that different groups of species could be associated with different depths. In the Guinea trawl survey 480 stations distributed along 63 transects at 40-mile intervals between Cape Roxo to the mouth of the Congo provided detailed data for analysis. The distribution and abundance of the 200 most common species were used in a statistical similarity analysis which separated

Table 1.4: The Productivity of the Guinea and Peruvian Upwelling Systems.

	Guinea Current	Peruvian Current Sub-area 1	Peruvian Current Sub-area 2
Area (x10^3 km^2)	100	288	191
Duration (days)	120	270	270
Primary Production			
mg C m^{-2} d^{-1}	600	325	1020
Tons C yr^{-1} (x10^6)	7.20	36.6	76.67
Secondary Production			
Tons C yr^{-1} (x10^6)	1.59 – 2.12	6.59 – 8.76	3.86 – 5.13
Efficiency (%)	22.1 – 29.9	17.9 – 23.9	5.0 – 6.6
Tertiary Production			
Tons C yr^{-1} (x10^4)	11.5 – 14.2	51.3 – 62.1	57.4 – 63.8
Tons wet wt. yr^{-1} (x10^6)	0.86 – 1.06	3.83 – 4.64	4.28 – 4.77

Source: Cushing (1971), Table IV.

out seven assemblages of the most common species which are shown in Figure 1.10 together with the physical characteristics of their respective zones. It was found that the type of bottom, closeness to the coast and, most of all, the position of the themocline, had the greatest influence on the associations of species found. The warm shelf waters (over 18°C) above the thermocline were dominated by the sciaenid community which occurred mostly over soft bottoms composed of sand or mud. The lutjanid community occurred most frequently over the relic outcrops of coral at the edge of the continental shelf where the bottom is sandy and hard. The zone on either side of the thermocline was dominated by two sections of the sparid community living mainly over soft muddy or sandy bottoms. Included in the deep element of the sparid community were three species of sub-tropical fish, the sparid *Boops boops*, the Japanese mackerel and *Trachurus sp*. The deepest assemblage on the continental shelf is similar to a group characterised by species of hake, and found at shallower depths in sub-tropical waters to the north and south of the tropical zone.

A similar zonation pattern was found by Lowe-McConnell (1962)

Figure 1.10: A Diagrammatic Cross-section across the Continental Shelf off Nigeria to Show Climatic, Oceanographic, Bottom Deposit and Fish Community Characteristics. See text for further details.

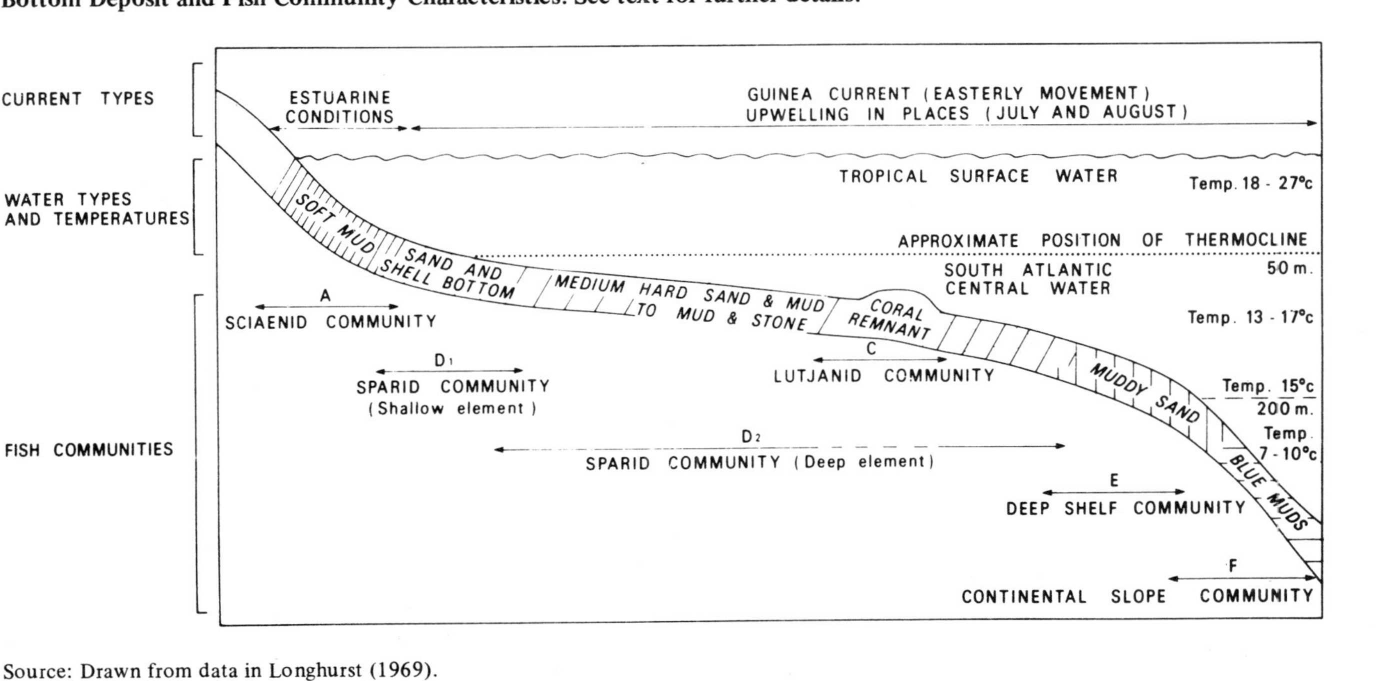

Source: Drawn from data in Longhurst (1969).

Figure 1.11: A Diagrammatic Cross-section across the Continental Shelf off Guyana, South America, to Show Climatic, Oceanographic, Bottom Deposit and Fish Community Characteristics. See text for further details.

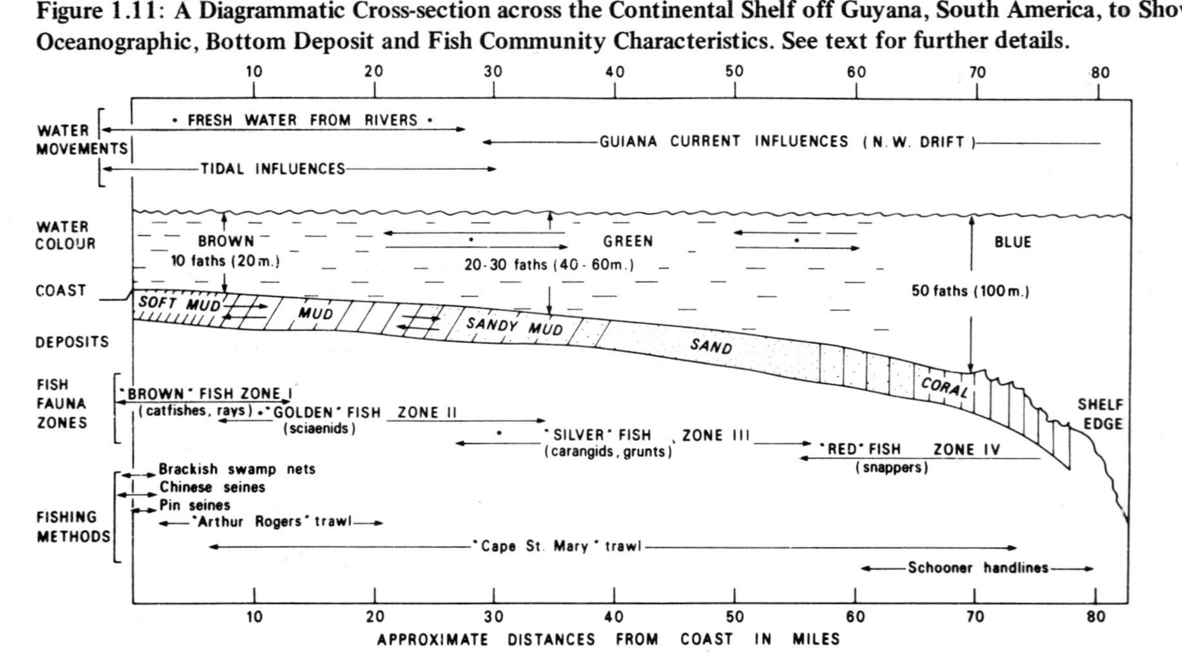

Source: Redrawn from Lowe-McConnell (1962) with permission from the Zoological Journal of the Linnean Society of London.

off Guyana. The results of this study were not treated in the same manner as were the data from west Africa, the different assemblages being characterised subjectively. Figure 1.11 shows the principal groups of species found in each of four zones on the continental shelf. The figure also provides information on the substrate and hydrography characteristic of each zone. Trawling could not be carried out in the most inshore region. Zone II yields the greatest catches, producing hauls of up to 251 kg per fishing hour of edible fish. As discussed in detail by Longhurst (1969), there are strong similarities on both sides of the Atlantic in the way the fish community is grouped. The biggest difference is the lack of the sparid community which is replaced by an assemblage dominated by the Pomadasyidae (grunts). Species of grunts may play the same ecological role in the west Atlantic as the sparids do in the east. In the tropical areas of the western Atlantic the greater occurrence of coral reefs is accompanied by a higher numerical and specific abundance in the lutjanid community, a fact exemplified by the lutjanid community found off Guyana. Throughout the tropical zone in the west Atlantic the specific composition of the lutjanid community is quite stable, which is not true of the sciaenid community which shows considerable local differences. Longhurst (1969) concludes his analysis with the observation that the communities found in the tropical Atlantic may also be found in the Indo-Pacific region but with certain changes in detail.

In the west Atlantic the dominant groups could be broadly divided into three groups according to feeding habits (Lowe-McConnell, 1977; see Figure 1.12). Bottom feeders are typified by a ventrally-placed mouth, a good acoustico-lateralis system and barbels around the mouth. An example is the croaker *Micropogon* which feeds on polychaetes living in the mud. The second group has fish with upturned mouths, allowing them to feed on pelagic shrimps. This type is exemplified by a carangid *Vomer*, and *Larimus* which is a croaker. The third group consists of species that feed mainly on fish and have evolved fusiform bodies enabling them to pursue fast-moving prey. Most of the sciaenids such as *Cynoscion virescens*, the so-called sea trout, belong to this group. Although diets seem to overlap considerably, competition is probably reduced by each species or different stage in the life history keeping separate either in space or time. For example, off Guyana many of the young stages can be found in Zone I, which means that adults and young do not compete for the same resources. Off West Africa most of the species are high in the food web with few exploiting the primary producers. The large rivers that reach the sea

Figure 1.12: Trophic Relationships of Some of the More Common Fish Found over the Continental Shelf off Guyana.

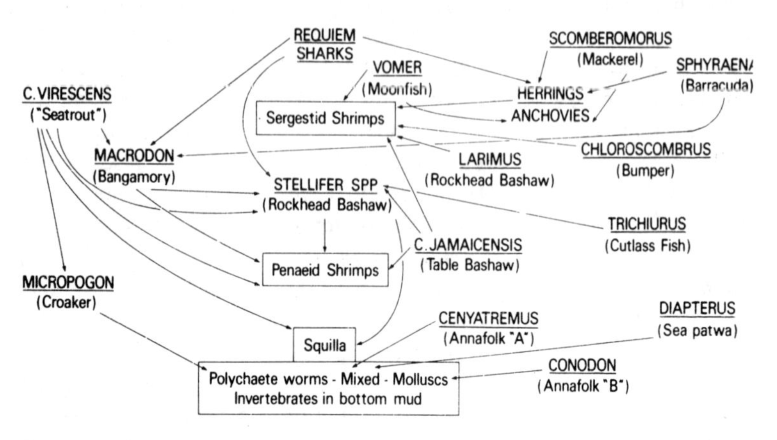

Source: Redrawn from Lowe-McConnell (1962) with permission of the Linnean Society of London.

in this area contribute much allochthonous detritus which probably supports many of the bottom-living invertebrates – a situation reminiscent of many lakes and rivers (see above). Most of the fish in the Gulf of Guinea have much shorter lives than is common in temperate areas. Fish aged four or more are rare, implying a rapid turnover and sensitivity to factors affecting year-class abundance. Production-to-biomass ratios are relatively high and although the communities are more affected by environmental fluctuations they can also recover more quickly from heavy exploitation.

We have described in this chapter the characteristics which distinguish the lives of fishes, and we have attempted to show how fish fit into the aquatic communities in which they live. The emphasis has been on the trophic interactions between the plankton, benthos and the fishes, as it is the dynamics of this system which ultimately determines how much fish is available to man. In the next chapter we describe the specific composition of the fish catch from the different parts of the world, the main methods used to catch fish and some of the factors which determine how fish are processed. We also briefly describe some processing methods.

2 WORLD FISHERIES

The fishery scientist is often confronted with problems that originate in the fishing industry, so that an appreciation of the character of this industry is necessary for a full understanding of fishery science. This chapter describes the growth of the fishing industry since 1945, the way in which the catch is used, the reasons why certain species rather than others are preferred as food, and considers the possibility of future increases in yield. We also describe briefly the main types of fishing gear, and the processing of the fish before they are marketed.

The Growth, Composition and Utilisation of the World Fish Catch

In the 15 years between 1955 and 1970 the total catch of fish taken by all nations more than doubled, increasing from 30 million to 70 million tons (Figure 2.1). Since 1970 the catch has fluctuated between 65 and 73.5 million tons. The levelling off has been mainly a result of the almost complete loss of the Peruvian anchovy fishing (see Chapter 7). Over the past six years the growth of fish catches used for direct human consumption has continued and it is instructive to examine the reasons for this trend.

The rapid increase in the quantity of fish caught was caused by the increasing world demand for protein, but this demand took two forms which stimulated two different types of fishery. As human populations expanded rapidly in the 1950s and 1960s there was an urgent need for more dietary protein at low cost. The capture of wild fish provided an attractive source that was economic and easy to obtain. Many socialist states, such as the Soviet Union, Poland, Bulgaria and Rumania and states with serious protein shortages such as Japan, invested in huge fleets of sophisticated long-distance trawlers which were able to rove the world seeking catches. A parallel trend occured on both sides of the north Atlantic where local fish stocks were unable to support continuously growing fishing pressure. The development of the plate freezer on board ships allowed distant stocks to be exploited and up until the early 1960s there still existed suitable, lightly exploited stocks to accommodate this trend.

Figure 2.1: The Catch of Fish by All Countries and How It Has Increased between 1950 and 1978. Also shown are the major uses to which the catch is put and the amounts coming from inland waters.

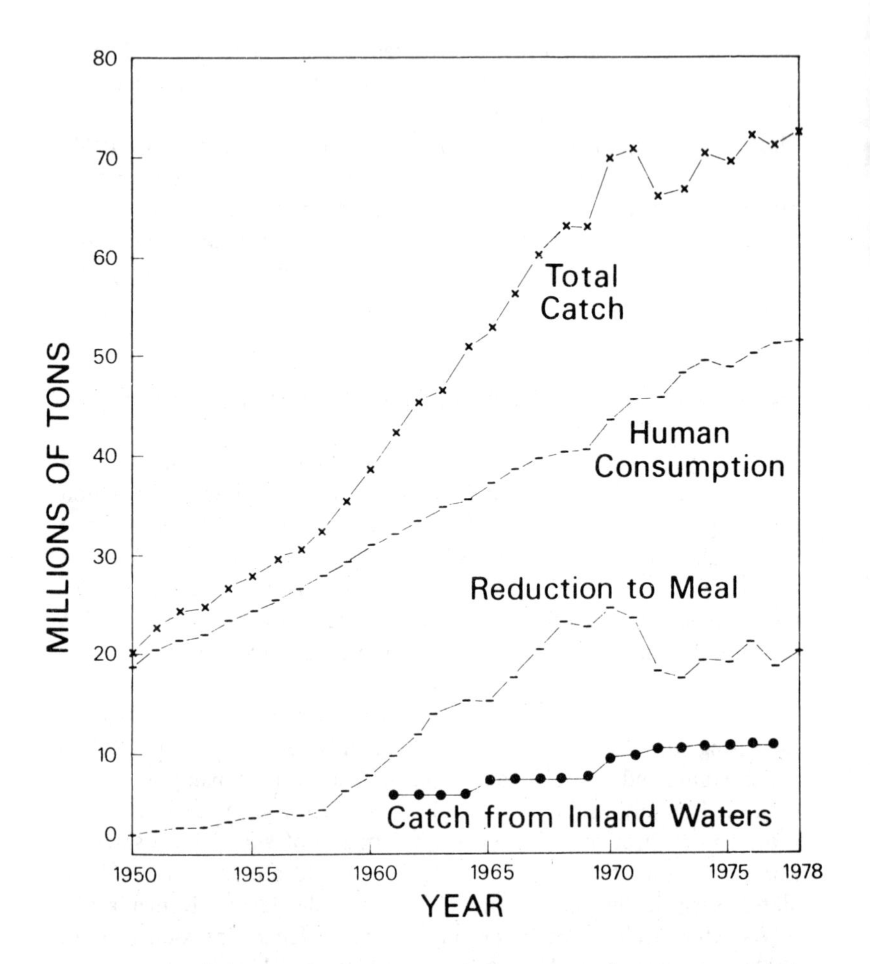

Source: Data taken from FAO Yearbook of Fishery Statistics (1979).

Figure 2.2: The Way in which the Catches of Six Groups of Fish Have Changed Since 1961. Each year's catch has been divided by the 1961 catch and one subtracted from the result. This gives a measure of how much the catch has grown over the period. The groups are as follows: 1 – flounders, halibuts, soles; 2– cods, hakes, haddocks; 3 – redfish, basses, congers; 4 – jacks, mullets, sauries; 5 – herrings, sardines, anchovies; 6 – tunas, bonitos, billfishes, mackerels, snoeks, cutlassfishes. (See FAO, 1979, for the taxonomic relationships of these groupings.)

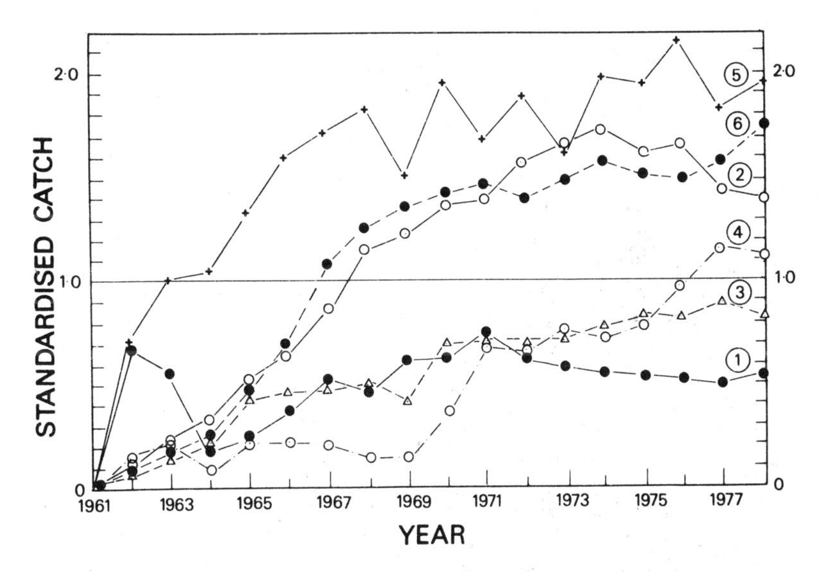

The second demand for protein came from increasingly industrialised pig and poultry farmers in the developed nations. Whilst the fish catch was expanding the output of poultry meat increased by 5.5 per cent a year while pork output increased by 3 per cent a year (Kolhonen, 1974). These industries used high-protein artificial diets and one of the chief sources of protein is fish meal. As a result fisheries on species suitable for reduction to meal, like the anchovy, were stimulated. Denmark, with an important pig-rearing industry, also developed a large fish-meal industry and even now about two-thirds of the fish caught by Denmark is used for reduction to meal (OECD, 1979). The rapid growth of the quantity going to reduction between 1960 and 1970 is shown in Figure 2.1, together with other processes.

After fish meal, the second fastest growing sector was the frozen fish industry. In the 1930s Clarence Birdseye produced a plan for the development of the frozen food industry based on advances he had made in the technology of freezing food (Farkas, 1978). Greater personal wealth in the developed nations in post-war years made it possible for more people to buy home freezers, so creating a market for frozen foods. Even Eskimos now have freezers! At the same time, the trend towards working wives and the increase in the number of people eating away from home enlarged the market for convenience foods. Fish prepared in a factory and sold deep frozen was, until recently, a cheap and easy source of protein. Further developments, such as 'boil-in-the-bag' products, enhanced the convenience of fish eating and sold more fish. The resulting acceleration in demand for white fish with low fat levels in the muscles (see later in this chapter), was supplied by the development of distant water fleets.

Since the declaration of 200-mile Exclusive Economic Zones (EEZs) by many countries, the traditional pattern of fishing by the developed nations has been radically altered. Suddenly, processing industries were cut off from reliable supplies of cheap raw materials. For example it has been estimated that the changed regime has deprived the countries of the European Economic Community (EEC) of 500,000 tons which had previously been fished off the coasts of non-EEC countries (FAO, 1979). Japan also has a huge market for frozen white fish which had previously been supplied by the country's large fleet of distant-water trawlers and factory ships. One of the most important species caught was the Alaskan pollock from the waters now controlled either by the USA or the USSR. In 1977 Japan's catch of pollock fell by more than 800,000 tons. However, since Japan exported much of the pollock to the USA, the fall in her catch did not greatly affect the home market. The USA is the third greatest market for frozen white fish and in the long run the new maritime regime must be to its benefit. In the meantime, the USA does not have a fleet capable of catching all the white fish available to it in its declared 200 mile EEZ. Consequently the country is still importing frozen fish (hake, Alaskan pollock, cod). Some of the imports, such as hake, come from new areas like Argentina, which provides a benefit to the economies of relatively poor countries.

The quantity of fish for the canning industry, which, like freezing, is another capital-intensive sector, has also grown rapidly in the past 20 years. In Europe the advent of EEZs and the depletion of stocks such as those of the North Sea herring has meant that many manufacturers have found it difficult to maintain raw material supplies.

Resource shortages have caused price increases which have resulted in a fall in demand (FAO, 1978). An example is the shortage of tuna for the US canning industry recently brought about by political conflict between the USA and Canada.

The quality of fish used in dried products has grown least over the past 20 years. Most of the trade is between Third World countries and there is poor information on the flow of goods. At one time Norway, Iceland and the Faroes produced large quantities of salt cod (stockfish), the biggest market being Nigeria. Nigeria now buys very little from Europe and the production has virtually ceased.

The changed legal regime at sea has established a fairer distribution of economic power. The processing industries in developed nations can no longer depend on regular supplies of cheap raw materials from off the shores of other countries. The developing countries can use their power to improve their own trade. For example, foreign ships may be allowed into a nation's waters to fish so long as the host country is given access to the markets of the visiting country. This could even apply to products not based on fish. Another tactic used by developing countries is to insist that foreign vessels land part of their catch in home ports. The host nation thus maintains the supplies to its factories and benefits from the value added to the raw materials during processing.

The rapid growth in fish catch did not affect all species in equal proportion. The gross view is illustrated in Figure 2.2 which shows the increase of six species groups (FAO, 1979). The largest increase of over 200 per cent was contributed by the herrings, sardines and anchovies, all species which are used in the fish-meal industry. The food fishes such as cods, hakes and haddocks and the tuna group also showed a 1978 catch which was about 1.6 times that of 1961. Fish from these two groups have been important to the expansion of the frozen fish and the canning industries in the developed countries.

This method of looking at the specific composition of fish catches is crude and obscures regional differences. The FAO collects and publishes statistics on a global scale. Poor reporting of catches is a perennial problem (Gulland, 1978b) so that data may be unreliable or out of date from areas with poor communication, a low level of education and small-scale artesan fisheries. With this qualification in mind, the FAO data provide a fascinating view of regional variations in the fishing industry. The data are conveniently grouped in the areas shown in Figure 2.3 and we will start with the North Atlantic.

The most important species in the north-west Atlantic area are the gadoids, such as cod, haddock, hake, pollock (called saithe in Europe),

Figure 2.3: A Map of the World Showing the Areas used by the FAO to Record Fishery Statistics. Each ocean is divided into different sectors in the following way. Pacific: north-west (nw), north-east (ne), west-central (wc), east-central (ec), south-west (sw), and south-east (se). Atlantic: north-west (nw), north-east (ne), west-central (wc), east-central (ec), south-west (sw) and south-east (se). Indian: (west) (w) and east (e).

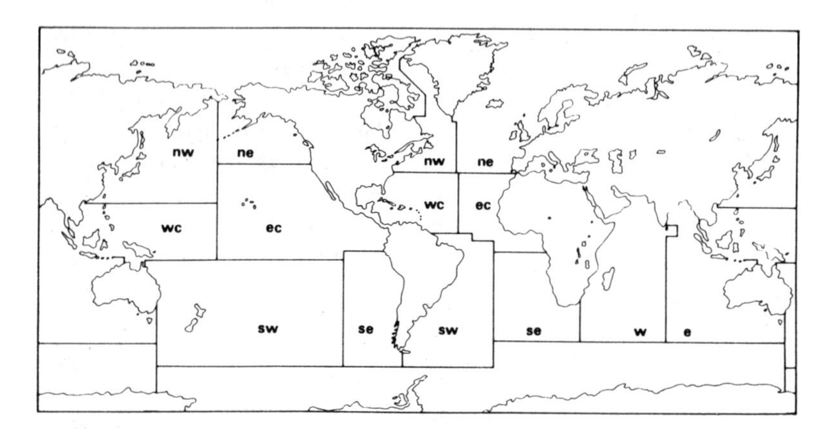

the redfish, several species of flatfish and herring, menhaden, capelin and the mackerel. There are also important stocks of shellfish such as lobsters, oysters and scallops. In the 1960s this area bore the brunt of the development of European distant-water fleets. The fishing pressure that developed greatly reduced the stocks of cod, haddock and herring so that at present they are considered to be depleted. Most other stocks are fully exploited and need careful management. Between 1970 and 1973 the total annual catch from the area was about four million tons. Since that date low total allowable catches (TACs) have been set, first in the International Commission for the North-west Atlantic Fisheries (ICNAF) and now by Canada and the USA who have had sole jurisdiction since 1977. The potential yield for the area is considered to be about the four million tons that was caught in the early-seventies. This can only be maintained if the stocks are carefully managed allowing the depleted ones to rebuild.

The north-east Atlantic, particularly the North Sea and continental shelf to the west of Britain, is one of the most productive sea areas in the world. In 1976 the total catch was more than 13 million tons. As on the western side of the Atlantic the main species come from the gadoids, clupeids and pleuronectids. Cod, haddock, saithe, whiting, Norway pout, blue whiting, lings and tusk are all significant contributors to the total. The stocks of the most important species such as cod and haddock are fully exploited. Blue whiting is a species that has recently received much attention as a stock that had been little exploited and is thought to be capable of yielding one million tons a year. The herring stocks of this area have supported fisheries for several centuries and have significantly contributed to economic growth and culture in the eastern United Kingdom, Denmark, Sweden and Norway, but the stocks are now so depleted that a total ban on heavy fishing has been brought into force. This has greatly increased the pressures on the mackerel stock in the English Channel to such an extent that it too is in danger of becoming depleted. The area was managed until recently by the North-east Atlantic Fisheries Commission (NEAFC), but since the declaration of 200-mile EEZs in all the countries concerned, management has fallen into disarray and will not recover until the EEC countries can agree on a Common Fisheries Policy.

The western-central Atlantic yielded 1.4 million tons of fish and shellfish in 1976. It is estimated that the yield could be raised considerably even though the present declared catch is only a lower limit. Many fish caught by sportsmen and as by-catch in the shrimp fishery are never declared. The species composition in the area is extremely diverse (see Chapter 1) which is a problem and much of the potential catch exists in remote areas. Increased catch could come from the pelagic fish such as the King and Spanish mackerels and various species in the carangid family. Demersal stocks of snappers, groupers, drums and croakers, porgies and grunts could all yield more, although in all groups the scattered distribution and poor knowledge of their biology makes it hard to estimate potential and realisable yields. At present most of the catch in the area is by the USA and other bordering countries. The area has been managed by the Western Central Atlantic Fisheries Commission.

The eastern-central Atlantic has a very different pattern of fishing although the structure of the fish community has strong similarities with that of the community in the west (see Chapter 1). In 1976 37 countries fished in the eastern-central Atlantic, the biggest catch being taken by the USSR followed by Spain. Although many of the bordering nations have fisheries in the area, they are small and simple and only take one-third of the catch. One of the consequences of this is a poor collection of statistics. As a result there are insufficient data for the Eastern-central Atlantic Fisheries Commission to make proper stock analyses. The total catch in 1976 was probably about 3.5 million tons and was mainly composed of pelagic species such as sardines, mackerels and various carangids. Some hake and sea breams are also caught and it is thought likely that the stocks of these species are over-exploited.

The catch in the Meditterranean and Black Sea in 1976 was 1.3 million tons. This level of production has been maintained for several years despite fluctuations in pelagic stocks like those of the anchovy. Most of the stocks are heavily exploited and increased pressure on the valuable demersal species has aggravated the problem.

The southern part of the Atlantic (excluding the Antarctic portion) is divided into two regions, each with similar fish faunas but with very different fisheries. The south-west Atlantic has been an underdeveloped area for many years and still contains stocks of hake, blue whiting, herring and anchovy which are only lightly or moderately exploited. In 1976 the total catch was 1.2 million tons, the most important single species being hake, although most of the catch was made up of mixtures of demersal and pelagic species. The region suffers from poor statistical reporting and also has a great variety of habitats ranging from nearly Antarctic conditions off Tierra del Fuego to the tropics in the north. Despite considerable growth in the fisheries over the past few years there is still room for expansion. The area has been managed by the South-west Atlantic Fisheries Commission. The contrast with the south-eastern part of the Atlantic could not be greater. Under the leadership of South Africa, with its advanced economy the area has been yielding 2.5–2.9 million tons of fish since the mid-sixties. The catch is composed largely of hake, kingclip, various carangids including the maasbanker or horse mackerel, mackerels, sardinella, pilchard and anchovy. The pilchard catch, together with that of anchovy and horse mackerel, has been the basis of a million ton per year canning and fish-meal industry in Namibia (Culley, 1971). However, the catch in 1978 was only 130,000 tons and only one out of the seven processing plants was still operating (Gulland, 1978b). The abundant fish production (especially from the hake stocks), together with the European culture of South Africa, has attracted many foreign fleets so that until recently the USSR, Spain and Japan featured significantly in the fishery. As a result of the heavy fishing many of the stocks are fully exploited or over-fished and there is little room for expansion. The area has been managed by the International Council for the South-east Atlantic Fisheries (ICSEAF), although the influence of this organisation has decreased now that South Africa has declared a 200-mile EEZ.

Like many of the sea areas that border developing countries, the

Indian Ocean has a potential for increased yield. Most of the bordering nations have small-scale fisheries which are labour-intensive and produce small quantities per man. The Indian Ocean is divided into east and west sections for statistical purposes. The Indian industry dominates both regions although in the east Burma, Indonesia and Bangladesh also play a part. In the last few years the Arab countries have begun to be more active in fisheries. The various fisheries are poorly documented but pelagic species such as the Indian oil sardine, Indian mackerel, tunas, and the demersal Bombay duck are important enough to have something known about their biology. Although most stocks are lightly to moderately exploited, large tunas are probably fully exploited. The areas as shown in Figure 2.3 extend far south, but most of the catch comes from a narrow strip bordering the land masses.

Like the Atlantic, the Pacific Ocean is heavily exploited in the north, but less so in the middle and south. This is a reflection of the distribution of the human population over the area and the productivity of the different parts of the ocean. In 1976 17.2 million tons of fish were caught from the north-west Pacific and during the last decade the catch from this area has made up about one quarter of the world fish catch. The principal fishing nation is Japan, with China, the USSR and North and South Korea all taking significant quantities. As in the north Atlantic, a number of different species are taken. Salmon caught at sea have always been important but many of the stocks are depleted. The biggest single species catch comes from the Alaskan pollock, which currently yields about 3.9 million tons. This species is used by the Japanese to produce Kamaboko, a type of fish sausage based on a minced raw material called surimi. Large catches also come from pelagic species such as saury, herring, jack mackerel, sardine, anchovy and chub mackerel. Most of these stocks are fully exploited; the herring stock around the north of Japan, like its North Sea counterpart, has nearly been fished to extinction. Most of the management in the area is carried out under bilateral agreement between the countries which exploit certain stocks.

The management of the north-east Pacific is now almost completely controlled by Canada and the United States. It is in this area that some of the first experiments in management were carried out on the Pacific halibut stock. The species, which was seriously overfished in the 1930s, was allowed to recover and now yields a small but valuable catch. The Pacific salmon species have also tested the ingenuity of North American fishery scientists, who have been as productive with management models for the stocks as the stocks have been with fish flesh. The halibut and the salmon have even had their own fisheries commissions to look after them, the International Pacific Halibut Commission and the International Pacific Salmon Fisheries Commission. The fisheries in the area are also managed by the International North Pacific Fisheries Commission (INPFC). The area yielded 2.4 million tons in 1976. This was composed of halibut, salmon, flatfish species, Pacific hake, sable fish, Pacific cod, ocean perch, tuna and herring. All these stocks are fully exploited with the ocean perch stocks being depleted. In the past, Japan and the USSR have been the major fishing nations in the area. Although 200-mile EEZs were declared in 1977, the pattern has not changed appreciably as Canada and the USA are not at present taking up their option of exclusive exploitation.

The western-central Pacific fisheries region includes most of the myriad Pacific islands and the chain of islands between Mayalsia and Australia. The area is productive with a catch of 5.4 million tons in 1976, made up of scads, Indian mackerel, skipjack tuna, ponyfish and milkfish. The main fishery is carried out by Thailand with the Philippines, Indonesia, Vietnam and Malaysia also making important contributions. Thailand has developed a modern fleet fishing in the Gulf of Thailand which is so efficent that some of the demersal stocks are showing signs of depletion. Elsewhere, especially in the deeper parts of the South China Sea, there are probably demersal stocks that could give higher yields.

The broad areas of continental shelf in the south-east Asian area are not mirrored on the eastern side of the central Pacific. The narrow shelf down the western margins of the Americas causes relatively poor fish production, especially of the more commercially valuable demersal stocks. The area still gives a good yield, being 1.4 million tons in 1976, although this was mainly composed of pelagic species such as tunas, Californian sardine, northern anchovy, central Pacific anchoveta, and thread herring. These fish were caught by fishermen from the USA, Mexico, Ecuador and Panama. It is considered that exploitation of demersal grunts could be expanded in this region.

The south-west Pacific is mostly open ocean, with only a small part of south-east Australia and New Zealand on it. The 1976 catch was 380,000 tons, consisting mainly of tuna, various demersal species, southern blue whiting, grenadier and small pelagic species. Local vessels only took about one-third of the total catch in 1976, the rest going to Japan and the USSR. At present New Zealand is working hard to develop its fisheries and has received much help in stock evaluation from the Japanese. There are some valuable pelagic and demersal species around New Zealand and most of the presently fished stocks are only partially exploited. Future exploitation by Japan and the USSR is likely to be limited now that New Zealand has declared a 200-mile EEZ.

The south-eastern Pacific is a complete contrast although the northern and southern sectors in the area have very different characters. The region has in the past been dominated by the anchovy fishery off Peru. Since the collapse of the fishery in 1971 Peru, with the assistance of the FAO, has tried to diversify the species used and catches of hake doubled between 1965 and 1973 but have since declined again. The hake stock is now moderately-to-fully exploited. Other species exploited are menhaden, bonito, jack mackerel, Spanish mackerel, sardine, the Chilean pilchard and the Chilean hake, and most stocks are now fully exploited. In 1976 the total catch for the area was 5.6 million tons. To the south of Valparaiso in Chile, from the Isles of Chiloe and beyond, lies a region where the land is sparsely inhabited and the seas barely explored. In the 1950s projects sponsored by the FAO investigated the demersal resources in the area. Recent trawling trials have confirmed that there are some commercially exploitable stocks such as the merluza de cola or grenadier. However, the area is a poor prospect for constant fishing because the bottom is too rocky in many places for trawling and the weather is often too violent for safety.

The Southern Ocean, which includes all sea areas bordering the Antarctic continent, has been of increasing interest to the fishing industry over the past ten years. Russian ships have been sporadically exploiting demersal stocks around South Georgia and Kerguelen. Species like the ice-fish and Antarctic cod have formed the basis of these catches. The productivity of fish on the shelves around these sub-Antarctic islands is low (Everson, 1977) and probably cannot support heavy fishing pressures. The Russian tactic is to use a type of fishing analogous to slash-and-burn agriculture in tropical rain forests – move into an area, fish it clean, and then leave it to recover for a number of years. The Antarctic resource that has really stimulated excitement is the krill *Euphausia superba*, a crustacean belonging to the Euphausiacea which was formerly the food of the huge blue and fin whale populations, now almost destroyed. Adult krill are planktonic and grow to about 6 cm. For the past ten years the Russians, the Japanese, and now the West Germans, Poles, Argentinians and others, have been carrying out trial fishing on the species. Krill can be detected relatively easily by sonar and exist in large sub-surface swarms that are convenient to capture (Eddie, 1977). There are logistic problems created by fishing in such a remote and boisterous sea area and the krill is not suitable for use in existing products unless it can be peeled whole. This is difficult to achieve because of the small size of the animal; it is also necessary to cook the krill soon after capture if they are to remain edible for any length of time. The prospect of a yield up to ten million tons a year has stimulated many to try to use the species in new products such as krill paste and krill sticks. The exploitation of the species may probably cause some problems because of its central place in the Antarctic food web. The problems of exploiting krill and the fish, squid, seals and whales which feed on it have recently been discussed by May et al. (1979).

The inland fisheries of the world produced about ten million tons in 1976 about 60 per cent of which came from aquaculture. Over the past decade the freshwater capture fisheries have shown a general decline as a result of competition from other users of waterways, such as navigation, dam building and water abstraction for human use. Increasing freshwater pollution and eutrophication is also an important factor. Statistics for inland fisheries are unreliable and sparse so that it is hard to assess their state accurately. The Asian region (see Figure 2.3) produces 70 per cent of all freshwater fish (7.4 million tons) followed by Africa (1.5 million tons), USSR (0.8 million tons), Europe (0.3 million tons), South America (0.2 million tons), North and Central America (0.1 million tons).

Aquaculture is growing in importance, the FAO estimating that there is a potential production of 30 million tons. We will return to the subject in Chapter 10. In the developed countries, freshwater fish are often more important as objects of sport than as the subjects of commercial fisheries. The initial value of the catch is mostly quite low but the industries supported by angling, such as tackle manufacture, make the sport of much greater commercial significance than would at first appear. In some cases, for example the Atlantic salmon, it costs the angler up to ten times the value of the fish as food to hook it out of some sought-after stretch of river. In Europe and the United States there is also much public money spent on managing rivers and reservoirs for the benefit of the sport fisherman.

Future Supplies of Fish

It has been estimated by the FAO that by the year 2000 there could be a requirement for 110 million tons a year of fish, a 57 per cent increase over current supplies. In this section we briefly consider ways in which potential global fish yields can be estimated. Gulland (1970) suggests that there are four routes to an estimate of potential yield and these are: (1) extrapolation from present trends in catch; (2) extrapolation to the world seas of estimates of fish yields per unit area in well-studied regions; (3) estimation of production at successive trophic levels starting with measured estimates of primary production; and (4) review of the state of all known and likely fish stocks and estimation of their potential sustainable yields, which are then summed.

As one might expect, extrapolations from present catches are not reliable. In the 1950s and 1960s, when world fisheries were growing at a steady rate of 7 per cent a year, it was possible to predict successfully world catch for two to three years ahead. Even then, this did not apply to individual fisheries which were all growing at very different rates. For example the Peruvian anchovy fishery expanded at up to 174 per cent a year in the 1950s. Using extrapolation nowadays would be very unreliable, as a steady trend no longer exists, catches since 1970 having fluctuated in an unpredictable fashion.

Predicting from a real estimate of fish yield is theoretically a sounder option, but its accuracy must depend on knowledge of the distribution of fisheries yields. For example Graham and Edwards (1962) used a figure of 22 kg ha^{-1} which was based on the average yields from fisheries in the north Atlantic in the late 1950s. This value was multiplied by the area of the oceans to give an estimated maximum potential catch of 55 million tons, slightly larger than the then world catch of 45 million tons. Subsequent developments in the north Atlantic fisheries showed that Graham and Edwards' figure was 3-4 times below the true potential of the area, thereby revising the world fisheries potential to 160-220 million tons. This assumes that all other fishing areas have the same level of productivity, which is not necessarily true.

Making estimates of potential by working up the food chain is an attractive approach, because it would allow one to include all potential resources without having to explore in detail all fishing areas. Extrapolation from present catch or from area estimates of fish production can never account for unexploited and unknown stocks. A good example of an estimate based on primary production rates and food chain interactions is by Ryther (1969) – see Table 2.1. The ocean was

Table 2.1: The Productivity of the World's Oceans. For a detailed explanation see text.

Variable	Open Sea	Coastal Zone	Upwelling Areas	Total
Percentage of ocean	90	9.9	0.1	100
Area ($km^2 \times 10^6$)	326	36	0.36	362
Mean primary prod. ($g\ m^{-2}\ yr^{-1}$)	50	100	300	–
Total primary prod. ($kg \times 10^9\ yr^{-1}$)	16.3	3.6	0.1	20.1
Trophic levels	5	3	1.5	–
Mean efficiency (%)	10	15	20	–
Fish production ($kg \times 10^9$ fresh wt.)	1.6	120	120	242

Source: Ryther (1969).

divided into three areas depending on level of primary production. Ninety per cent of the area is open ocean with the lowest mean rate of primary production. The coastal zones, where most of the worlds fisheries are sited, have twice the productivity but only 10 per cent of the area. The very productive upwelling zones are small and are also sporadic in occurrence. Ryther made assumptions about the number of levels in the food chains of each zone and used the findings from research on trophic ecology to estimate the degree of efficiency with which food is transferred from one level to another (see Table 2.1). His final estimate of potential production was 242 million tons of fish a year, which would include not only presently exploited species but also many such as the myctophids, which are hard to catch and difficult to use. From this potential level of production one might expect about half to be available to fishing fleets, giving 121 million tons a year. This estimate and the means by which it was calculated was criticised by Alverson *et al.* (1970), who consider that this type of food chain estimate depends on three important assumptions. First, one assumes that the complex dynamic food web can be reduced to a simple chain of fixed length; even for fish such as the herring this is not the case, as is shown in Figure 1.7. Secondly, Ryther assumed that a single value of ecological efficiency could be applied to all levels in the food chain. This is rarely the case (Steele, 1974; and see Figure 1.6). The third assumption is that one can accurately assess how much of the total fish production is available to man on a sustained basis. As explained in Chapters 7 and 8 this is not at all easy. Alverson *et al.* (1970) pointed out that Ryther's method of estimating fish production is very sensitive to these assumptions and that the estimate could be wrong by as much as two orders of magnitude!

Adding the estimated potential catch for all known fish stocks gives a projected catch of 100 million tons of conventional species. If certain squid and lantern fish (myctophid) stocks are added then the potential would be raised to 200 million tons. The estimate of 100 million tons assumes that all stocks are at their optimum level of abundance and, as explained in Chapter 11, it is extremely unlikely that a multispecies fishery can be managed in such a way that all stocks are yielding optimum quantities. Gulland (1970) assumed that only 80 per cent of the potential yield can be realised, which means that we are already very near to catching all the traditional species we can. Management of these existing stocks by developed nations is as yet far from rational. The development of fisheries for squid and other unusual species will depend more on economic incentives and the willingness of people to try new foods than on stock abundances.

Fishing Methods and Fishing Fleets

There are probably as many different fishing methods as there are species of fish caught by man, and so we can describe only the main kinds, together with an account of the composition of fishing fleets in the developed countries. Despite the wide variety of methods there are only three main techniques: (1) hooking individual fish, as used by anglers and longliners; (2) tangling fish in netting, as found in gill nets, trammel nets and drift nets; and (3) actively catching fish in a net such as a seine, a trawl or a trap. The increased power of large fishing ships has made trawling and seining by far the most efficient method, so we will concentrate on these after a brief description of hook-and-like fishing.

Cod, haddock, redfish, ling and torsk which are often caught in trawls or seines, can also be captured by hook and line. Some species of tuna are nearly always caught on hook and line. The method is centuries old but has lost favour during the last fifty years because it is labour-intensive and more fish can be caught in trawls. The ships required for hook and line fisheries can be small, which makes the method popular with inshore fishermen. Bare hooks are used to catch species such as mackerel which will strike at anything. This type of hook-and-line fishing is called jigging. However, most hook-and-line equipment uses bait or some form of lure. Baited hooks can be used most simply on hand lines, but are now more often mounted in groups on lines that are either anchored to the bottom if benthic fish are to be caught or suspended from buoys to catch pelagic species. Skipjack and yellowfin tuna are caught by lines suspended over the side of the ship on the ends of stout bamboo poles. Some attempt is now being made to machanise hook-and-line fishing.

The hook and line is, of course, the mainstay of sport fishing. The basic equipment of rod, line and reel is the same for all species caught, but the variations on this theme are numerous. Likewise, there is tremendous variation in the types of bait and lure used, depending on the species to be caught, the nature of the habitat and the whim of the angler.

The mechanised seine net is made in two main forms, the Danish seine and the purse seine, both derived from ancient draft and beach nets which were used from the shore to encircle fish. Both types are set as the ship steams in a circle starting from a marker buoy and paying out the net over the stern of the vessel. The Danish seine, designed by Jens Vaever in 1848, is used on the bottom like a trawl. The net and attached ropes rest on the sea bottom immediately after laying. As the ropes are hauled in they frighten fish inwards where they are eventually caught by the advancing net which is held open vertically by floats on the headline and lead weights on the foot-line. The purse seine is used mostly for capturing schooling pelagic fish, such as anchovy, mackerel, tuna, capelin and herring. Once the school has been detected on sonar it is surrounded by the net which is designed to drop very rapidly in the water off the ship. It then hangs in the water (Figure 2.4b) supported at the surface by floats and pulled down at the bottom by weights. These nets can be up to 1.5 km long and over 100 m deep and as a result are very costly (around £40,000). Once the school is encircled, a line running through the eyes on the net bottom is hauled in, closing the bottom of the net and creating a bag rather like an upside-down umbrella. The net is then gradually taken on board using a powered pulley (power block), so reducing the size of the bag and concentrating the fish. When they are sufficiently concentrated the fish are either lifted out with a scoop or pumped on board to be stored in refrigerated holds or sometimes frozen. Purse seine ships do not need to be as large as ships whose engines have to tow a trawl and many of the American tuna purse seiners look more like luxury cruisers than fishing boats (Figure 2.4a).

The trawl is primarily an instrument for pulling a sock-like net along the sea bed, the weighted foot-line being designed to keep the net on the bottom. When encountered, fish at first keep station with the front

Figure 2.4: A. A 190 foot (57 m) Tuna Seiner as Used by American Tuna Fishermen Fishing in the Pacific. Note the helicopter used for searching for tuna schools. B. Drawing to Show the Way in which a Purse Seine Net is Set around a Fish School. See text for further details.

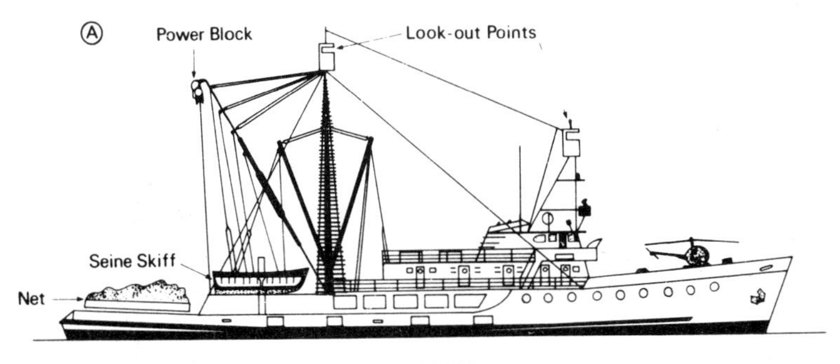

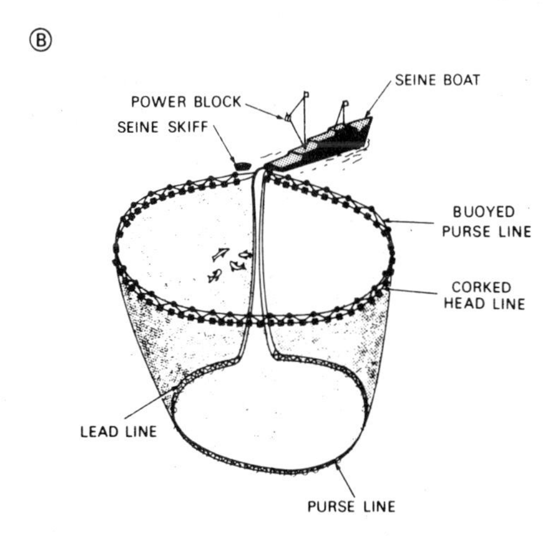

Source: Figure A redrawn from the Encyclopedia of Marine Resources (1969), with permission.

of the trawl and its ropes. The towing speed must be fast enough to bring the white muscle of the fish into use, so that they soon fall back exhausted into the sock. For this reason the development of the steam-powered ship greatly enhanced the efficiency of the method.

For the first part of the twentieth century all ships using the trawl

Figure 2.5: Top and Side Views of a Side Trawler to Show the Arrangement for Winches, Trawl Cables and Other Features. See text for further details.

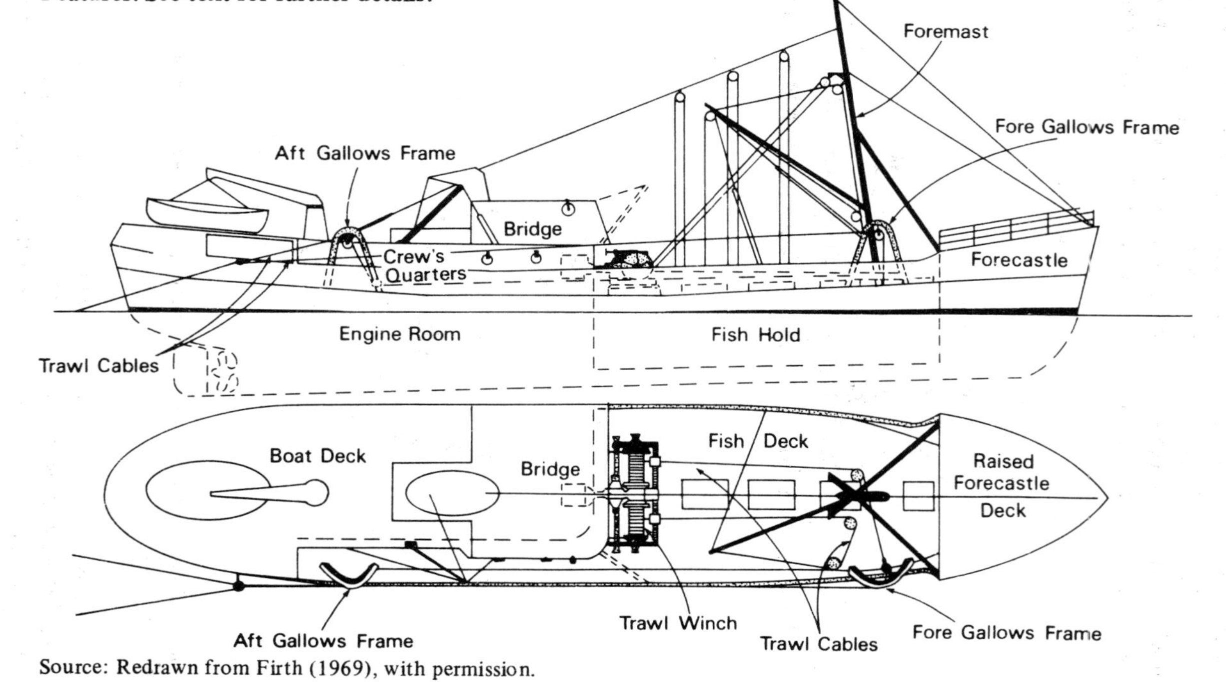

Source: Redrawn from Firth (1969), with permission.

Figure 2.6: Diagram of a 250 foot (75 m) Stern Trawler with Views of the Stern Section from Above and Astern to Show the Principal Features. The two dotted lines at the stern on the overhead view represent the trawl warps.

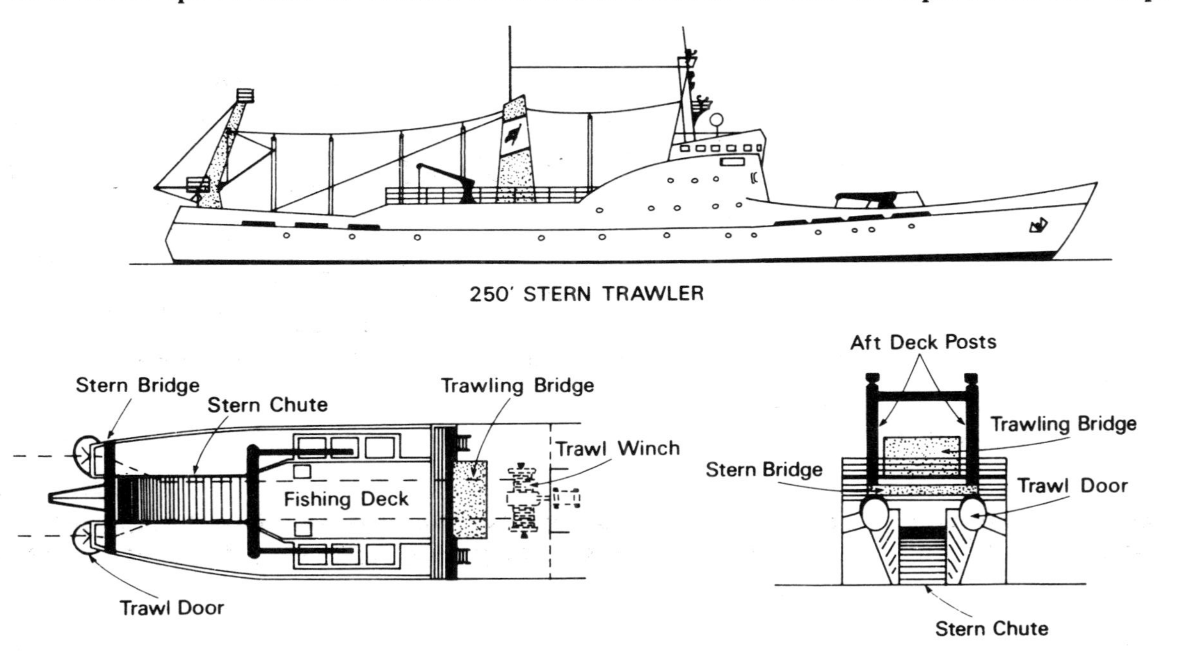

Source: Redrawn from Firth (1969), with permission.

Figure 2.7: Parts of a Trawl When being Towed along the Bottom. The gear shown is of the Vigneron Dahl type, the otter boards being separated from the net by a length of warp (wire), so increasing the amount by which the mouth of the net is spread open.

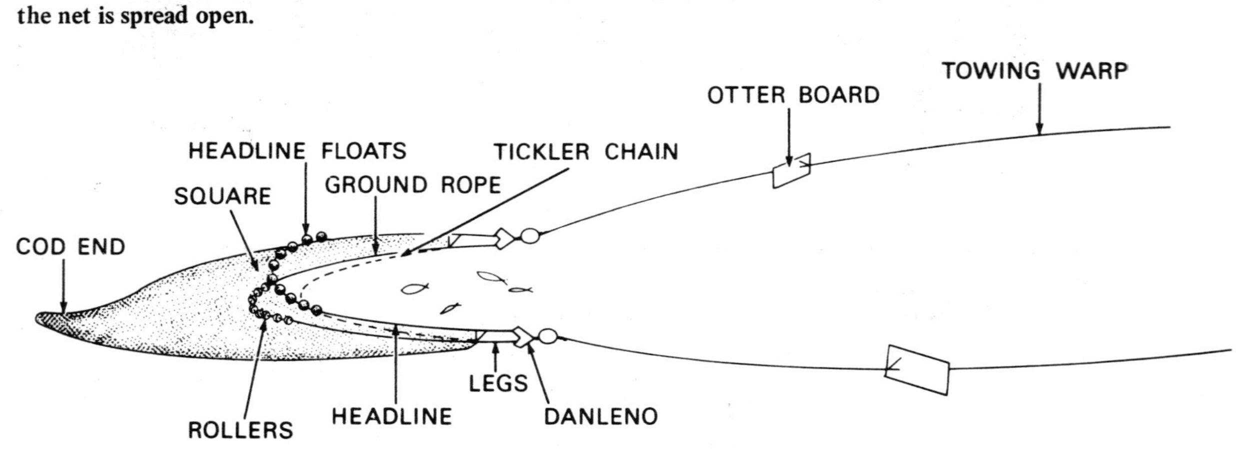

launched and hauled the net over the side of the ship and such a side trawler (Figure 2.5) usually has A-shaped frames fore and aft to take the warps (ropes) and trawl doors (see below) and an appropriate winch just forward of the wheelhouse to haul the net in. The warps pass from the winch through a system of pulleys and fair leads (guides) to the net. All side trawlers must lie broadside to the sea when hauling the net in, which sets a limit to the type of weather in which they can safely work. The development of the stern trawler (Figure 2.6) has improved this aspect of a trawler's performance. The net and associated gear is launched down a ramp at the stern of the ship and is hauled back the same way. During both operations the ship keeps head on into the sea. The otter trawl, designed for commercial use in 1894 by a Mr Scott of Granton, Scotland (Graham, 1956), is now the chief type of net used. At the end of the last century it replaced the beam trawl which had its entrance held open by a stout beam about 50 feet (16.5 m) long. The beam trawl is awkward to handle but is still used by Dutch fishermen to catch sole in the southern North Sea and in many prawn fisheries throughout the world. The innovation that led to the success of the otter trawl was the removal of the beam and the attachment of the otter boards or trawl doors. These are two large wooden or steel rectangles typically about 3 x 1.5 m in size and weighing about one metric ton. One of these is attached on each end of the trawl arm (see Figure 2.7) and the inside of the door is attached to the towing wire just forward of its centre. As a result of this eccentricity the doors when pulled forwards are pushed outwards by the water pressure, so spreading the mouth of the net. A modification that leads to a wider spread is a piece of wire attached between the trawl doors and the net. Such trawls are of the Vigneron Dahl type and were introduced in the early 1920s. The footline (Figure 2.7) of the net is often equipped with rubber or wooden rollers called bobbins to allow the net to pass easily over a rough bottom. The headline is kept up by means of floats. More recently the bottom trawl has been adapted for use in mid-water to catch pelagic shoaling species of fish or organisms such as the Antarctic krill. Some of these nets can have staggering proportions being up to 100 m long with a mouth opening of 400–500 m^2. These large nets have trawl doors that weigh about three tons each. As the shaft power required in a vessel to pull a large net through the water is proportional to the mouth area, very large vessels of the order of 2,000–3,000 tons are needed (Eddie, 1977). These large ships can develop 2,500–3,000 shaft horse power. Mid-water trawls can also be towed by two ships. Mid-water trawls are remarkably precise instruments

Table 2.2: The Composition of a Selection of Fleets from OECD Countries.

Size of Vessel (grt)	Denmark 1977	Canada 1977	Country Japan 1977	Norway 1977	UK 1978
With engines					
4			341,068		
5–9	6,752	26,220	13,267	27,700	6,367
10–29		5,245	9,564		
30–49			1,842		
50–99	230	408	4,058	270	331
100–199			1,350	544	283
200–499			1,664		
500–999	3	352	77	67	40
Over 1,000	0		176	5	30
Vessels without engines	2,943	4,060	35,785		14
Total vessels	10,274	32,225	408,851	24,847	7,067
Total catch	1,782,927	1,207,895	5,384,000	3,217,504	916,356
Catch per vessel (tons)	174	37	13	129	130

Source: OECD (1979).

when used together with advanced electronic fish-finding gear. Using sonar the skipper is able to monitor the position of the fish shoal and the depth of the net which can be altered accordingly. The design of trawls has very recently become an exact science and engineering formulae are now available for predicting the speed and required power for a given mesh size and type of net filament.

Each country or geographical region has its own characteristic type of fishing vessel. This regional variation has been reduced in modern times, but the nature of each country's fishing industry is still reflected in the particular structure of each national fleet. A selection of data from member countries of the Organisation for Economic Cooperation and Development (OECD) is shown in Table 2.2. In all the countries shown, and probably throughout the world, the majority of a national fleet is made up of small vessels. The important differences lie at the other end of the scale where there is greater or lesser emphasis on large ships. Japan for example has 176 vessels greater than 1,000 tons. The last row of the table shows the mean catch per vessel and this reflects differences in the structure of the industries and fleets. Denmark has a high value because the emphasis in the industry is on catching fish for reduction to meal. This means large catches of small fish which are bulk-stored in the ship's hold. The dominance of small vessels is indicated by the mean catch per vessel for Canada and Japan. The table is misleading in this instance. The OECD statistics show that in Canada 352 vessels are greater than 100 tons but the largest category is 150 tons and over, and it is unlikely that there will be many really large vessels as would be found in the Japanese fleet (Macdonald, 1979). Japan has many small ships but also an extensive fleet of distant-water trawlers that fish the world's oceans. Norway and Britain show close similarities, although Norway has a greater number of small vessels and Britain until recently has had greater emphasis on distant-water fishing. This pattern is fast changing as a result of the 200-mile EEZs.

The Storage and Handling Properties of Fish

Fish flesh is a good quality food for humans. The chemical composition of a selection of species is compared with the composition of other common sources of protein in Table 2.3a. Fish protein contains all the ten amino acids essential to humans and is particularly rich in the amino acids lysine and methionine which are generally lacking in cereal protein (Table 2.3b). The fat content of fish is highly variable depending on the species, season and state of maturity. In fatty fish such as the

Table 2.3a: Comparison of Proximate Composition, Energy, Mineral and Vitamin Content in Some Fish Species with Other Foodstuffs.

Species	Proximate Composition			Energy MJ/kg	Minerals μg/g			Vitamins				
	Protein	Fat	Carbo-hydrate		Ca	P	Fe	Vit. A	Thia-mine	Ribo-flavin	Niacin	Vit. C
	%	%	%					IU	μg/g	μg/g	μg/g	μg/g
Salmon	23	13	0	8.9	79	186	1	310	0.1	0.2	7.2	9
Cod	18	1	0	3.3	10	194	1	0	0.1	0.1	2.2	2
Herring	17	11	0	7.4	4	256	1	110	0.1	0.1	3.6	–
Mackerel	19	12	0	8.0	5	239	1	450	0.1	0.3	8.2	–
Beef	16	34	0	15.9	9	142	2	70	0.1	0.1	3.7	0
Pork	17	25	0	12.5	10	193	3	0	0.8	0.2	4.4	0
Egg	13	12	2	6.8	54	–	2	1,180	0.3	0.3	0.1	0
Cheese	17	25	2	12.5	105	–	1	1,010	0.1	0.7	0.8	0

Table 2.3b: Essential Amino Acids in Three Species of Fish Compared to Beef. Figures are per cent of total protein.

Amino Acids	Salmon	Herring	Mackerel	Beef
Argenine	5.8	7.1	5.8	5.3
Histidine	2.6	1.9	3.8	5.7
Isoleucine	4.9	6.2	5.2	4.7
Leucine	7.3	7.1	7.2	7.2
Lysine	8.0	8.3	8.1	8.3
Methionine	3.0	2.6	2.7	2.8
Phenylalanine	3.7	3.6	3.5	3.5
Threonine	4.4	4.1	4.9	4.5
Tryptophan	0.9	0.8	1.0	1.0
Valine	5.6	5.4	5.4	5.1

herring and mackerel the fat is not always distributed evenly. Fish flesh is also a good source of minerals which are present in an easily accessible form. Vitamins in fish are not always present in the parts that are eaten and the vitamin content can vary with the season. For example most of the vitamin A and D in cod occurs in the liver while in the eel they occur in the muscle. The chemical composition of fish is also a strong constraint on the ways in which fish can be stored and processed.

In the tropics large quantities of fish are wasted because they rot or are eaten by pests before they are sold and eaten. In temperate climates spoilage is not such an acute problem but it still occurs. Spoilage can be prevented by suitable methods of handling and preservation. Smoking, drying and salting have been practised since civilisation began, but it is only in the last 50 years that the problems have been taken up by science (Burgess, 1979). In the developed countries establishments such as the Torry Research Laboratory in Aberdeen, UK, the US National Marine Fisheries Service (various centres), and SIK – the Swedish Food Institute in Göteborg, Sweden – have developed ways to handle and preserve fish so that less is wasted and more used for direct human consumption. The large processing companies such as Findus in Europe, and Birdseye in Europe and the USA, have done much to improve the quality of frozen fish products. The developing countries are only just beginning to get the benefits of better handling and processing, although FAO has done much to try to increase the technical contacts between industries in developed countries with developing fisheries.

When a fish dies, metabolism is interrupted and the immediate effect is that waste products cannot be removed (see Tarr, 1969; Sikorski, 1979). If the fish is exhausted, the glycogen in the white muscle will be depleted. Immediately after death the sarcomeres are relaxed and the actin and myosin molecules in the muscles are unlinked. Fish cooked at this stage has a characteristic fresh taste, the texture depending on how much the muscles have been shortened in cooking. Beyond this point (a few minutes at normal temperatures, or about a day in ice), rigor mortis sets in as ADP and Ca^{++} accumulate. Rigor is a linking of the actin and myosin molecules to form actomyosin; when this happens some of the membranes (myocommata) between the sarcomeres rupture. Fish cooked at this stage is rigid and tough. However, the accumulated calcium ions then activate lysosomal enzymes, which resolve the rigor after an hour or so (about half a day in ice) by disrupting the muscle structure, and the muscle proteins break up, becoming partially hydrolysed. Fish cooked at this stage tastes fresh and tender; most fisheries would aim to market their

Figure 2.8: Diagram Showing Increase in Bacterial Numbers, Ammonia from Breakdown of Protein and Trimethylamine in Haddock Fillet Stored in Ice.

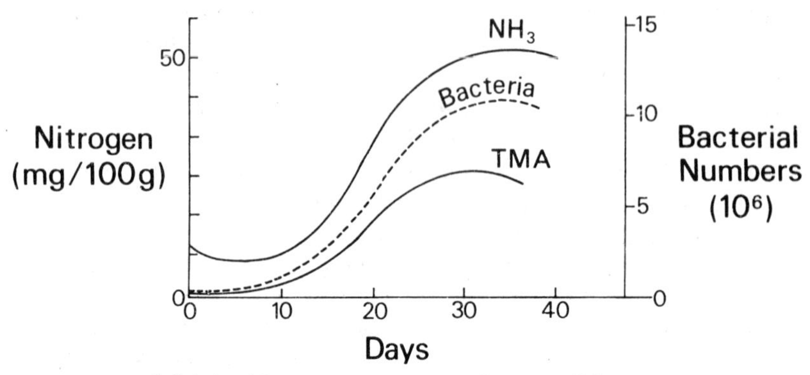

unprocessed fish in this state, or as near to it as possible.

As decomposition proceeds, both endogenous and bacterial enzymes break up the protein further, the pH rises, and more bacteria invade. Ammonia is released at first, especially from gadoid and elasmobranch fish, soon followed by the characteristically fishy-smelling methylamines, usually trimethylamine (TMA, see Figure 2.8). From this stage on the fish turns stale if on ice, or begins to decompose visibly at ambient temperatures. In the gadoid fishes, dimethylamine (DMA) and formaldehyde produced by enzymatic degradation of TMA lead to toughening because the proteins become insoluble (Castell *et al.*, 1973). This problem is accentuated by freezing on capture, followed by unfreezing to process the fish (Hiltz *et al.*, 1977). For this reason it has been found best to store freshly-caught gadoid fishes destined for processing on ice. Blood and red muscle tends to increase this problem of the build-up of DMA, but at least in fish we do not normally have to bleed red meat as in mammal flesh, although in some countries fish are laid out on boards to 'bleed'.

The key to good quality products is first-class raw materials. To obtain these it is necessary to have good knowledge of the handling and storage properties of fish backed up by good practice on board ship and in the factory. We will see in Chapter 4 that fish either store fat reserves in the muscle or in the liver, the location making a big difference to the storage properties of fish. Species like herring or mackerel with up to 25 per cent fat in the muscle do not keep well, either in fresh or frozen form, because the fats quickly oxidise and go rancid. These species are better preserved using smoke, salt or vinegar, or canned, and these are the most common methods used, giving products such as kippers, rollmops, Scandinavian-type pickled herrings, and canned pilchards and sardines. Species such as cod, haddock, Alaskan pollock and hake have between 0.5 and 5 per cent fat in the muscle and are ideal for freezing. They have formed the basis of the expansion in the frozen fish industries.

When fish is kept fresh from the moment it is caught to the time it reaches the auction, it is termed wet fish. Handling wet fish at sea and on land needs special care if quality is to be maintained. As soon as the fish are landed on deck they are gutted and thoroughly washed. This removes bacteria and digestive enzymes. The fish are then stored in the hold either on shelves or in boxes; in either method the fish are well mixed with ice. If white fish are gutted and washed immediately after capture and stored at 0°C with plenty of ice then they should keep fresh for five to six days (Horne, 1971). After 10–12 days they would be stale and would become uneatable after 16 days. The fish holds in most modern trawlers are usually insulated and have a temperature which is just above freezing. Fatty fish are usually smaller than white fish and are not often gutted at sea. Once iced they should be on land within one to two days.

Some of the larger trawlers have the capacity to process and freeze fish on board. The simplest system is for the fish to be gutted, perhaps by machine, and then frozen whole. The more sophisticated ships will have the full range of gutting, filleting and skinning machines so that frozen blocks of fillets can be produced. Investment is greater, but such ships can stay at sea longer because better use is made of hold space. Factory trawlers will also have a fish-meal plant to make profitable use of the offal. Freezing fish at sea can mean a higher quality product for the final consumer.

Once landed, most fish in Great Britain and in many other European countries are sold at auction. In Britain, the largest centres are Aberdeen, Hull and Grimsby. In Europe, important centres are Nantes, Boulogne, Esbjerg, Bergen, Bremerhaven and Hamburg. A proportion of fish is bought by wholesalers who have small establishments near the fish-market where the fish are hand-filleted, packed into cartons and loaded into insulated trucks for distribution to inland markets or to retailers. In Grimsby market quite a large proportion of the fish is bought by the large frozen food companies. Soon after purchase the fish are transported to the factory where they are size-graded and put through the filleting and skinning machines. The fillets are hand-trimmed and then packed into waxed cartons held rigid in a metal frame. The cartons are then loaded into a plate freezer and frozen. The blocks of fillet are then used as a raw material for fish fingers and other ready-made fish products. One of the key practices in such an operation is to keep a stock of frozen fish which can be used when supplies at the auction are low or expensive (the two often go together).

In the western USA, Canada and Alaska, the canning of salmon is an important industry, sockeye salmon being the most important species. The first canning plant was set up in 1864 by the Hume Brothers at Sacramento, California. As the industry grew it spread northward until it reached Alaska. The fish are mostly caught as they move inshore to migrate up river to spawn. They are captured by trolling, seining and gill nets and are usually sold directly to the processing factory.

Other than the salmonids, freshwater fish are not used on any large scale in the processing industry, but the fish are usually sold fresh or smoked for local consumption. Most commercial freshwater fisheries are on large lakes and depend to some extent on local tastes. For example, there are large fisheries for tilapia in the African lakes, for carp and bream in Eastern Europe, for coregonids in N. America, Ireland and the USSR, and for several cyprinid species in south-east Asia. Eels are fished extensively in rivers and lakes in temperate regions.

One problem in freshwater bream is a muddy flavour caused by the fish ingesting a blue-green alga, *Oscillatoria agardhi*. The alga produces geosmin in the fish flesh, and correlations have been noted between the timing of peaks of algal growth and a muddy taste appearing in the bream catch (Persson, 1979).

In most countries there is inadequate infrastructure to support the capital-intensive industry that now exists in developed nations. Chilling or freezing fish at sea or in the processing factory puts stringent demands on the distribution system that takes the product from the dock or factory to the consumer. In developing countries the emphasis is on devising processing methods and techniques which can be operated without expensive technology. Smoked, salted, dried and canned products are suited to such economies. Such products are already common in developing countries but the task for the future is to improve the quality and keeping properties.

In the next chapter we begin to examine the biological basis of this industry. The first task will be to look at how the structure of fish populations affects production. This will lead into the study of fish resources under exploitation.

3 THE STRUCTURE OF FISH POPULATIONS IN SPACE AND TIME

The Concept of a Unit Stock

It is important for the fishery biologist to know the limits of the fish population with which he or she is working. Within a species, growth differs from one population to another just as other characteristics do, and this affects yield. Recruitment of young fish also varies between different populations. Differences may be genetic, as discussed in Chapter 5, or the result of the differing environments, or usually a mixture of both. The cod stocks of the north Atlantic provide examples of our point. Each major stock is in the main isolated from every other (Harden-Jones, 1968) and has its own particular characteristics as shown in Table 3.1. Over the past 70 years much research has been done to find out how each stock maintains its integrity in what appears to us to be a featureless environment.

In fisheries science an exploited fish population is termed a 'stock'; as far as we can see this is synonymous with 'population' as precisely defined by other ecologists – that is 'a group with unimpeded gene flow' – so we use the standard fisheries term in this book. For a discussion of the practical problems of defining a population see Krebs (1979).

The Arcto-Norwegian cod stock is found within the area shown in Figure 3.1. The predominant currents originate with the north-easterly North Atlantic Drift which divides into the east-flowing North Cape current and the north-flowing West Spitzbergen current. Year after year cod spawn on the east side of the Lofoten Islands where there has been a line-and-net fishery for hundreds of years for the spawning cod, called skrei by the Norwegians. The fish spawn in February and March in middle waters at a depth of about 100 m, where there is a boundary between fresher light water at the surface and denser saltier water deeper down. Fish leave the spawning ground in April and individuals that have been tagged at this time have been recaptured in the Barents Sea, around Bear Island, and to the south-east of Spitzbergen (Cushing, 1968). These areas are the feeding grounds where the spent adults replenish their energy reserves by feeding through the summer on the rich food sources such as benthic crustacea and molluscs and the pelagic capelin. Cushing suggests that the distribution within

Table 3.1: Characteristics of the North Atlantic Cod Stocks.

Stock	Estimated Average Recruits (2-yr-olds) x 10^6	Growth Rate k^c	Ultimate Max Lt $L_\infty{}^c$ cm	Ultimate Max Wt $W_\infty{}^c$ kg	Mean Wt at 5 yrs Old	50% Mature Age (yrs)
North Sea, south	75	0.33	111	17	6.1	3
north/central	50	0.27	119	15	5.2	3
West Scotland	14[b]	0.24	108	14	6.6	3
Irish Sea/Bristol Channel	4	0.33	101	12	6.5	2/3
Faroe, Plateau/Bank	16	0.19	115	14	4.5	3/4
Iceland	300	0.16	118	15	3.4	7/8
Baltic	250	0.20	90	6	1.5	4
Arcto-Norwegian	1250	0.12	130	17	1.5	7/8
East and south-west Greenland	80	slow			1.3	8/9
West Greenland	180	0.20	100	8	1.9	7/8
Labrador	2000	0.15	81	5	0.8	7/8
East Newfoundland		0.11	114	12		
Flemish Cap	50	a	a	a	a	
Grand Bank	200	0.10	152	25	1.4	4/5
St Pierre Bank	80	0.10	137	18	1.1	4/5
West Newfoundland	a	0.05	143	20		4/5
South Gulf of St Lawrence	125	b	b	b	1.0	
Banquereau	75	0.14	105	11	1.4	4/5
Brown's/Lehave Bank	16	0.20	117	16	2.1	4/5
George's Bank	a	0.12	142	20	3.0	3/4

Notes: a. Not Known. b. Has varied widely 1967–77. c. See Chapter 4 for the meaning of these parameters.
Source: Garrod (1977).

Figure 3.1: A Map of the North of Norway and the Barents Sea to Show the Annual Migrations of the Arcto-Norwegian Cod Stock. The thin southward-pointing arrows show the late winter movements of the mature cod to the spawning grounds. The north-pointing arrows show the spring movements of the spent cod back to the summer feeding grounds. The major area of spawning is inside the Lofoten Islands which are marked.

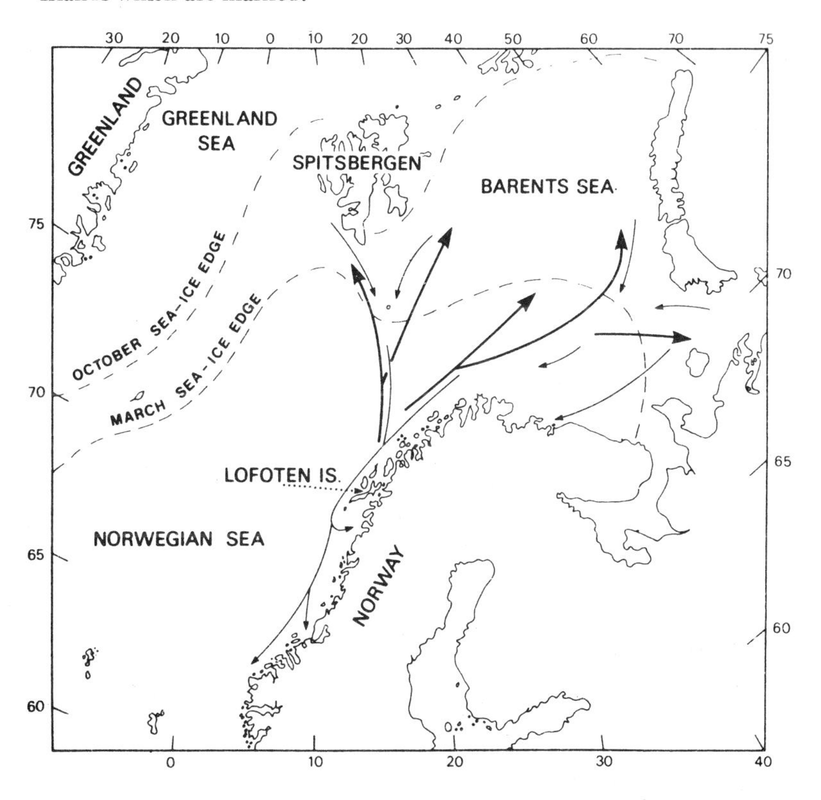

Source: From maps given by Harden-Jones (1969).

the Barents Sea is limited by the two degree isotherm; the cod stay in water warmer than 2°C.

Meanwhile the eggs drift north around the Norwegian Cape carried by the North Cape current. The eggs hatch as they go and the larvae grow fast on the rich Arctic zooplankton which increase in abundance as spring progresses. When the young fish reach the Barents Sea they are large enough to descend to the bottom where the juveniles live in

Figure 3.2: A Map of North Britain and the Norwegian Sea to Show the Annual Migrations of the Blue Whiting. The shaded area includes most of the spawning grounds, the south-pointing continuous arrows show the early spring movements whilst the north-pointing broken arrows show the early summer migration back to the summer feeding area in the Norwegian Sea.

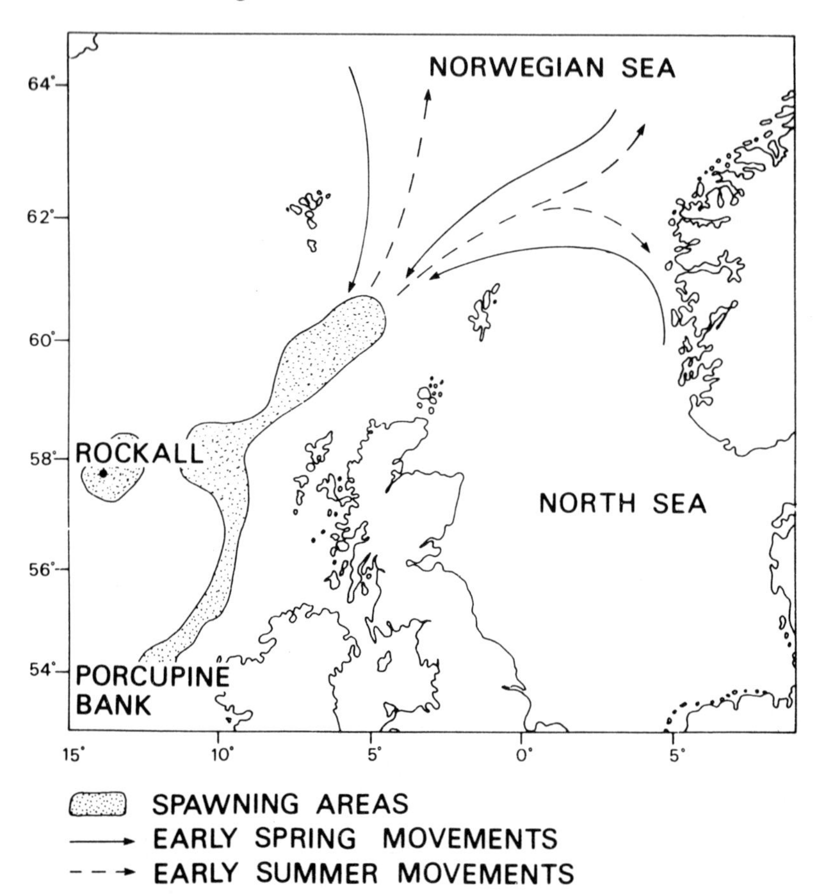

shallow inshore regions for 2–3 years. The young cod do not mature until they are 7–8 years old, but carry out a partial migration from about the age of three. By August the feeding adults are at their northernmost limits, reaching north to Spitzbergen and east to Goose Bank. In October the cod begin to move south and east, congregating south of Bear Island and off the North Cape. Soon after this the cod move south along the Norwegian coast, some moving into fjords and bays as the movement continues, but the largest part of the population continuing on to the Lofoten spawning area.

The blue whiting, a mid-water gadoid, is another species that illustrates an annual cycle of movements which maintains the integrity of the stock. Spawning occurs in March and April over Porcupine Bank (to the west of Ireland), Rockall bank and on the slopes of the continental shelf to the west of Scotland (see Figure 3.2). After spawning the adults move north and north-east, some moving between Iceland and the Faroes while others keep further south to go between the Faroes and Scotland. During the summer the adults disperse in the Norwegian Sea where they feed and recover condition, some penetrating as far south as the Norwegian Rinne on the north-eastern side of the North Sea. The eggs and then larvae are carried by the North Atlantic Drift around the top of Scotland into the North Sea and Norwegian Sea. The cycle is completed when adults migrate back to the spawning grounds in early spring.

Many other temperate species show similar patterns of behaviour, a further example being the plaice in the Southern Bight of the North Sea (see Chapter 5 for a description). Many of the examples that can be found show the same pattern of movements. Spawning occurs each year in the same place, usually up current from the feeding grounds of the adults. The prevailing current carries the planktonic eggs and larvae to nursery areas where the young fish feed and grow until they are large enough to join the main stock when they are said to have been 'recruited' to the fishery (see Chapter 6). This stage is not always as distinct as it is in plaice. The characteristic cycle of events ensures that the integrity of the stock is maintained, ensuring that members of the species remain in a favourable environment. This is probably also true of stocks which mix during the feeding stage of the cycle. Cushing (1967) demonstrates that the Downs, Dogger and Buchan stocks of autumn-spawning North Sea herring most probably separate out to spawn, so that each stock shows slight but consistent morphological differences and can be identified by its pattern of serum proteins.

The Abundance of Fish and its Estimation

The abundance of a commercially important fish population is of vital interest to the fisherman. With data on abundance and weight, the biologist can make estimates of production which provide the basis for developing management strategies (see Chapters 4, 7 and 8). Data in Table 1.3 show that in the North Sea only 12 species make up the bulk of the commercial catch and part of the reason for this is that the species are amongst the most abundant. Similarly, a few species of freshwater fish numerically dominate British rivers and lakes. The abundance of a fish species is not a fixed quantity and it varies from one place and one time to another, producing spatial and temporal patterns which the fisherman usually learns by experience (see Chapter 11).

Whatever the relative abundance of a fish species, it is a common fact that all populations show change in abundance through time. Fluctuations are controlled by a changing balance between death and birth rates and by the availability of resources. The simplest equation which we usefully can study is the logistic which describes the way in which the size of a population approaches an asymptote and fluctuates about it as the relationship changes between births and deaths. The equation is:

$$\frac{dN}{dt} = rN\left(1 - \frac{N}{N_{max}}\right) \tag{3.1}$$

where N is population abundance, r is the intrinsic rate of increase and N_{max} is the maximum abundance that can be supported by the environment, the carrying capacity. (In most ecological literature N_{max} is labelled as K; we have used N_{max} because K is frequently used elsewhere in fisheries.) The rate of increase, r, is a composite parameter reflecting the difference between birth rate b and death rate d, such that $r = b - d$ (note that these are instantaneous rates). The shape of the curve describing the growth of N, assuming that the logistic equation is valid, is shown in Figure 3.3. Starting from a low N, abundance increases slowly at first, then faster until the rate of increase drops away as N approaches N_{max}. The carrying capacity is the population abundance towards which the population converges whenever equilibrium is disturbed. The logistic model is unrealistic in several respects. No animal population would stay at equilibrium for all time, but would be more likely to wander about the equilibrium point, which would itself move as environmental change occurred. The model also assumes that as population density increases, birth and death rates adjust and have an instantaneous effect on the rate at which the population increases. The existence of finite development times ensures that this last assumption is never true, so that adjustment of r to density always lags behind the initial stimulus.

Despite these limitations the logistic model is still a useful aid to understanding population dynamics. MacArthur and Wilson

Figure 3.3: Rate of Change of Population Size and Growth of Population Size with Time as Described by the Logistic Model.

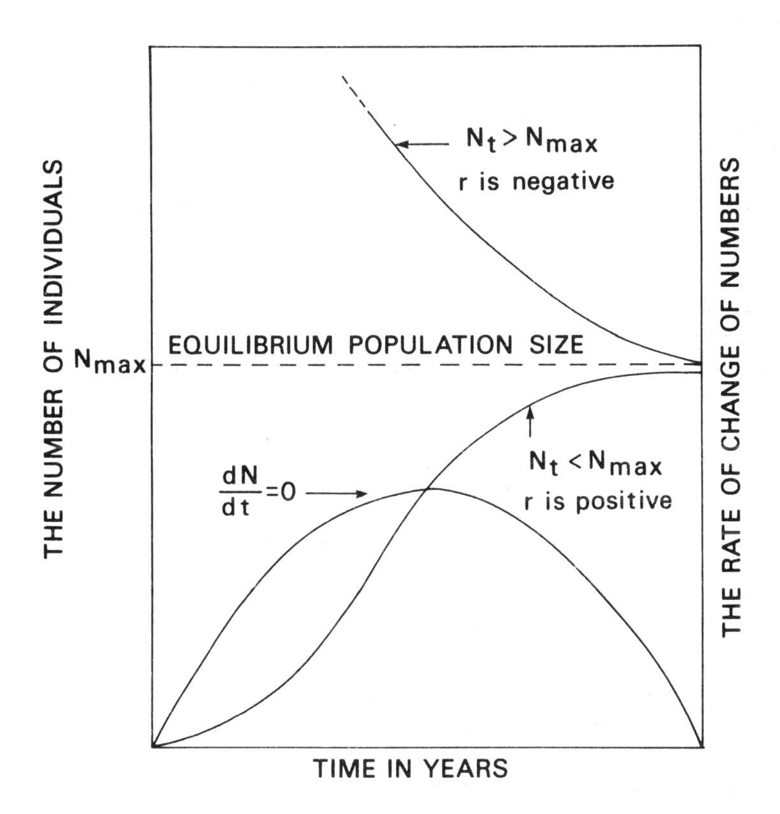

(1967) drew attention to the possibility of classifying organisms by the main process that controls their population; *r*-strategists are species that rely for their persistence on the ability to colonise new habitats and increase rapidly to make use of short-lived resources. The principal control factors are environmental and unpredictable. *K*-strategists ($K \equiv N_{max}$ in this book) live in stable environments where it is important for organisms to persist and to out-compete rivals by subtle behavioural means; in other words the main controlling factors are biological. Pianka (1970, 1978) extended this idea and drew up a list of characteristics which is associated with each type of strategy; this is reproduced in Table 3.2. Like many simple classifications, this one is an oversimplification and few animals are pure *r*- or *K*-strategists

Table 3.2: Some of the Characteristics That Are Found with *r* and *K* Selection.

Characteristic	*r* selection	*K* selection
Habitat	variable and/or unpredictable: uncertain	fairly constant and/or predictable: more certain
Niche	broad	narrow
Mortality	catastrophic, density-independent	density-dependent
Population size	variable in time; usually below carrying capacity; ecological vacuums; recolonisation each year	fairly constant in time; at or near carrying capacity; no recolonisation necessary
Intra- and inter-specific competition	variable	intense
Selection favours	1. rapid development 2. high r_m 3. early reproduction 4. small body size 5. semelparity	1. slower development 2. Low r_m 3. delayed reproduction 4. large body size 5. iteroparity 6. greater competitive ability
Longevity	short	long
Outcome	productivity	efficiency

Source: Adapted from Pianka (1970).

but lie somewhere in between. Life history strategies in space and time also have to take account of reproductive investment. Generally, those that lie near the *r*-strategy end of the spectrum have shorter generation times than those at the *K* end of the range. Few fish species are pure *r*-strategists even though some produce massive numbers of eggs and have huge potential rates of increase (see Chapter 5). Fish such as the anchoveta which are largely herbivorous have short generation times (about three years) and rapid rates of increase could be an exception. Fish at the *K* end of the spectrum are not uncommon; a particular example are the mouth-brooding African Cichlids in the Mbuna group of rock fishes from Lake Malawi, which produce very low numbers of eggs and live in communities densely packed with species (Fryer and Iles, 1972).

Changes that can be expected in the abundance of fish populations are illustrated in Figure 3.4. One of the most striking facts is that year-to-year variations can be huge, so that one year class can be up to 400 times the size of another (LeCren *et al.*, 1977). As shown for the data

Figure 3.4: The Recruitment and Spawning Stock Size of North Sea Haddock between 1905 and 1970 Showing Apparent Large Increase in Abundance towards the End of the 1970s.

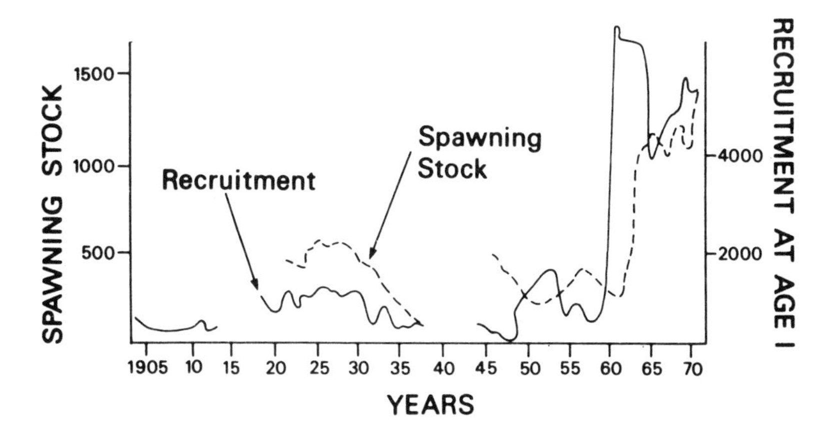

Source: Redrawn from Sahrhage and Wagner (1978), with permission.

on North Sea haddock in Figure 3.4, there can also be long-term trends in abundances, so that haddock were more abundant at the end of the 1960s than they were at the end of the 1950s. Similar long-term trends were shown in the Windermere perch population between 1941 and 1966. In this case the change resulted from documented human intervention so that a partial explanation is possible. For the changes found in the North Sea over the past 70 years there are as yet no convincing explanations (Cushing, 1980).

Within its area of distribution the abundance of a species will vary with season. This is a result of the seasonal migrations described above. This seasonal variation is often reflected in the pattern of fisheries. For example, there is a fishery in December and January on the ripe cod moving south along the north Norway coast as they move to the spawning grounds. Then there is the intense fishery, many centuries old, on the fish as they spawn off the Lofoton Islands. At these times the fish are concentrated into small parts of their total area of distribution and this makes it easier to exploit them. The blue whiting concentrates at spawning time in vast shoals over Rockall Bank and the west Scottish continental shelf, whilst for the rest of the year it is mainly dispersed throughout the Norwegian sea. Sometimes shoals collect along oceanographic discontinuities, such as the front which forms to the east of

Iceland where water of the East Icelandic current meets the North Atlantic Drift. Fish congregate at such boundaries because of the greater supplies of planktonic food to be found there. Similar aggregations occur in the North Sea where the diatom population increases in spring as the water column stabilises and light increases. The *Calanus* populations then begin to increase too as they feed on the diatoms. In spring, *Calanus* is the preferred food of herring which congregate on the copepod, creating local patches which can be exploited by fishermen (Cushing, 1973).

Points of upwelling also lead to the concentration of fish. Cushing (1973) describes how yellowfin-tuna catches in the Pacific are highest in the zone of divergence which straddles the equator. The divergence of water in this zone draws to the surface cool water from a depth of about 100 m. This water is also richer in nutrients than the tropical surface water so that the zone is characterised by increased plankton production. Small pelagic fish, such as lantern fishes, feed on the plankton and in turn provide food for the tuna. Other upwelling areas are similarly characterised by congregations of fish; examples are anchoveta off Peru, and maasbanker and pilchard off Namibia.

Good estimates of abundance are hard to make yet essential for a proper analysis of population dynamics. To estimate population density or numbers reliably one must be able to sample all age groups without bias and this is difficult in most aquatic habitats. There are now several statistical techniques that can be used to estimate abundance and these are thoroughly reviewed by Seber (1973), Ricker (1975), Youngs and Robson (1978), and Southward (1980). Marking and tagging is usually an integral part of abundance estimates; practical methods are reviewed by Laird and Stott (1978). We discuss next some of the assumptions made when making abundance estimates and give an outline of two frequently used methods.

The simplest model to use is the 'mark and recapture' method attributed to Petersen (1896). The method relies on the capture of a sample, which is then marked and released. At a later time a second sample is taken and the number of recaptures recorded. Let n_1 be the number captured and marked on the first occasion and n_2 the number captured on the second occasion of which m_2 will be marked. It is assumed that the proportion of fish that are marked in the second sample is representative of the proportion that are marked in the whole population. We can therefore write

$$\frac{m_2}{n_2} = \frac{n_1}{N}, \quad \text{so that} \quad \hat{N} = \frac{n_1 n_2}{m_2} \tag{3.2}$$

where $\hat{N}$ is the estimate of total population size. The following six assumptions underlie the method:

(1) The population is closed so that N is constant. A closed population is one in which there is no movement of individuals either in or out by means of death, immigration or recruitment.
(2) All fish have the same probability of being caught on the first sampling occasion.
(3) The catchability of the fish does not change after marking.
(4) The second sample is a random sample of the population.
(5) No marks are lost between sampling occasions.
(6) No marks go unrecorded in the second sample.

A small lake or pond may well be the only habitat where these constraints are likely to hold, and then only for a short period of time. The marking method will possibly be a cause of death but this can be controlled. Over any time greater than a few hours, predation is likely to cause mortality and marked fish may be more vulnerable. Although marks should all be released at once, the recaptures can be made over a longer period of time (Ricker, 1975). As a result, recruitment can dilute the number of marks present leading to a change in the ratio of n_1/N. There are methods of correcting for recruitment (Seber, 1973; Ricker, 1975).

It is always difficult to ensure that both the initial and subsequent samples are random with respect to the whole population; the estimates usually only apply to a subset of individuals. This is a result of the different properties of each type of sampling apparatus. Nets have fixed mesh sizes which cannot retain small fish, traps do not allow large fish in and it is too easy for small fish to swim out. Size selection also occurs with hooks. Stunning with electricity is a useful method of capturing fish in small ponds and streams, but even this method is size-selective. The method depends on a voltage difference of about 2 volts between the head and tail of the fish. The electric field dissipates exponentially with the distance from the electrode, so that bigger fish are shocked first, smaller ones only being stunned when the electrode is quite close. As the electric fishing procedure usually frightens fish away from the apparatus, big fish are more likely to be caught than are small ones. Size selection can mean that a biased sex ratio is recorded because female fish are often larger than the males (see below). Sampling gear will also miss parts of the population for other reasons. For example the very young fish will only be found in shallow water or amongst weeds where the gear cannot reach. In the sea, juveniles and adults can be separated by large distances. A classic case is plaice in the southern North Sea where the further one travels from the north European coasts, the older the fish become.

The fifth assumption is that no marks are lost between sampling occasions. The truth of this can be tested by keeping marked fish in tanks and recording the rate of tag loss (Pitcher and Kennedy, 1977), or by statistical methods (Seber, 1973). A close examination of fish can also sometimes show scars where tags have fallen out leaving a wound. Unreliable reporting of recaptures results in overestimated population size. In ecological work in small bodies of freshwater the investigator is most likely to make all tag recaptures, so that misrecording is reduced. Tagging programmes on marine fish will more often rely on tag recovery by commercial or sport fishermen. This is more likely to lead to errors of recording as a result of tags being missed or not returned to the investigator. Good financial rewards can help but increase the cost of the investigation. It is often recommended (e.g. Ricker, 1975) that it is worthwhile ensuring that a high proportion of recoveries are made by trained observers. Seber (1973) provides methods for testing whether the proportion of tags returned by fishermen is different from the proportion return by trained workers.

Few natural populations are in fact closed so that methods have been devised which can be used when individuals are being gained and lost during the course of the work. A population in such a state of flux is said to be 'open'. Gains can be from immigration or recruitment of young, whilst losses will be from emigration and mortality. The most general method for estimating the abundance of an open population is that due to Jolly (1965) and Seber (1965), now called the Jolly-Seber method. The underlying model is stochastic and produces estimates of survival and mortality as well as the numbers present on each sampling occasion. The Jolly-Seber method can be used when recapture data are available from a number of occasions. A special case when $n = 3$, and depending on a deterministic model, is equivalent to the Bailey Triple Catch method (Bailey, 1951, 1952). We will use it to illustrate the procedures underlying models for use with open populations. For greater detail see Seber (1973) and a recent clear review by Begon (1979).

To carry out Bailey's method, three samples are taken (see Table 3.3). On the first occasion all the fish are marked and returned. After a relatively short while (a week in the example) the habitat is sampled again and the recaptures recorded. All unmarked fish are given a new

Table 3.3: An Example of the Use of Bailey's Triple Catch Method Applied to Data on Roach from River Nene. (N = 1, 128; $s_{1,2}$ = 0.65; $r_{2,3}$ = 5.24.)

	Sampling Occasions (One Week Apart)		
	Day 1	Day 2	Day 3
Number of fish caught	$n_1 = 449$	$n_2 = 303$	$n_3 = 214$
Number recaptured, marked on day 1	–	$m_{1,2} = 77$	$m_{1,3} = 9$
Number recaptured, marked on day 2	–	–	$m_{2,3} = 6$

Source: Hart and Pitcher (1973).

mark. On the third sampling occasion the recaptures from the first and second occasions are recorded as are the number of unmarked fish. Using this data set it is possible to calculate the number of fish on each occasion, and the survival and recruitment between them. Abundance on the second occasion is calculated from:

$$\hat{N}_2 = \frac{(n_2 - m_{1,2})(n_2 + 1)m_{1,3}}{(m_{1,2} + 1)(m_{2,3} + 1)} \tag{3.3}$$

where the subscripts 1, 2 and 3 refer to the three sampling occasions and the other symbols have the same meaning as before. Survival from occasions 1 to 2 is calculated from:

$$\hat{s}_{1,2} = \frac{(n_2 - m_{1,2})(m_{1,3})}{(m_{2,3} + 1)n_1} \tag{3.4}$$

and recruitment is calculated from:

$$\hat{r}_{2,3} = \frac{m_{1,2}n_3}{n_1(m_{2,3} + 1)} \tag{3.5}$$

The value of $\hat{N}_1$ and $\hat{N}_3$ can be calculated by applying $s_{1,2}$ and $r_{2,3}$ assuming that these rates are the same for the period between occasions one and two and two and three. Variances for the three estimates can be calculated (Ricker, 1975).

Although $s_{1,2}$ and $r_{2,3}$ are called survival and recruitment, they need to be freshly interpreted for each survey. Their interpretation will depend on the time interval between sampling occasions. In the survey illustrated in Table 3.3 the interval was only a week, so that s and r are best interpreted as immigration and emigration rates. If the interval was several months, then survival and recruitment would be more appropriate, although a separate analysis might have to be done to disentangle the influence of movements into and out of the sampled area.

Mark-recapture techniques are often too slow or too difficult to carry out on commercial marine fish populations. Modern management needs quick, cheap methods with which reliable estimates of numbers can be obtained. The short-cut method used in fisheries is based on the relationship between catch and effort. In its simplest form it is assumed that a given unit of fishing effort will catch fish in proportion to their abundance (Beverton and Holt, 1957). The argument can be developed in the following way. The number of fish (N) alive in a year class at time t is given by

$$N_t = N_R \exp(-Zt) \tag{3.6}$$

where N_R is the number of young fish of the year class recruiting to the commercial part of the stock, and Z is the instantaneous rate of mortality. From this we can calculate that the number dying at the end of t years is

$$N_R - N_t = N_R - N_R \exp(-Zt) = N_R\,(1 - \exp(-Zt)) \tag{3.7}$$

Now Z can be split up into mortality resulting from fishing, F, and mortality caused by natural processes, M, such that $Z = F + M$. The proportion of the fish that die in t years that are caught is then

$$N_R \left(\frac{F}{Z}\right) (1 - \exp(-Zt)) \tag{3.8}$$

This is the yield in numbers to the fishery from one year class. If it is assumed that the population is in a steady state, then all age classes will be subject to the same F and M and the yield in numbers from the whole population in a year will be the same as the yield in numbers from one cohort over its whole lifetime. Therefore we can write

$$Y_n = N_R \left(\frac{F}{Z}\right) (1 - \exp(-Zt)) \tag{3.9}$$

Expressed in terms of yield per unit of fishing mortality we have

$$Y_n/F = N_R/Z\,(1 - \exp(-Zt)) \tag{3.10}$$

It is now usually assumed that F is proportional to the fishing intensity exerted by the fishing gear. This is most often labelled f and can be expressed for example as the number of boats fishing per year per unit area, or the number of hooks set per year per unit area. We can write the relationship between F and f as

$$F = qf \tag{3.11}$$

where q is the catchability coefficient – the degree to which the fish are vulnerable to the gear. It is this factor which is altered by improvements in ship power, in nets or by the advances in electronic fish-finding gear. We can now write the equation for catch in numbers per unit effort as:

$$\frac{Y_n}{qf} = \frac{N_R}{Z}(1 - \exp(-Zt)) \tag{3.12}$$

This is the relationship between catch per unit effort and stock abundance. In application it will always be necessary to estimate q, and check for trends in its value (see Chapter 8).

There is a known relationship between the size and age of a female fish and the number of eggs she is likely to produce (see below). From these data a mean absolute fecundity can be calculated which can be used with an estimate of egg abundance to derive estimates of adult abundance. It is now quite common for fishery biologists to sample in spring the eggs of a particular species, and to use the data to estimate the abundance of the parent stock (see Chapter 6). The method only works if certain conditions are fulfilled. The eggs should be spawned over a relatively short period of time and at a known site. The usual path of drift should be known, as should the stages of development through which the eggs pass. The relationship between the number of eggs and female length and age should be known along with the sex ratio. One should also be sure that only one stock is contributing to the eggs being sampled.

Electronic fish-finding equipment has also become an essential tool in stock estimation (Cushing, 1973, 1978). Echo-sounders, originally designed to find submarines and measure depth, have been adapted to detect fish shoals. Fish flesh is not much more reflective to sound waves than is water, but the air in the fish's swim bladder is and this produces a good sonar echo. Techniques have been developed to pinpoint fish shoals and to make estimates of the abundance of fish in them, although the species must still be determined by sampling some of the fish. Large individual fish can also be detected and counted. This approach makes it possible to make rapid estimates of distribution and abundance, being particularly useful in areas where no fishery exists. During the 'Eureka' programme off Peru, echo-sounding was used to obtain a synoptic survey of the anchovy stock on a single night (Cushing, 1973). About 20 fishing vessels were co-opted, each one starting from a fixed point off the coast and steaming out to sea for an agreed distance. The ships then turned and steamed back to a pre-arranged point further up the coast. This way the whole population was surveyed in a night giving an almost instantaneous estimate of abundance. Had the survey been done in stages using nets, fish movements between areas would have biased the results.

Population Age Structure, Mortality Schedules and Fecundity

Most fish of commercial importance spawn only once a year but live for more than one year so that several ages are present in the population at any one time; these populations have an age structure composed of several cohorts, usually equivalent to year classes in fish populations. The most obvious point is that there are usually more young fish than old in the population, but this is not necessarily so at any one time. However, in one cohort, numbers are steadily reduced by mortality as the fish get older. Fishing often kills a limited range of age classes (see Chapter 8), so that the age structure of the population is directly affected. Mortality and fecundity are also related to age. In combination with age structure, the schedule of fecundities for each age determines the reproductive capacity of the population. To understand the full implications of these comments we will first look briefly at the theory of life tables. The theory will then be applied to fish populations when we have discussed the characteristics of fish fecundity.

For a full discussion of life tables and how they may be used in population ecology the reader is referred to Southwood (1978), or Ricklefs (1979). Since this major field of ecology has been rarely applied to exploited fish populations and we consider that a number of important explanatory points can be made, we are including a detailed discussion here.

The Theory of Life Tables

The schedules of survival, mortality rate and fertilities (to which are sometimes added estimates of further expectation of life) are known as life tables, and are similar to the actuarial calculations used by insurance companies for human populations. In theory, the full set of information contained in the life tables for a given species-population enables one not only to make a complete description of its population ecology but also to deduce the controlling factors which determine its population dynamics.

There are two types of life tables. In *time specific* (equivalent to static, current or vertical) life tables, the structure of the population at an instant of time is examined, and the calculations are based on the mix of cohorts that may be present. There could easily be more older fish than younger, from the survivors of a particularly successful year class. Although this is an extreme possibility, there will inevitably be some imbalance between the cohorts and so this type of life table is not always an accurate way of assessing the population dynamics unless there is very little change from year to year. However, information for *age-specific* life tables (equivalent to dynamic, cohort, general or horizontal) is much more difficult to gather. In this type of table we follow the fate of each individual cohort (represented by year classes in most fish populations). Provided that we have information on enough cohorts, the major advantage is that changes in mortality, fecundity, etc., in each equivalent stage can be followed and related to the numbers or densities in that stage. In this way we can locate the mechanism determining population numbers ('key' factors), and population regulation (density-dependent 'regulating' factors). In practice when information on numbers is not very accurate or is incomplete, a single composite life table may be composed of the averages from each source. In this case we have a third type of life table which gives the average picture for the years included, but is neither exactly time- or age-specific.

We now come to the process of constructing a life table; we will be concerned only with age-specific tables. Following an age class through its life will help to introduce the concepts. Assume that at birth there are 1,000 individuals (see Table 3.4). At birth (age 0) all the young are alive. The column labelled l_x gives the probability of an individual born alive reaching the start of age x. When $x = 0$, $l_o = 1$ as would be expected intuitively, but as time passes the chances diminish of newborn reaching greater ages. In the example in Table 3.4 the pattern of survival is quite fish-like, being low in the early years, unchanging in the middle years and falling away sharply as the organisms grow old.

In life tables the fertility, or number of births, is expressed as the number of females produced and the label for this parameter is usually b_x or m_x. Mammal studies would most likely record the number of live

Table 3.4: A Hypothetical Life Table. (Net productive rate, R_o = 25.552.)

Lower Bound of Age Class	Number	Survival	Number of Young (Age Specific Fertility)	Actual Young Produced
x	n_x	l_x	b_x	$l_x b_x = v_x$
0	1,000	1.000	0	0
1	800	0.800	0	0
2	500	0.500	0	0
3	300	0.300	10	3.0
4	250	0.250	20	5.0
5	200	0.200	30	6.0
6	180	0.180	30	5.40
7	150	0.150	25	3.75
8	100	0.100	20	2.00
9	40	0.040	10	0.40
10	2	0.002	1	0.002

Figure 3.5: Graphical Representation of Hypothetical Life Table Data shown in Table 3.4. (See text for further details.)

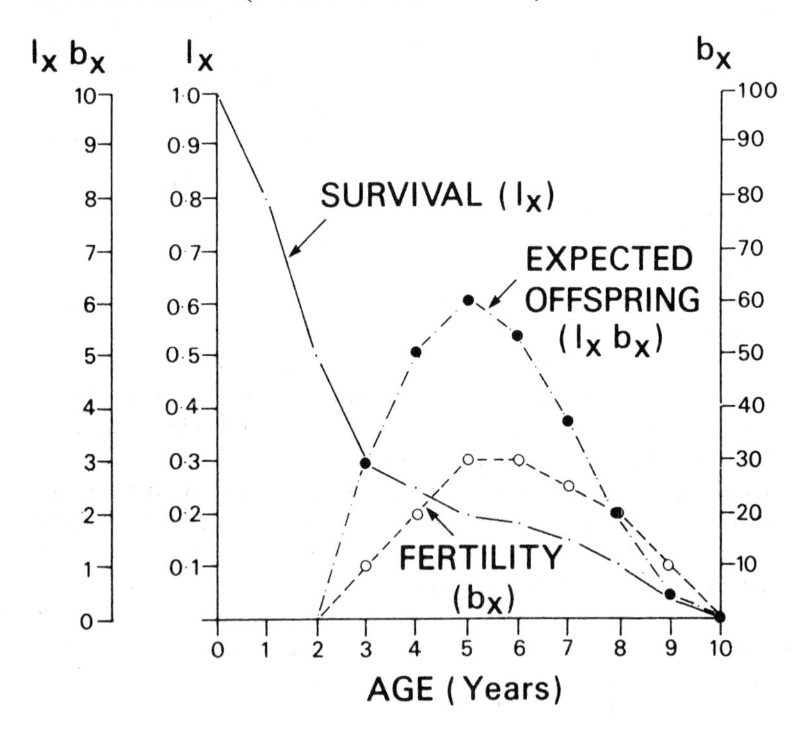

births and the temporal pattern might look as shown in Table 3.4 and Figure 3.5, with a peak in middle age. As we will see, the temporal pattern of fish reproduction does not follow this example. For species that do not bear live young the b_x values will be the numbers of fertile female eggs laid. One must assume that the sex ratio at birth is one-half unless one has more specific information (see below).

Multiplying b_x by l_x calculates the expected number of eggs or off-spring to be contributed by a female at each age. The sum of the expected number of offspring is a balance between decreasing survival with age and the curve of reproductive capacity and gives the number of female offspring that each famale is likely to produce during her lifetime. This value, 25.5 in the example in Table 3.4, is called the net reproductive rate and is conventionally labelled R_0. A value of R_0 less than 1 would be found in a population in decline. A stable population has an R_0 of 1, each female replacing herself exactly. In an expanding population R_0 is greater than 1, as in the example given. From this it is intuitively easy to see that R_0 should have an influence on the way in which the population grows. To obtain a complete picture it is necessary to take generation time into account. This can be done by using the exponential model for population growth. (Note that we will use power notation, e^r, not exp(r) from here to Equation 3.26, since other powers are involved.) So we have

$$N_t = N_0 e^r m^t \tag{3.13}$$

where N_t is the population abundance at time t, N_0 is the abundance at time $t = 0$, and $r_m = (b - d)$, The latter is the intrinsic rate of increase and is the difference between birth and death rates. The equation can be rewritten as

$$\frac{N_t}{N_0} = e^r m^t \tag{3.14}$$

In species where the parents die immediately after their offspring have been born, the net reproductive rate is the expected number of female offspring that will be reproduced from the number of females in the population. The time between the birth of the parents and the birth of their offspring is called the generation time and is labelled T by convention. Under such conditions the population will have expanded R_0 when T has elapsed so that

$$R_0 = \frac{N_T}{N_0} = e^r m^T \tag{3.15}$$

Taking the logarithm of this expression gives

$$ln\,(R_0) = r_m T \tag{3.16}$$

so that

$$r_m = \frac{ln\,(R_0)}{T} \tag{3.17}$$

In many animal populations the generations are not discrete but overlap; this is particularly true of many fish populations which contain a number of age classes at any one time, each of which is able to reproduce. As a result, Equation 3.17 no longer holds exactly, giving only an approximation to r_m. Following Laughlin (1965), it is better to rewrite the equation as

$$r_c = \frac{ln\,(R_0)}{T_c} \tag{3.18}$$

where r_c is called the capacity for increase. As such it is an estimate of r_m and so long as generations overlap r_c is always less than r_m. The size of the differences between the two is a function of R_0 and the length of time during which reproduction takes place (Laughlin, 1965). T_c in Equation 3.18 is not the same as T in 3.17. As generations overlap, T is no longer a discrete time interval but is replaced by a number which is defined as the mean age at which females produce their young. It is calculated from

$$T_c = \frac{\Sigma\, x l_x b_x}{\Sigma\, l_x b_x} = \frac{V_x}{R_0} \tag{3.19}$$

So if one has life table data it is possible to make a reasonable estimate of r_m.

Strictly, r_m can only be calculated from life tables when the age distribution is stable, a condition towards which a population will converge if not disturbed (see Ricklefs, 1979, for an exposition of this result). Assuming that a stable age distribution exists, a relationship can be derived, involving l_x and b_x, from which r_m can be calculated as accurately as we please. This is done in the following way. We shall derive the equation using discrete time intervals (designated by Δx) and will sum them where necessary. The equation can equally well be expressed in integral notation.

Call N_0 the number of individuals born at time t originating from all the age classes existing in the population. Then

$$N_0(t) = \Sigma N_x(t) b_x \Delta x \tag{3.20}$$

where $N_x(t)$ is the number of individuals in age class x. But we can write

$$N_x(t) = N_0(t-x) l_x \tag{3.21}$$

where $N_0(t-x)$ was the newborn number of individuals now aged x. This can be substituted into the first equation to give

$$N_0(t) = \Sigma N_0(t-x)\, l_x b_x \Delta x \tag{3.22}$$

A further substitution can be made if it is assumed that the population has been growing by a factor e^{r_m} each year. This is that

$$N_0(t-x) = N_0(t)\, e^{-r_m x} \tag{3.23}$$

This means that the number of newborn at time $t - x$ is equal to the present number of newborn decreased by $e^{-r_m x}$. Substituting this gives

$$N_0(t) = \Sigma N_0(t)\, e^{-r_m x} l_x b_x \Delta x \tag{3.24}$$

Both sides can be divided by $N_0(t)$ to give

$$\Sigma\, e^{-r_m x} l_x b_x \Delta x = 1 \tag{3.25}$$

If Δx is made extremely small then Equation 3.25 becomes the integral equation

$$\int_0^\infty e^{-r_m x} l_x b_x dx = 1 \tag{3.26}$$

which is called the Euler-Lotka equation (Wilson and Bossert, 1971), or just the Lotka equation. This equation is central to ecological theory and it has been used extensively in theoretical work on life history strategies (e.g. Gadgil and Bossert, 1970, and Law, 1979a, b). A few points from this theory will be mentioned at the end of the next section.

Equation 3.25, which is an approximation to Equation 3.26, can be used with life table data to estimate iteratively the exact value of r_m (see Southwood, 1978, p. 290). A preliminary estimate to start the procedure can be obtained from $r_c = ln(R_0)/T_c$.

One further point about R_0 needs to be mentioned. We have already said that the net reproductive rate is the amount by which the population expands each generation. In Equation 3.15 we saw that $R_0 = e^{r_m T}$ so that if we set $T = 1$ the amount by which the population will increase per generation is e^{r_m}. This multiplicative factor is usually expressed as λ_m or in the case of Equation 3.18, λ_c.

The Characteristics of Fish Fecundity

It is now necessary to outline the characteristics of the reproductive capacity of female fish, their fertility. A few teleosts, for example the mouth-brooding cichlids such as *Tropheus moorii*, may produce only five to seven eggs (Fryer and Iles, 1972). In such species, as in the mammals and birds, the eggs and young are painstakingly cared for by the parent during their growth, so reducing the chances of mortality; parental investment is therefore high. Most teleosts produce enormous numbers of eggs, taking an alternative strategy to ensure that they are represented in the next generation; they produce so many eggs that even though the majority die, enough remain to replenish the stock.

Fertility, i.e. the number of live fertilised eggs produced, is very difficult to measure in teleost fish because external fertilisation of thousands of eggs makes it impossible to determine the level of success. We usually have to be content with fecundity, i.e. the potential number of offspring present before fertilisation. Fecundity is easily estimated by opening the fish, preserving the ripe ovaries and subsequently counting the numbers of eggs (see Bagenal, 1978, for methods). Fecundity is bound to give something of an overestimate of fertility, but in most species with lots of eggs this is not likely to be too serious for the maintenance of the species. In elasmobranchs the distinction between fertility and fecundity is likely to be more important since the energy invested in one yolky shelled egg is very large for the female; eggs which are ovulated may go to quite an advanced stage before having to be reabsorbed. In many ways the elasmobranch reproductive system is as advanced and specialised as that of the more familiar birds and so we are mainly considering the teleost system as 'typical' of fish.

Unlike mammals and birds, the absolute number of eggs produced per individual (the absolute fecundity) increases with age in teleost fish. This is a consequence of the indeterminate growth found in fish (see Chapter 4). Generally the fecundity is related to some power of the length although there is considerable variability (see Bagenal, 1973). The most usual relationship can be written as

$$F = aL^b \tag{3.27}$$

where F is the fecundity, L the length of the fish and a and b are fitted constants. If logarithms of both sides are taken the equation becomes

$$ln(F) = ln(a) + bln(L) \tag{3.28}$$

which is an equation describing a straight line. In this linear form it is easier to fit the relationship to data. Although a, the intercept, will vary considerably, the value of b is often close to 3, indicating that a direct volume (cube) relationship exists with length. An example of this type of relationship is shown in Figure 3.6.

There are various other definitions of fecundity which are sometimes mentioned and these will be briefly defined. Relative fecundity is the mean number of eggs per unit weight of fish. Such a value is useful for comparing fecundities when fish are different in size. It is assumed that relative fecundity is constant with fish total weight (see below). Population fecundity is the number of eggs produced by all the females in the spawning population, and is given approximately by

$$F_p = \Sigma N_x \bar{F}_x \tag{3.29}$$

where N_x is the number of females aged x and $\bar{F}_x$ is the mean fecundity per female aged x or the age-specific fecundity. A more accurate method allows for the normal distribution of sizes in each age class and an algorithm for this integral is given by Pitcher and Macdonald 1973b). When life tables are being constructed the age specific fecundity is often calculated from the relationship between absolute fecundity and length using a mean length for each age.

An illustration of these relationships is given in Figure 3.7, which contains data from the pike population of the River Stour, UK (Mann, 1976a). The increment in absolute fecundity increases up until age five after which it decreases. If the exponent relating length to weight is equal to the exponent relating fecundity to length, then the eggs per gram will remain constant as the fish grows. This is shown by the following argument. Relative fecundity is given by

$$RelF_w = \frac{F_t}{w_t} \tag{3.30}$$

But

Figure 3.6: Fecundity of Minnows (*Phoxinus phoxinus*) Plotted against Fish Length. Fitted regression is log F = 3.42 log L – 3.56. Dashed lines indicated the 95 per cent confidence limits.

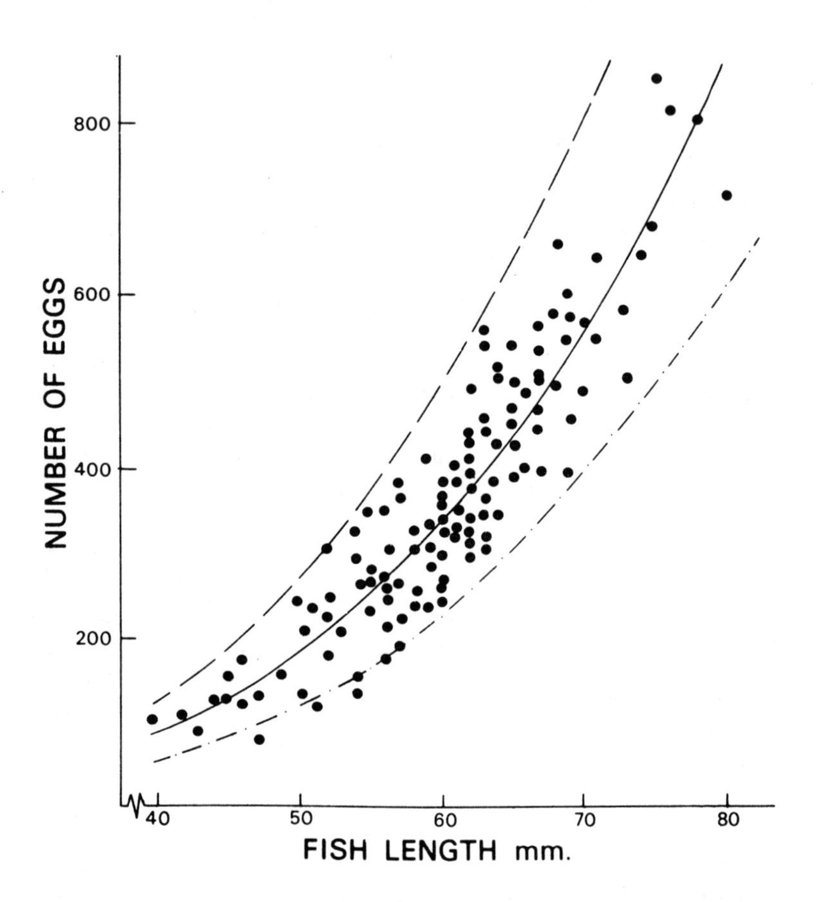

Source: Pitcher (1971).

$$F_t = a\ l_t^{\,b \simeq 3} \tag{3.31}$$

and

$$w_t = a_1 l_t^{\,b_1 \simeq 3} \tag{3.32}$$

so

Figure 3.7: Relative Fecundity and the Change in Absolute Fecundity, both Plotted against Age, for Pike from the River Stour, Dorset, UK

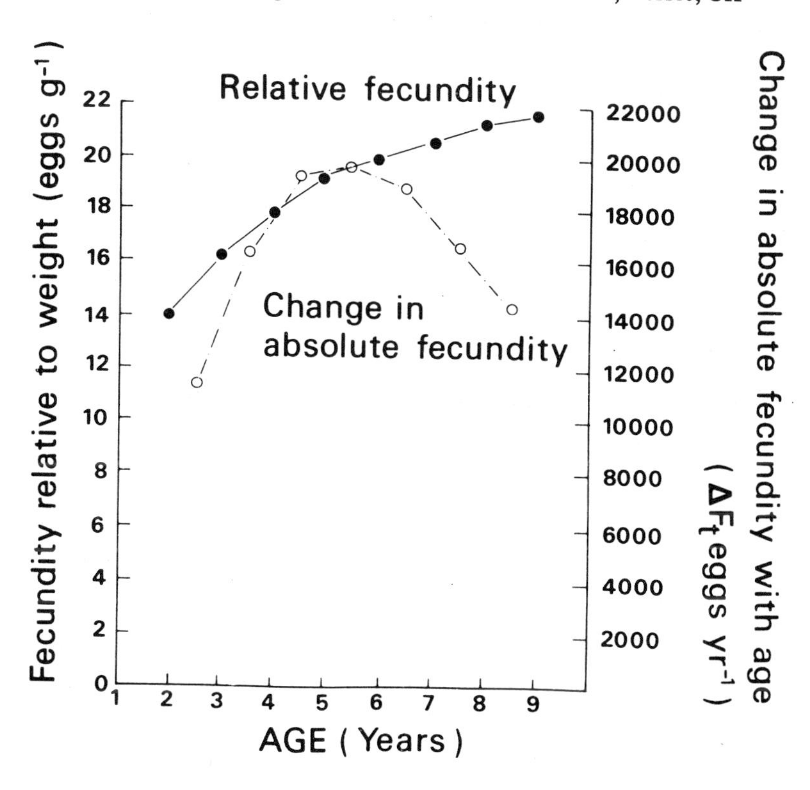

Source: Plotted from data in Mann (1976a).

$$RelF_w = \frac{al_t^{\,b}}{a_1 l_t^{\,b_1}} = aa_1^{-1} l_t^{\,b-b_1} \qquad (3.33)$$

Because $b \simeq b_1 \simeq 3$, $b - b_1 \simeq 0$ and $l_t^0 = 1$, so $RelF_w = aa_1^{-1}$ which is constant.

This unchanging relationship between F_t and w_t is dependent on constant values of b and b_1. This condition is not always met, as in the Stour pike where $b - b_1 = 0.51$. The condition, or fatness, of a fish (see Chapter 4), is reflected in the value of b; a high b means a heavier fish for a given length. The value of b_1 is often calculated using the weight of the soma plus gonad. As fish ripen the gonad will increase in size and this increase can come from a gain in water from the environment or by a gain in materials which could come either from the environment or the somatic tissue. So relative fecundity will remain nearly constant only if most of the gonad increase is transferred from somatic tissue. In our example $a/a_1 = 0.68$ and so the equation relating relative fecundity to length is $0.68 l_t^{0.51}$ (Figure 3.7). In practice, beyond age 5 in our example, the increase in relative fecundity is negligible, so the assumption of constant relative fecundity is not unreasonable.

Absolute and relative fecundities can be influenced by environmental factors (Bagenal, 1973). A clear inverse relationship has been found between population abundance and relative fecundity in plaice, pike, witch and Norway pout. The cause is not known, but changing relative abundance of food could be partly responsible. Changes in relative fecundity will mean changes in absolute and population fecundity which imply higher egg production at low population density.

Relative fecundity can be used to compare reproductive strategies in fish. A high relative fecundity usually means numerous small eggs, each with a poor chance of successfully producing a mature adult. The female achieves success by swamping the elements so that at least a few escape destruction. Cod and plaice have this type of strategy. Fish such as salmon have a lower relative fecundity and make each egg more valuable with sufficient food reserves to maintain the young for some days after hatching. There are also fish like the cichlids and the sea horse which have low relative fecundity, producing few eggs, which are then, however, cared for, as are the young, so reducing the chances of death. The effort is put into the parental care rather than into the numbers of eggs.

It is also possible that fish can change their reproductive strategies according to the mortality schedule to which they are subjected. Gadgil and Bossert (1970) devised a model that showed how the pattern of reproduction employed depended on the balance between costs and benefits. Where the curve of costs against reproductive effort is concave, the best strategy is to reproduce in several bursts throughout life, a pattern termed *iteroparity* (e.g. cyprinids). When the cost curve is convex the best strategy is to wait and reproduce only once in a lifetime – called the 'big-bang' strategy by Gadgil and Bossert. This style of life history is said to be *semelparous* (e.g. the Pacific salmon). The gains from reproductive effort are expressed in terms of offspring born. The costs are measured as mortality caused and offspring thereby lost in the future as a result of reproductive effort that takes place now. Law (1979a) has studied the way age-specific predation can induce changes in reproductive strategies. Using a model which is in principle the same as that used by Gadgil and Bossert (1970), both being based on the Euler-Lotka equation, Law has shown that reproductive effort is increased in the age group i of a population if predators select prey from age groups greater than or equal to i. In these circumstances the cost of reproduction is decreased as the chances of living to ages greater than i are already lowered by predation. Future mortality induced by a unit of reproductive effort at age i is devalued. If predators specialise on prey younger than age i, then reproductive effort at i is decreased. The predictions of the model are made more plausible by the work on Windermere pike. Older pike have been removed by a gill-netting programme since 1944 (see Frost and Kipling, 1967, and Chapter 5); fecundity studies have shown that in 1963 a three-year-old female pike produced about 50,000 eggs. In 1975 the mean value was 80,000. The three-year-old pike are at the lower end of the range removed by gill netting. This evidence could be equivocal because the change in fecundity could have resulted from a change in mean size at three years, but it seems that this was not the case. Law's study makes it clear that a fishery could have important selective effects on the fished stock and the subject is expanded in Chapter 5.

Life Tables for North Sea Haddock

In this section we discuss the application of life table theory to fish, bringing out the features special to the group. The example we have chosen is for North Sea haddock, and we have taken data from Sahrhage and Wagner (1978). Only the data for the years 1921-30 and 1957-67 are used. The haddock population in the North Sea during the two periods did not show a stable age distribution, so that any values of r_c or r_m calculated are not generally applicable. A life table was constructed for each of the year classes born between 1921 and 1930 and between 1957 and 1967, including seven ages in the first period and six in the second. From these a composite life table has been calculated, one for each period, and these are shown in Table 3.5. The numbers (n_x) are the mean number of fish caught per ten hours of trawling and are only an index of stock abundance. The number of fish alive at the start of year zero is the estimated number of eggs laid. All the numbers in column two of Table 3.5a, b and c, refer to the number of females. To arrive at this estimate we assumed a sex ratio of 0.5 at all ages. As a result all other values are with reference to females only. As explained earlier, fecundity is used as an estimate of fertility and the values for fecundity in the column headed b_x have been adjusted to account for different proportions of each age group being mature. For example the

Table 3.5: Composite Life Tables for North Sea Haddock.
a. All Year Classes Born Between 1921 and 1930.

Age	l_x	b_x	l_xb_x	xl_xb_x
0	1.000	0	0	0
1	2.128×10^{-5}	0	0	0
2	1.713×10^{-5}	1,690	0.029	0.058
3	5.881×10^{-6}	37,650	0.221	0.664
4	2.337×10^{-6}	75,923	0.177	0.710
5	1.109×10^{-6}	110,992	0.123	0.615
6	6.984×10^{-7}	139,000	0.097	0.582
7	1.060×10^{-7}	175,000	0.019	0.130
			R_0 = 0.667	2.760

$T_c = 4.140, r_c = -0.098, r_m = 0, \lambda_m = 1.$

b. All North Sea Haddock Born Between 1957 and 1967.

Age	l_x	b_x	l_xb_x	xl_xb_x
0	1.000	0	0	0
1	1.145×10^{-4}	0	0	0
2	5.853×10^{-5}	3,815	0.223	0.447
3	1.777×10^{-5}	47,062	0.836	2.509
4	6.195×10^{-6}	95,500	0.592	2.366
5	6.702×10^{-7}	136,263	0.091	0.457
6	4.559×10^{-7}	200,000	0.091	0.547
			R_0 = 1.833	6.326

$T_c = 3.451, r_c = 0.176, r_m = 0.180, \lambda_m = 1.$

c. Age-specific Life Table for the 1962 Year Class of North Sea Haddock, One That Was Extremely Abundant and Had a Strong Influence on the Stock

Age	n_x	l_x	b_x	l_xb_x	xl_xb_x
0	11,650,000	1.000	0	0	0
1	14,076	1.208×10^{-3}	0	0	0
2	4,551	3.906×10^{-4}	3,815	1.490	2.980
3	1,868	1.603×10^{-4}	47,062	7.546	22.638
4	715	6.133×10^{-5}	95,500	5.857	23.428
5	28	2.361×10^{-6}	136,263	0.322	1.610
6	31	2.661×10^{-6}	200,000	0.532	3.192

$T_c = 3.420, r_c = 0.806, r_m = 0.873, \lambda_m = 1.$

mean number of eggs expected from a female aged two is 70,000 for the years 1957–67. But only 10.9 per cent of females of this age were mature, so 70,000 has been multiplied by 0.109 and divided by two to give the average number of female eggs expected, this being 3,815. All females aged six and over were mature.

The life tables in Table 3.5 illustrate very clearly the dramatic mortality rates in the first year of life. The averaging done to produce the tables hides the fact that in some years the numbers per ten hours of trawling dropped from millions to only a few tens in the first year. The fecundity column shows the characteristic rise with age. Comparing the two composite tables is instructive as the haddock population was decreasing between 1921 and 1930 and increasing between 1957 and 1967. This is reflected in the life table parameters T_c and r_m. In the twenties the mean generation time was 4.1 years, whilst it was only 3.4 in the sixties; there was a shorter turn-over time. r_m in the early period was zero, meaning that the population just replaced itself ($\lambda_m = 1$). In the sixties $r_m = 0.18$ ($\lambda_m = 1.2$) which would be expected from the expansion observed.

The spawning stock of any one year is composed of representatives of many different year classes. In most fish, year class strength varies greatly from one year to the next and as a result one good year class can dominate the population for a number of years. This can be shown using the life table data. For both periods the age-specific net reproductive rates were assigned to the years in which they occurred. For example, in Table 3.6b age 3 of the 1962 year class was alive in 1965, so 7.546 was assigned to a column labelled 1965. In the same year the spawning stock would also contain representatives of the 1963, 1961, 1960 and 1959 year classes. The age-specific R_0 for these year classes were 7.546, 0.004, 0.234, 0.066 and 0.022 respectively, giving a total of 7.872. But 96 per cent of this R_0 was provided by the 1962 year class. Table 3.6 shows the annual R_0's calculated in this way for both the twenties and sixties. There is no doubt that the higher mean value of r for the sixties was almost completely a result of the 1962 year class which dominated the population from age two to six.

The analysis of year class net reproductive rates also gives valuable insight into the effects of fishing on the spawning success of the stock. Increased fishing means a reduced chance of a newborn fish getting to a particular age. It is true that egg and larval survival fix the approximate size of the year class, but heavy fishing will reduce the abundance of the sexually mature age groups and directly influence the composition of the spawning stock. Heavily exploited populations are strongly

Table 3.6: R_0 for Spawning Stocks Calculated from Year Class Life Tables.
a. Years 1928–33.

Age	Year					
	1928	1929	1930	1931	1932	1933
0	0	0	0	0	0	0
1	0	0	0	0	0	0
2	0.029	0.004	0.064	0.008	0.010	0.020
3	0.146	0.144	0.040	0.234	0.097	0.058
4	0.128	0.105	0.073	0.043	0.388	0.068
5	0.081	0.052	0.044	0.037	0.024	0.088
6	0.013	0.022	0.015	0.015	0.015	0.003
7	0.090	–	0.005	0.005	0.013	0.004
Total	0.487	0.327	0.241	0.342	0.547	0.241

b. Years 1963–9.

Age	Year						
	1963	1964	1965	1966	1967	1968	1969
0	0	0	0	0	0	0	0
1	0[a]	0	0	0	0	0	0
2	0.351	0.490[a]	0.004	0.0005	0.0001	0.003	0.554
3	0.289	0.354	7.546[a]	0.060	0.0005	0.002	0.065
4	0.039	0.047	0.234	5.857[a]	0.003	0.002	0.001
5	0.033	0.026	0.066	0.107	0.322[a]	0.032	0.001
6	0.018	0.011	0.022	0.023	0.006	0.532[a]	0.229
Total	0.730	1.928	7.872	6.048	0.332	0.571	0.850

Note. a. 1962 year class.

influenced by a few year classes. One very good year class can give the impression that the stock is in good condition. Concentrated fishing on that year class will seriously affect the way in which it can influence the recruitment of future generations; this point will be returned to in more detail when we consider fishery models in Chapter 9.

The Sex Ratio in Fish

In calculating the life tables it has been assumed that the sex ratio is 0.5 at all ages. This assumption needs examining in more detail. Breder and Rosen (1966) describe how many species of fish spawn in large shoals within which males and females associate in pairs or in

Table 3.7: The Numbers of Male and Female Chub from the River Stour, Dorset, UK

Age	Males	Females	Ratio
1	44	6	7.3 : 1.0
2	23	9	2.6 : 1.0
3	8	3	2.7 : 1.0
4	8	3	2.7 : 1.0
5	8	2	4.0 : 1.0
6	5	3	1.7 : 1.0
7	1	1	1.0 : 1.0
8	3	1	3.0 : 1.0
9	3	3	1.0 : 1.0
10	11	15	0.7 : 1.0
11	7	5	1.4 : 1.0
12	6	13	0.5 : 1.0
13	7	6	1.2 : 1.0
14	3	6	0.5 : 1.0
15	1	2	0.5 : 1.0
16	1	3	0.3 : 1.0
17	0	2	–
18	0	2	–
19	1	1	1.0 : 1.0
20	0	2	–
21	0	0	–
22	0	1	–
		Overall ratio	1 : 1

Source: Mann (1976a).

polyandrous groups. The formation of pairs is consistent with a 1:1 sex ratio, but unless the same group of males travels about the shoal fertilising a succession of females, a polyandrous association would imply a surplus of males. This type of spawning is typical of many clupeids and salmonids and is quite common in members of the Order Cypriniformes, although the rule is distinct pairing. Recent demographic studies on British freshwater fishes (Frost and Kipling, 1967; Banks, 1970; Mann, 1973, 1974, 1976a, b) including data on pike, perch, roach, dace and chub show that often in the early years of a year class's existence males are either more numerous or equal in number to females. As the fish age, the males suffer a higher mortality rate so that older fish are nearly always female. An example is illustrated in Table 3.7. These findings could be typical of many temperate freshwater and marine fish.

Sex Changes in Fish

Some fish species have a sex change part way through their lives. For example the bluehead wrasse which lives on coral reefs in the western Atlantic (Warner *et al.*, 1975) has individuals that are born female and only become male on reaching a certain size. Studies of the mating system have shown that males, which are brightly coloured, establish spawning territories on reefs. Females visit these sites and spawn with the largest and most colourful of the males. This lek-type system, where males must be big to gain territory, means that small young males are at a great disadvantage. Warner *et al.* (1975) propose that in such circumstances selection will favour a sex change. Early in life it pays to be the sex where fertility increases only slowly with size and then to switch to the opposite sex when a sufficient size has been reached. In many species female fertility increases most with size. As a result individuals are born males and become female at a later age. The sex change may be genetically controlled or triggered by some environmental factor, such as the loss of a dominant male from the harem. This type of sexual system is not common in commercially important species, but where such species are fished, stock management must take the unusual system into account.

In this chapter we have described the structure of fish populations and how this may be measured, and have briefly analysed its consequences in adaptive life-history terms. The next chapter looks at how individual fish grow in size.

4 FISH NUTRITION, GROWTH AND PRODUCTION

To grow, fish need a proper diet. An individual fish in nature has to take care of the problem itself, but as soon as fish become farmed animals then we need to have information about their nutritional requirements. In the first part of this chapter we discuss fish nutrition. In the wild, the first task a hungry fish has to face is the collection of food; in the second part of this chapter we discuss some of the strategies that fish employ to do this. Once caught, the food must be digested and how this is done depends on gut structure. It is also important in growth studies to know how fast it takes food to pass through the gut. We discuss the way energy from food is partitioned in the fish, since assimilated energy can be used for a number of purposes. Part of this energy is used for growth; the efficiency of utilisation and the growth rate produced are influenced by both biotic and abiotic environmental factors. The final sectors of the chapter discuss ways in which growth can be modelled and how production can be estimated, both predictions being needed in fishery management.

Nutrition

Halver (1979) has pointed out that all animals need an energy source, sufficient essential amino acids, essential fatty acids, a range of specific vitamins and minerals. However, there is more to nutritional research than the discovery of the substances necessary for fish to survive. As Cowey and Sargent (1972) write: 'the ultimate aim of nutritional research is the provision of a balanced diet, which will meet the requirements of the animal with respect to any one of a number of physiological functions ranging from growth to reproduction'. In this section we divide the subject of nutrition into qualitative and quantitative parts. First we discuss the basic nutritional requirements of the fish and then we consider the balance between the different food inputs needed to achieve optimal growth performance. These features are important to an understanding of the ecology of fish and are directly relevant to the practice of fish farming (see Chapter 10). Recent progress in fish nutrition has been reviewed by Cowey and Sargent (1979).

The Substances Required

Protein. Proteins are necessary to maintain and repair tissue wear and tear, to allow the reconstruction of depleted tissues and to allow growth to take place by the addition of new proteins. However, the fish's need for protein cannot be supplied from any source. For all fish which have been examined the following ten amino acids are essential and cannot be omitted from the diet (Halver, 1979): arginine, histidine, isoleucine, leucine, lysine, methionine, phenylalanine, threonine, tryptophan and valine. These must be present in nearly equal amounts. They are essential in that the fish cannot synthesise them from simpler substances in quantities sufficient for normal growth, or synthesise them from non-essential amino acids.

Further work has determined the necessary proportions of each amino acid in the diet of three species of fish. This was first done for chinook salmon by de Long *et al.* (1962) who carried out a set of experiments to determine the proportion of threonine required in the diet. The experimental results showed that threonine must form about 0.9 per cent dry weight of the diet irrespective of temperature. This proportion means that if the diet contains 40 per cent protein then 2.3 per cent of this should be threonine. Similar experiments with the other indispensable amino acids have shown the necessary proportions for each. The results for chinook salmon, Japanese eel and carp are shown in Table 4.1, together with the requirements of rats to provide a comparison. The percentages given are all in terms of the total dry diet of which 40 per cent is protein in the chinook salmon, 37.7 per cent in the eel, 38.5 per cent in the carp and 13.2 per cent in the rat. Further work has shown that certain nitrogen supplements, including some amino acids, can reduce the proportion necessary of some of the indispensable amino acids (Cowey and Sargent, 1972; Halver, 1979).

Not all dietary protein is assimilated. Eighty per cent of animal protein fed to trout or channel catfish was assimilated, whilst with proteins of vegetable origin, for example from soybean meal, wheat middlings or rice bran, 70 per cent or less was assimilated. Factors such as the feeding rate, nutrient level in the diet, fish size and water temperature, may alter the quantities assimilated (see also the section on energy budgets).

In fish farming, sparing of protein can be achieved by including in the diet sufficient sources of energy other than protein. Fish fed low-energy diets break down proteins to obtain energy; consequently protein, which is expensive to provide in a farming enterprise, is used in a wasteful way. Experiments have shown that addition of lipids and

Table 4.1: The Amino Acid Requirements of Three Species of Fish and the Rat.

Amino Acid	Chinook Salmon	Japanese Eel	Carp	Rat
Arginine	2.4	1.7	1.6	0.2
Histidine	0.7	0.8	0.8	0.4
Isoleucine	0.9	1.5	0.9	0.5
Leucine	1.6	2.0	1.3	0.9
Lysine	2.0	2.0	2.2	1.0
Methionine	1.6[a]	1.9[a]	1.2[b]	0.6[b]
Phenylalanine	2.1[c]	2.2[c]	2.5[c]	0.9[c]
Threonine	0.9	1.5	1.5	0.5
Tryptophan	0.2	0.4	0.3	0.2
Valine	1.3	1.5	1.4	0.4

Notes: a. Methionine plus cystine. b. In the absence of cystine. c. In the absence of tyrosine.
Source: Cowey and Sargent (1979).

carbohydrates to protein-poor diets increases the growth rates of the fish compared with those not given the energy supplement. It is concluded from such experiments that more of the protein goes towards tissue construction when sufficient alternative energy sources are available. Fish are not quite as efficient with ingested energy as are chickens (Cowey and Sargent, 1979). The ratio of energy required for daily growth to energy required for maintenance of body weight is 27 per cent in the carp but 30 per cent in the chick. In terms of nitrogen the ratio of N retained to N ingested is about 30 per cent in carp but over 50 per cent in chickens. It is suggested that fish farming is advantageous because it optimises the use of manpower rather than because it is an efficient converter of energy.

Carbohydrates. In many fish species carbohydrates are not an important item in the natural diet and are not missed if absent. Certain vegetarian species can use large quantities of carbohydrate, for example the grass carp. Despite the ease with which certain species can do without carbohydrate it can be incorporated into diets in quite large amounts; in this way a cheap source of energy can be provided in artificial diets. All fish have a sufficient complement of enzymes to cope with starches and disaccharides. The latter in the form of sucrose or lactose can be fully assimilated by rainbow trout when forming from 20 to 60 per cent of the dry diet (Cowey and Sargent, 1972). However, with potato starch and dextrin, assimilation decreases with increasing amounts in the diet. Carbohydrate in the diet can lead to a saving of protein and Buhler and Halver (1961) showed that chinook salmon use protein most efficiently when dextrin formed between 20 and 30 per cent of the dry diet.

Lipids. Teleosts can be regarded as lipid specialists; many, like the pilchard, capelin and salmon even concentrate oils within their body. Dried fish were used as torches by the American Indians! The fats are used in two ways, either as a rapidly-mobilised energy reserve (an important adaptation in the fish's patchy environment) or as structural phospholipids.

Lipids themselves can be divided into two groups; the polar lipids and the non-polar lipids. Lipids from both groups contain fatty acids linked to different substances. If the group to which the fatty acid is linked is an alcohol, then the combination forms a neutral fat. Fish have no adipose tissue like mammals have. There are two main sites where lipid is stored, composing what is known as depot fat. In some fish lipids are stored in an adiposite liver, for example in cod, whilst others possess adiposite muscle where the major store of lipids is in the muscle tissue, an example being the herring. The bulk of fatty acids used for energy are triglycerides.

Unlike mammals, the herring, sardine, and other pelagic fish are able to tackle a diet high in oil and wax, especially the copepods which are packed with wax esters. The wax esters are metabolised to glycerol, but at the expense of a little dietary protein used for energy. Freshwater fish like the trout do not seem to have the same ability. The use of oils to spare protein in farmed fish diets has been mentioned. This practice has its limits; too high a level of oil can cause deposition in the flesh, making the fish unmarketable and hard to freeze.

Polyunsaturated lipids are used to build the structural phospholipids whose function is to keep living membranes pliable. To do this fatty acids in the diet are desaturated and their carbon skeletons elongated. In terrestrial homeotherms like mammals, where the great majority of animal nutrition studies have been carried out, the $\omega9$ and particularly $\omega6$ series of fatty acids predominate. For example 18:2$\omega6$ linoleic acid is essential in man's diet and undergoes chain elongation and desaturation to form 20:4$\omega6$ arachidonic acid. By contrast, in aquatic poikilotherms temperatures lower than 20°C favour metabolisms based on the more unsaturated ω3 series of fatty acids as these remain fluid at low temperatures. Fish convert dietary ω6 to ω3, although most fish cannot cope with a large proportion of ω6 lipids. The ability to do this is greater in freshwater than in marine teleosts and the bodies of trout and carp contain more ω6 fats than those of marine fish, probably because their diet includes food from the land.

Freshwater fish which can survive at high temperatures, for example farmed catfish at 30°C, can in fact survive on a diet containing wholly ω6 fatty acids from beef tallow but cannot do so at lower temperatures, illustrating the link between the ω series and the animal's metabolic temperature. The marine biosphere is characterised by 22:6ω3 linoleic acid, although shorter-chain polyunsaturates abound in the lower trophic levels. In freshwater, the shorter C_{18} chains are also the rule. The ability of different teleost species to elongate C_{18} chains to the longer molecules like C_{22} which can be incorporated into membrane phospholipids, shows a fascinating correlation with position in the food web. Fish at higher trophic levels, like turbot and some gadoids, cannot elongate 18:3ω3 to the 22:6ω3 required and so in fish farming these species require such long-chain lipids in the diet; usually marine oils provide the most economic source.

There is a parallel here with the felids, who have similarly lost the ability to synthesise arachidonic acid from linoleic. Such top carnivores have dispensed, by means of evolution, with the need to use energy to elongate lipid chains. Red sea bream can elongate 18:3ω3 to 22:6ω3, but grow better in culture if fed on the latter, demonstrating the metabolic costs of the pathway. Turbot can taste the difference, and prefer to feed on food containing the longer-chain lipid. Nutrition experiments have shown that plaice, herring, capelin, menhaden and most salmonids can elongate C_{18} fatty acids successfully, and the herbivorous fish such as mullet and anchovy probably have the best ability to do this.

Vitamins. The term vitamin covers a large heterogeneous group of substances classified by what they do rather than by what they are. Cowey and Sargent (1972) consider that modern definitions of the term vitamin usually include the following points: the organic nature of the substances concerned; the presence of these substances in natural food in extremely small amounts; the distinctive differences between these substances and the major components of the food such as proteins; the absolute dietary requirement which is displayed by animals (they cannot be synthesised); specific deficiency diseases occur when these compounds are totally absent from the diet.

The water-soluble vitamins, which include vitamin C and the B complex vitamins, are, as in mammals, part of the co-enzyme systems in biochemical pathways and play a basic role in fish biochemistry. Deficiencies in water-soluble vitamins induce biochemical failure, resulting in changes in the form or behaviour of the animal. The process is gradual starting with a steady depleting of the co-enzyme. Next follows a decline in activity of those enzymes dependent on the missing co-enzyme. As a result of this decreased enzyme activity there is a general decline in the well-being of the animal shown by loss of appetite and excessive irritability. Finally diagnostic tissue pathology may develop leading to permanent damage and death.

Halver (1979) lists the quantities of water-soluble vitamins required by fish in milligrams per kilogram of body weight per day. At present there is no knowledge on how the requirement of a specific vitamin can be reduced if another vitamin is included in large amounts or if a vitamin precursor is present. There is also little work to show whether or not vitamin requirements are altered by different proportions of protein, carbohydrate and fat in the diet. It is known that thiamine requirements are related to carbohydrate intake in certain herbivorous species such as the mullets. Likewise the requirement for pyroxidine is increased with increasing dietary protein in certain carnivorous fish. Other factors, such as the water temperature and stress, can also affect the requirement for certain vitamins.

Little quantitative information is available on the requirements for fat-soluble vitamins. It is known that in salmonids vitamins A, K and E are required for normal growth and maturity (Halver, 1979). In water with a high level of calcium, vitamin D is not required by salmonids, but vitamin D_3 can promote calcium uptake in the gut in certain water systems and has produced better growth under experimental conditions. Fish can store fat-soluble vitamins and if too much is accumulated their health can suffer. It is clear then that fish have requirements for specific quantities of fat-soluble vitamins and these requirements need to be determined before a proper artificial diet can be made up.

Cowey and Sargent (1972) define the vitamin dietary requirement as 'that . . . level of a vitamin which permits optimal activity of all those enzymes for which the vitamin serves (possibly in a modified form) as co-enzyme'. To determine the level of vitamin required one would have to measure the biochemical activity of each enzyme affected.

Minerals. The needs for common minerals such as calcium and phosphorus have been determined for a few species (Halver, 1979). Calcium can be absorbed through the gills and through the gut given enough vitamin D_3 in the diet. Analysis has shown that in many diets presently used for fish, calcium levels are adequate but phosphorus is often not present in large enough quantities. Fish seem particularly sensitive to the type of phosphorus so that gross chemical analyses do not provide sufficient information.

Many trace minerals act as co-enzymes in biochemical systems (Cowey and Sargent, 1972; Halver, 1979), but not much is known about their requirements. It is considered that the classical trace elements are required to activate certain enzyme systems and so they are included in most diets. Other elements known to produce deficiencies if absent are iodine and cobalt. Although it has been shown that cobalt cannot be used by the metabolism of the fish to synthesise vitamin B_2, it is still added to diets because it is thought that gut bacteria in fish can incorporate the cobalt into the vitamin.

The Economics of Feeding

Fish have to expend energy to gain energy. There are four main problems a fish has to solve in finding and capturing food. First, patches of food must be located. Once found, a decision must be taken as to whether the patch is worth exploiting and, if so, for how long. In the patch the fish needs to decide which items of food are worth selecting, assuming that the food organisms are not all the same size; this can be called diet selection. Research on the economics of feeding is at present fashionable (see Pyke *et al.*, 1977), and contributes towards a better understanding of the factors that influence growth. For example, Werner (1980) has argued that 'what is critically needed (for a better appreciation of production biology) is an understanding of the mechanisms by which a fish chooses to utilise specific resources, with what efficiency, and how this relates to the capabilities of other species (competitors and predators) in the system'. Only with this information can the fisheries ecologist take proper account of the amount of food a fish eats, often the starting point in many studies of fish production. For reasons of space we concentrate in this book on the economics of diet selection in fish.

Individuals that are most efficient at some activity will produce more offspring and as a result the genes that control the activity leading to success will become more numerous. It is thought that the ability to forage for food optimally has been selected for in many animals (although it may not apply to strongly opportunist predators). Any function or structure can be analysed in the light of this theory, but we will look specifically at the analysis of optimal feeding behaviour and ecology. Feeding behaviour is composed of the four different activities mentioned above linked by a decision chain. This produces a foraging strategy which is composed of a cluster of behaviours. The fitness of each variant could in theory be measured, but this is difficult in fish. Lifetime reproductive success has only been measured directly in a few species of birds and insects by means of accurate long-term field studies. Consequently a common currency must be chosen that is positively correlated with fitness, and which is preferably easily measured over a short period of time.

Energy is the currency most often chosen (Pyke *et al.*, 1977) and in fish this choice is judicious. Since higher energy gain is associated with fast growth, then a fish which has had a consistently high energy intake will be bigger at a given age than one of equal years that has had a lower energy intake. Also, in fish the number of eggs produced increases with size (see Chapter 3), so the well-fed and larger fish will produce more eggs earlier and will contribute more offspring to the next generation. Consequently the behaviour that led to the bigger fish obtaining more energy will be favoured and the genes which enabled it to forage more efficiently will spread through the population. For these reasons the currency chosen in optimal foraging studies is energy, and net energy gain per unit time is the item usually optimised. This is assumed to happen either by maximising the net gain or minimising the energy loss during feeding. Several studies (see Pyke *et al.*, 1977) have made a further assumption about the utilisation and gain of energy. Direct estimation of energy consumption during pursuit is not always practical, so it is assumed that time is positively correlated with energy consumption. Similarly it is assumed that a larger food item will contain more energy than will a smaller one. Following from these assumptions results are expressed as grams caught per unit time and this is taken as an index of joules caught per joule expended. In some instances an animal might be best optimising joules caught per unit time, as time saved in food gathering means more time left for doing other things.

Each potential food item has a food value to the fish which can most simply be expressed as the number of joules per gram. In obtaining each item there are costs incurred by the fish. First the prey must be pursued, and with fish bigger items will often have to be pursued longer as the speed at which fish can swim increases with their length (Webb,

Figure 4.1: A Hypothetical Graph Showing the Net Energy Gained per Food Item Plotted Against the Food Particle Size. The average gain is marked by the horizontal line dividing the figure into the upper section A and the lower section B. The optimal diet includes all those items that yield a net gain that is greater than the average gain.

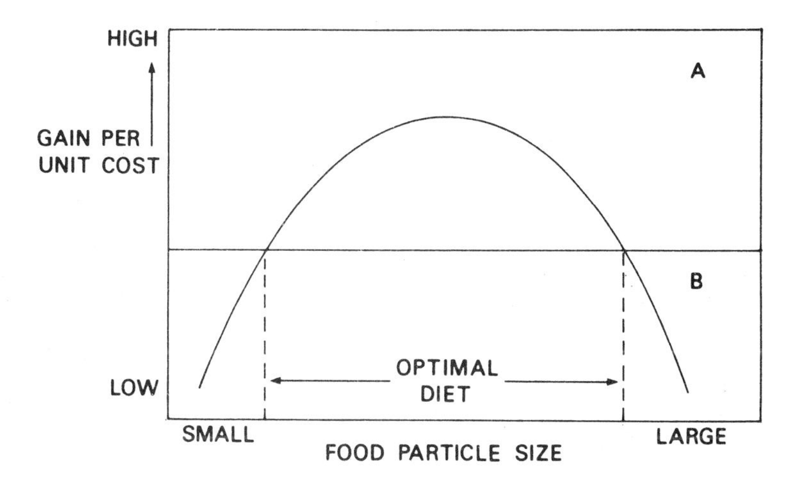

1975). Once caught the item must be manipulated or chewed before it can be swallowed. This takes longer for large items which could mean increased energy consumption, depending on the degree to which the prey struggles. A 100 g pike can swallow a 1 g minnow in 2 seconds but it takes 14–15 s to swallow a 4 g minnow (Hart, unpublished data). Handling times can be difficult to measure in fish like cyprinids, which engulf prey by protrusion and chew it up with pharyngeal teeth at the back of the pharynx. When pursuit and handling times are added we have the total time spent obtaining the item. The net gain can then be expressed as grams or joules gained per unit time. The prey items can now be ranked according to their profitability; theory predicts (Krebs, 1978) that an item should be added to the diet so long as the net gain contributed by it is greater than the average net gain received from all the items previously selected. As soon as this inequality no longer holds the new item is rejected and those items selected constitute the optimal diet. The principle is illustrated in Figure 4.1. A further prediction is that the decision whether or not to include an item in the diet is independent of the abundance of the item. The choice only depends on the relative abundance of the more profitable items (Pyke *et al.*, 1977).

This means that as the abundance of the most profitable food decreases the diet should widen to include less and less profitable foods. The reverse is true when the abundance of the preferred food item increases.

Some of this theory has been tested using the North American sunfishes (Werner, 1979). The predictions about optimal diet was tested using blue-gill sunfish feeding on three different sizes of *Daphnia*. The experimental results compared very favourably with the predicted outcome; the optimum prey size (the largest) was chosen more often than were the other two sizes when prey densities were high. At low prey densities all three size groups offered were chosen equally. O'Brien *et al.* (1976) provided a causal explanation for this result. They suggested that the sunfishes will take the apparently larger prey, so that at low densities small prey would be taken more often because the chances of closely encountering a large prey would be small. Gibson (1980) tested this 'apparent size hypothesis' with sticklebacks feeding on *Daphnia*, and found that it did not correctly predict the observed choice. The fish's choice of food was still consistent with optimal diet theory; even at high densities the large prey were not common enough to make it worthwhile for the fish to ignore small prey.

In any fish species, bigger individuals eat bigger prey (Werner, 1972). In general small individuals of a species eat a more restricted range of prey sizes, for reasons of mouth morphology. Werner (1979) analysed the consequences of mouth size in the bluegill sunfish. The costs per unit weight of prey caught for different sizes of bluegill are shown in Figure 4.2. The results show that small fish have a narrow range of prey size in the diet, whilst larger fish can profitably take food from a wider size range. The dotted line in Figure 4.2 shows the diet width for different sizes of consumer when costs in time per unit return are 4 s mg^{-1}. An interesting point is that all fish greater than 50 mm are competing with each other for food. Werner suggests that this partly explains why bluegills often become stunted and why different size classes tend to occupy different habitats.

Using these results on the size range of food, Werner was able to calculate expected mean prey sizes for different sizes of bluegill. These are shown in Figure 4.3 along with mean prey sizes from field studies and it can be seen that agreement between the two is good. The optimum prey size for each size class is also indicated and it is clear that the bigger the fish, the larger the difference between mean prey size and optimum prey size. Does this large difference have an influence on growth rate? In other words, could the older bluegills grow faster if their mean food size was closer to the optimum? Data from sunfish

Figure 4.2: The Cost of Prey Capture, as Handling Time per Unit Capture, Plotted against Prey Weight for Bluegill Sunfish of 20, 30, 40, 50 and 70 g. The dotted horizontal line represents the average handling time for an environment and defines diet width for the differently-sized sunfish (see Figure 4.1).

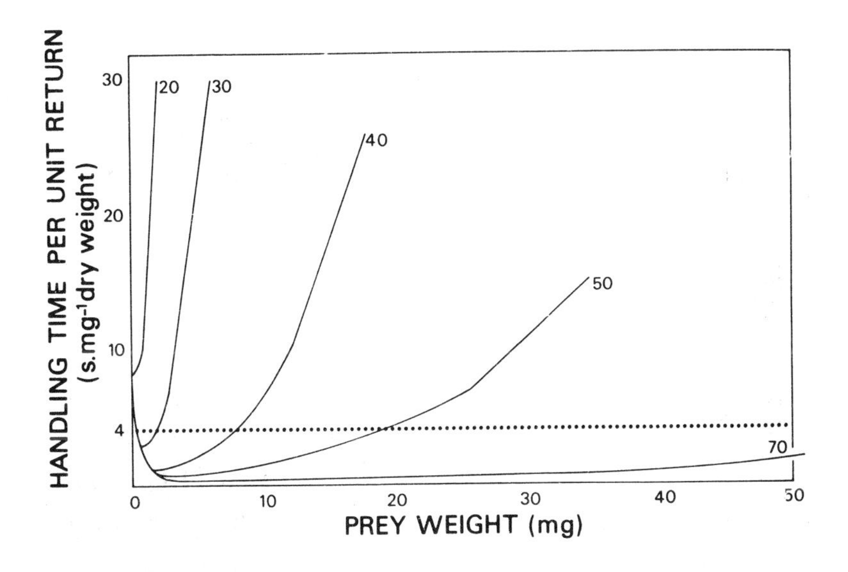

Source: Redrawn from Werner (1979), with permission.

show a positive correlation between prey size and growth rate (Werner, 1979), but at present there is little further evidence. Jones (1976) has suggested that Arcto-Norwegian cod grow to a larger size than the Faroe cod because of the presence in the Barents Sea of the capelin, although this size difference could just be the result of greater food abundance in the Barents Sea. In freshwater old roach can increase their growth rate if they get large enough to tackle big water snails; similarly perch begin to grow faster at four or five years old when they become large enough to switch to small fish as prey.

As well as selecting the optimal diet, fish are capable of searching for the most profitable patches and can switch their hunting effort as the relative rewards levels in the patches alter.

Figure 4.3: The Optimum Prey Size and the Predicted Mean Prey Size Plotted against Different-size Bluegill Sunfish. The optimum prey size is calculated from the data shown in Figure 4.2. The black squares show actual mean prey sizes.

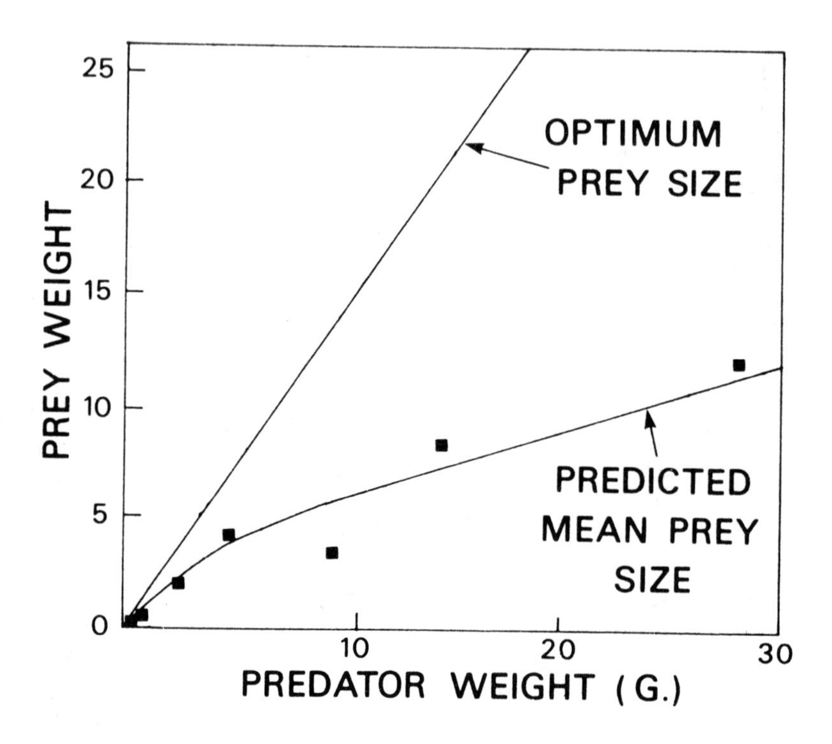

Source: Redrawn from Werner (1979), with permission.

Fish Guts and their Digestive Capacities

The form and function of the fish alimentary canal determines the types of food that can be eaten and the rate at which it can be processed. The alimentary system includes the mouth, jaws and teeth, but we will not discuss details of these here. Primitive forms such as the lampreys have the simplest type of system which merely consists of a straight-through tube. Elasmobranchs and most teleosts have a pharynx, esophagus, stomach, intestine and rectum although some teleosts, such as the cyprinids, do not have a true stomach. Associated with the

Figure 4.4: The Form and Structure of the Alimentary Canal. A. Of an Elasmobranch as Illustrated by a Shark. B. Of a Teleost as Illustrated by a Perch.

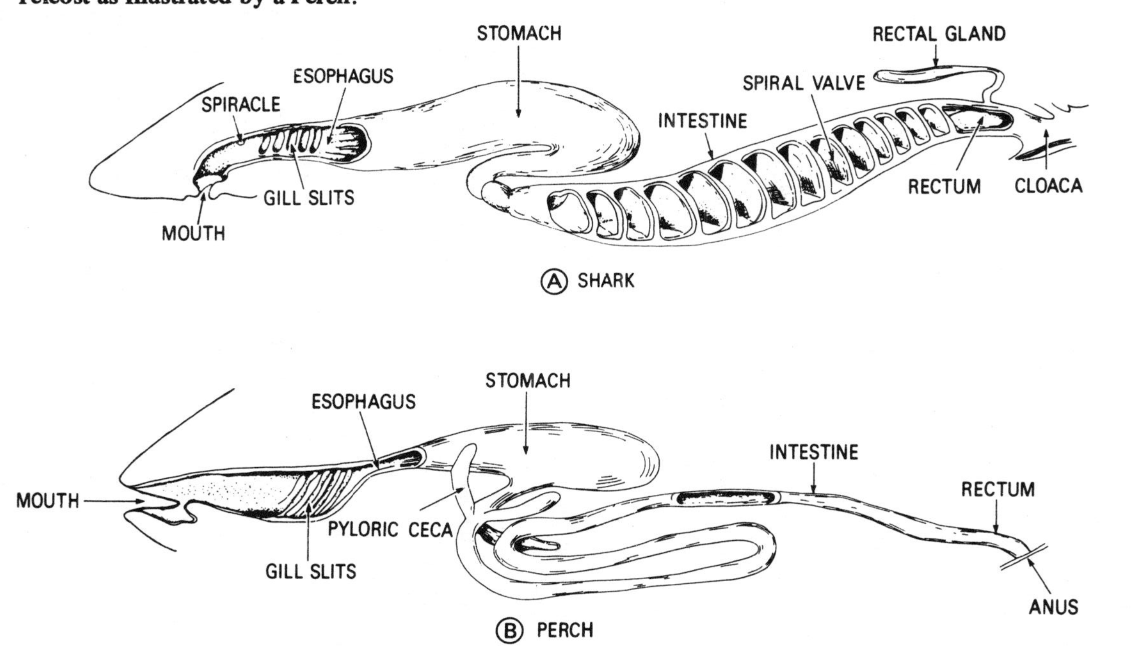

Source: Redrawn from Romer (1962), with permission.

system is the liver, pancreas and spleen. Figure 4.4 shows a teleost- and elasmobranch-type gut.

Fish have a rigid tongue and no salivary glands, although mucus is produced in the mouth and esophagus. Some fish, such as the cyprinids with their pharyngeal teeth, masticate their food but most tear food apart and swallow small chunks or take in the food whole. The stomach is both a storage device and a site for some predigestion. The cells of the stomach wall produce pepsin, a protein-digesting enzyme and hydrochloric acid which gives the correct pH for the functioning of pepsin. Most fish stomachs are thin-walled but in some species, for example the detritus-eating mullets, a part has become more heavily muscled and is lined by a horny layer so that it acts to grind up the food. Some species have a number of pyloric cecae attached near the exit of the stomach (Figure 4.4). Their number varies from species to species and their function is not entirely clear although they probably help to increase the digestive area (Greenwood, 1975).

The intestine is the area where digestion is completed and food absorbed. In piscivorous teleosts it is quite short but vegetarian species have long, highly-coiled intestines which allow a much larger absorptive surface. Elasmobranchs and primitive fish have a spiral valve (see Figure 4.4) which acts to increase the surface area in an otherwise short intestine. A range of digestive enzymes are excreted into the intestine either from the intestinal wall or from the pancreas. Many fish have only a slight distinction between parts of the intestine, but some have a short straight section joining the coiled absorptive part to the anus; this is analogous to the large intestine in mammals. In elasmobranchs the rectum opens into a cloaca which also receives the exits from the urino-genital system. Teleosts have a separate anus.

The ecologically important feature of digestion is the rate at which food can be processed, as this determines the upper limit to the intake of energy and hence the upper limit to the growth rate. Methods of estimating daily rates of food consumption have been reviewed by Elliott and Persson (1978), Elliott (1979) and Fänge and Grove (1979).

The manner in which food is evacuated from the stomach has been modelled in at least four different ways (Wirjoatmodjo, 1980). The simplest approach assumes that food is evacuated from the stomach at a constant rate leading to a linear relationship between stomach content and time after feeding. This type of model fits data from channel cat-fish and garfish. The second model assumes that stomach content declines exponentially with time, meaning that the rate of emptying is proportional to fullness. The exponential model fits data from sockeye salmon and brown trout and can be fitted to data by plotting the natural log of fullness against time. A third model assumes that the amount of contraction exerted by the stomach on the contained food, and hence the quantity of food evacuated, is proportional to the amount by which the stomach is stretched. The radial distension model fits data from large-mouth bass and plaice. Plotting the square root of stomach fullness against time gives a straight line. For the fourth model it is proposed that the rate of digestion is proportional to the surface area of the food bolus; digestion works from the outside in. If this model is appropriate then a plot a stomach fullness raised to the power of 0.67 against time is a straight line. The surface area model has been fitted to data from plaice, dab and turbot. For all four models the rate of evacuation will vary with temperature, meal size, food quality and fish size. For example, the rate usually increases exponentially with temperature.

Since all four models have been fitted successfully to data from different fish species, no single model can be the best. Wirjoatmodjo (1980) studied evacuation rates in the flounder and fitted all four models for comparative purposes. The results showed that the radial distension model gave the best fit if the stomach was scored as empty when 1 per cent of food eaten was left. The exponential model provided the best estimate of the starting weight of the food in the stomach. It also gave a better estimate of emptying time if the stomach was scored as empty when 5 per cent of the original meal was left. Overall, the exponential model was favoured because it was easy to use and to fit to data from digestion experiments on three natural food organisms of flounders.

Elliott and Persson (1978) have developed an equation which describes the amount of food in a fish's stomach at a particular time, from which an estimate of food consumed can be obtained. It is assumed that the instantaneous feeding rate F is constant and that R, the instantaneous rate at which the stomach is emptied, follows the exponential model. We can then write

$$\frac{dS}{dt} = (F - R)\,S \qquad (4.1)$$

which describes the rate at which the stomach content S changes with time. The amount of food in the stomach at time t is obtained by integrating the previous equation to give

$$S_t = S_o \exp(-Rt) + \frac{F}{R}\,(1 - \exp(-Rt)) \qquad (4.2)$$

where S_o is the amount of food in the stomach at the start of the time period. Rearranging this to get an equation for $F_t = C_t$, the amount of food consumed in the time period, gives

$$C_t = F_t = \frac{(S_t - S_o \exp(-Rt))Rt}{(1 - \exp(-Rt))} \tag{4.3}$$

To use the equation for fish in the field one would need estimates of S_o, S_t and R. Elliott and Persson (1978) tested the model against data from brown trout and perch, and found a close agreement except when large fish were used which fed close to their level of satiation. With these, C_t seems to fall away as the fish becomes satiated and S_t reaches a peak approximately at the point where the rate of food intake is at a maximum.

Allocation of Ingested Food

The food has now been selected and ingested. Once digested the materials will be used for repairing or building tissue whilst the energy will provide the fuel for these processes and for essential behaviours such as swimming, territorial defence and reproduction. In this chapter we are principally concerned with factors affecting growth; we look in this section at how much of the ingested energy is available for growth. Energy for growth is greatly affected by the amounts of energy needed for metabolism and we will discuss the amounts required. Materials are also used for growth but most research work on the internal allocation of materials and energy expresses its results in terms of the total energy equivalent.

A simple energy budget can be written in the following terms (Brett and Groves, 1979),

$$I = E + M + G \tag{4.4}$$

where I is ingested food, E is excretion, M is metabolism and G is growth. All biological systems obey the laws of thermodynamics so that the energy ingested must be accounted for by one of the three processes on the right-hand side of the equation. A variable amount of the ingested energy is excreted and we will briefly discuss the factors which influence this. Energy required for metabolism also changes with the activity of the fish. Consequently growth can only occur if there is an excess of energy after excretion has taken place and metabolism has been satisfied.

For a given type of fish the amount of ingested food which is excreted as faeces is a measure of the digestibility of the food. Winberg (1960) proposed that 80 per cent of most foods was digested by carnivorous fish, and that this figure could be taken as a general one for ecological calculations. Recent work (Brett and Groves, 1979) confirms that carnivorous fish digest and assimilate the largest proportion of food taken in. Even within this group the efficiency of digestion varies with the nature of the diet; carnivores feeding on invertebrates with a hard exoskeleton lose 17 per cent of the ingested energy compared to only a 5 per cent loss when feeding on soft-bodied invertebrates like polychaetes. Piscivorous species too are thrifty with the food eaten losing approximately 6 per cent of ingested energy. In contrast, the herbivorous fish have a harder time, like their counterparts in terrestrial ecosystems. Species such as grass carp and milkfish can lose 50 per cent of ingested energy in the faeces.

The proportion of food energy lost in the faeces is not static, but is a function of environmental and biological conditions. Elliott (1976) studied the effects of temperature, ration size and body weight on faecal energy loss in brown trout fed on *Gammarus pulex*. It was found that at a fixed temperature the energy loss increased from 14 to 23 per cent as the diet size approached the maximum. For a given ration, faecal loss decreased from 29 per cent at 4°C to 20 per cent at 19°C. Fish weight had no significant effect on the percentage. Presumably these results reflect increased enzyme action with temperature and increased enzyme production with ration.

Protein in the diet is usually highly digestible unless bound up inside plant cells. For example, sunfish feeding on mealworms digest about 97 per cent of the protein irrespective of diet size (Gerking, 1955). If the proteins digested are in excess of those required then they can be deaminated and the nitrogen excreted as ammonia or urea across the gills. This excretion represents a loss of ingested energy and must be included in the budget.

Metabolism takes a significant but variable share of the ingested energy (Webb, 1978; Brett and Groves, 1979). Variations in metabolism are caused by changing activity in the fish. The resting rate, when no activity is observed, is called the standard metabolic rate. At the other extreme is the upper limit of metabolism which accompanies intense activity; this is called active metabolism. In between the two extremes there are two other levels of significance: routine metabolism which accompanies spontaneous low-level activities, and feeding metabolism

Figure 4.5: The Active and Standard Metabolic Rates for Fingerling Sockeye Salmon as a Function of Temperature. The stippled area shows the metabolic rates associated with feeding at maintenance and at maximum ration. The stars show the maximum oxygen consumption rates at 15°C associated with aggression, excitement and lakeward migration.

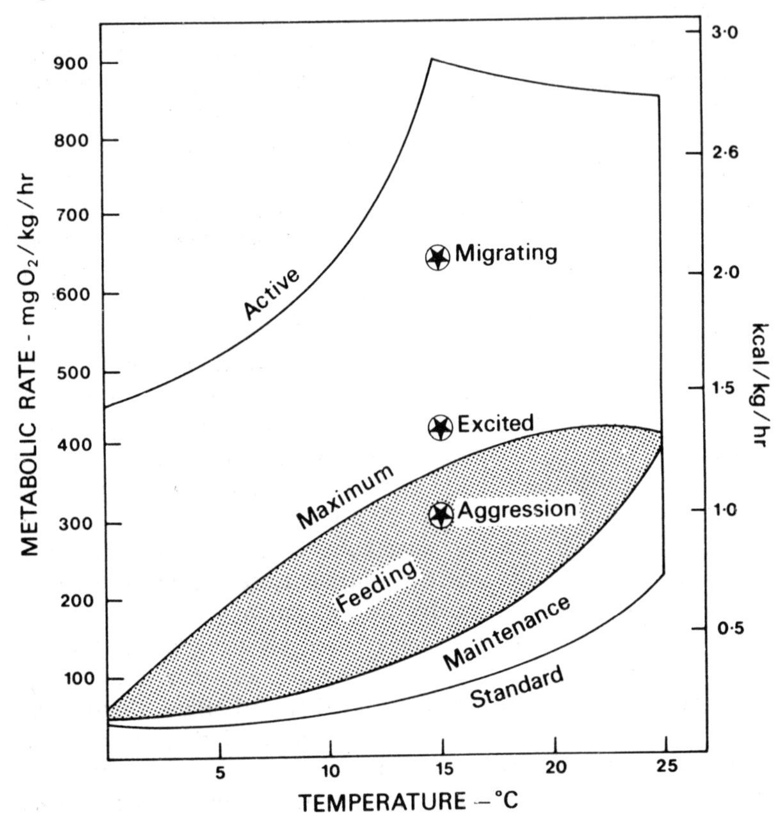

Source: Redrawn from Brett and Groves (1979), with permission.

which is the metabolic activity associated with food processing. An example of the range of these metabolic rates is shown in Figure 4.5.

The standard metabolic rate is influenced by both temperature and fish size. Most species are adapted to adjust their standard metabolic rate according to the temperature range normally experienced. Fish living in the polar regions have a standard metabolic rate ranging between the same limits as does the rate for temperate species. Tropical species tend to have rather higher standard rates. The size effect is described by the power function

$$M_s = \alpha W^{\gamma} \tag{4.5}$$

where α and γ are fitted constants and W is fish weight. The mean value for γ, calculated for a number of species, is 0.86 ± 0.03 (SE) (Brett and Groves, 1979). This relationship means that bigger individuals have a relatively lower standard metabolic rate than do small individuals. For example a 3 kg salmon has a standard rate which is approximately one-fifth that of a 1 g fry.

Active metabolic rate is caused by sustained swimming, such as cruising or migration. The rate is the upper limit of aerobic oxygen consumption. Some sudden burst of activity could go above the active rate but would be driven by anaerobic respiration which could not be sustained for long. In some species the active metabolic rate is influenced by temperature in the same way as is the standard rate, increasing according to a power curve. In other species the active rate increases up until a certain temperature and then falls away (e.g. in Figure 4.5). The scope for increase in metabolic rate above the standard rate can be eight to ten times in streamlined fast-swimming fish. Even a few short daily bursts of fast swimming can double the metabolic costs to the individual fish. Size also influences the active metabolic rate so that bigger fish have rather lower values. An interesting practical outcome of studies on active metabolic rate is that nervous, continuously-moving fish will use a lot of energy in metabolism which cannot, as a result, go to growth. The message seems clear, farmed fish should be kept calm.

Studies on cod by Saunders (1963) showed that the routine metabolic rate of 75 mg $O_2\,kg^{-1}\,hr^{-1}$ rose to 112 mg $O_2\,kg^{-1}\,hr^{-1}$ after feeding and took seven days to return to the routine rate. Subsequent work on other species of fish has shown similar results. A summary of results on fish that have been fed a high ration (Brett and Groves, 1979) shows that the ratio of this feeding metabolism to standard metabolism is 3.7 ± 1.2 (SD). The feeding metabolism is strongly influenced by ration size and temperature. At a given temperature feeding metabolism increases with ration size. Temperature changes raise or lower the metabolic cost of food processing. In fingerling sockeye salmon it has been found that the metabolic cost of feeding on a high ration at a high temperature is some eight times the cost at low ration size and temperature (Brett and Groves, 1979).

Figure 4.6: Flow Diagram to Show the Average Partitioning of Dietary Energy for a Carnivorous Fish. 100 calories (419 J) is equivalent to about 2 per cent dry weight per day for a 1 kg fish. The figure should be examined in conjunction with Table 4.2.

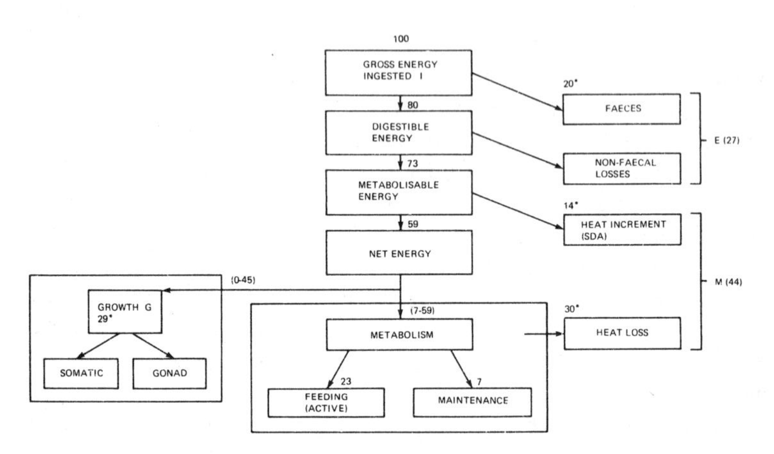

Source: Adapted from Brett and Groves (1979).

Table 4.2: Energy Budgets for Fish Species from a Range of Trophic Types.

Species	Diet	Energy Budget E%	M%	G%	Reference
Perch[a]	*Gammarus*	24.3	62.5	15.8	Solomon & Brafield (1972)
Brown trout[a]	*Gammarus*	31.5	53.3	15.8	Elliott (1976)
Grass carp	*Egeria*	41.7	8.3	50.0	Stanley (1974)
Atlantic cod	Chopped fish	5.3	72.5	22.2	Edwards *et al*. (1972)
Bluegill[a] sunfish	Mayfly nymphs	24.0	44.3	31.7	Pierce & Wissing (1974)
Carnivorous fish	–	27	44	29	Brett & Groves (1979)
Herbivorous fish	–	43	37	20	Brett & Groves (1979)

Note: a. Figures given in this table are averages of those in the original.
Source: Simplified from Brett and Groves (1979).

Part of the energy used during feeding metabolism goes towards the biochemical breakdown of proteins. This has been called the specific dynamic action (SDA) by many workers. Brett and Groves (1979) recommend that this energy component should be called the heat increment, as SDA is a mistranslation of the original German term and the process is not only specific to protein deamination. The heat increment is usually between 12 and 16 per cent of the ingested food energy; this proportion does not change significantly with fish weight.

The energy remaining after faecal energy loss and metabolism has been accounted for can go to growth. This can either be in the form of somatic or gonadal growth; the balance between the two will be discussed in a later section. The partitioning of energy is summarised in Figure 4.6 and some energy budgets are given in Table 4.2. In the next section we will discuss how growth rate and efficiency respond to factors such as ration size and temperature.

Growth Dynamics

Increase in fish mass or length slows with age but never ceases altogether as fish have indeterminate growth. (The typical form of fish growth curve is shown in Figure 4.12.) The rate at which fish grow depends on the difference between I and $M + E$. M and E vary little for fixed conditions and so growth rate is most dependent on the size of I. This was shown very clearly by Brett *et al.* (1969) working with fingerlings of the sockeye salmon. An example of their growth response to ration size is shown in Figure 4.7a. Three key ration sizes are defined as shown in the figure. Growth rate is expressed as the gain in weight per gram per day times 100 (the percentage specific growth rate, see below). The important ration sizes are the maintenance ration, R_{maint}, which just suffices to keep the metabolism of the fish ticking over with nothing spare for growth; the optimum ration which produces the maximum rate of growth per unit of food; and the maximum ration which allows the highest growth rate. Experiments have shown that each of these levels occured at a greater ration size as the temperature increased from 0°C to about 16°C, after which growth fell away as the temperature approached the lethal level at 25°C. In other words, increased temperatures caused the curve in Figure 4.7a to move to the right and upwards.

A further important feature of growth is the efficiency with which food is converted into new tissue. This can be divided into gross and

Figure 4.7: A. Growth Rate of Sockeye Salmon Fingerlings as a Function of Ration Size at 10°C. The tangent to the curve defines the point where the maximum rate of growth lies. B. Gross Conversion Efficiency of Sockeye Salmon Fingerlings as a Function of Ration Size at 10°C.

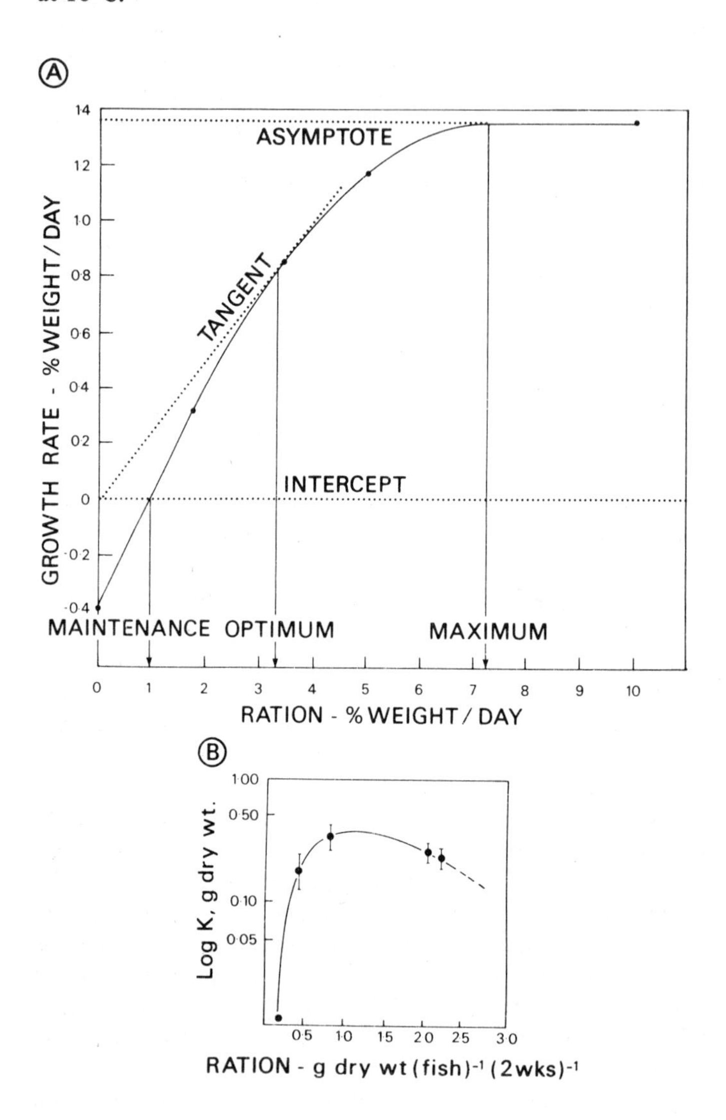

Source: Redrawn from Brett *et al*. (1969), with permission.

Figure 4.8: The Relationship between Specific Growth Rate in Per Cent Weight Gain Per Day (*G*), Ration Size (*R*) Given as Per Cent Body Weight Per Day, and Four Abiotic Factors. A. The Influence of Temperature, a Controlling Factor; B. The Influence of Salinity, a Masking Factor; C. The Influence of Light, a Directive Factor; D. The Influence of Oxygen, a Limiting Factor. For each factor growth at three levels is shown. See text for details of the categories of limiting factors.

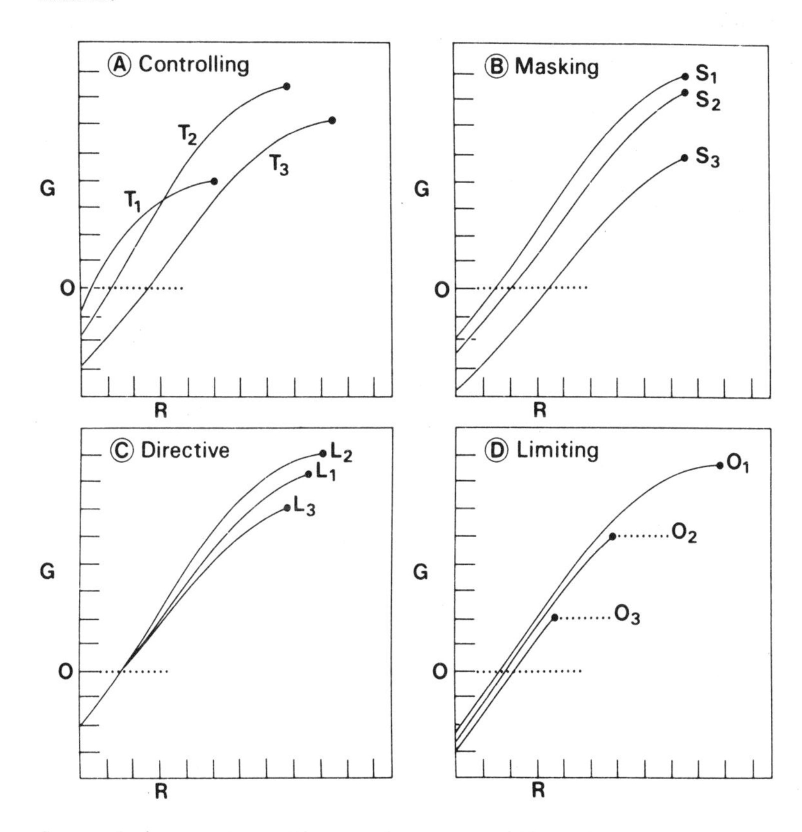

Source: Redrawn with simplifications from Brett (1979), with permission.

net growth efficiencies, and at least in the sockeye salmon, both change with temperature and ration size (Brett *et al*., 1969). We can define

$$\text{gross efficiency} \quad K_1 = \left(\frac{G}{R}\right) \times 100 \tag{4.6}$$

Figure 4.9: The Way in which Specific Growth Rate (G) in Relation to Ration Size (R) is Influenced by Four Biotic Factors. A. The Effect of Diets of Decreasing Nutrient Value and the Effect of Reduced Diet on Growth (G_1–G_3); B. The Influence of Increasing Weight (W_1–W_3) on the Growth Rate; C. The Way in which Dominance Order Affects Growth Rate Either by Increasing the Amount of Energy Consumed or by Decreasing the Amount of Food Available; D. The Case Where Size Differences Between Groups of Fish is Only Significant When the Ration is Restricted (R_{res}). See text for further details of the categories of limiting factors.

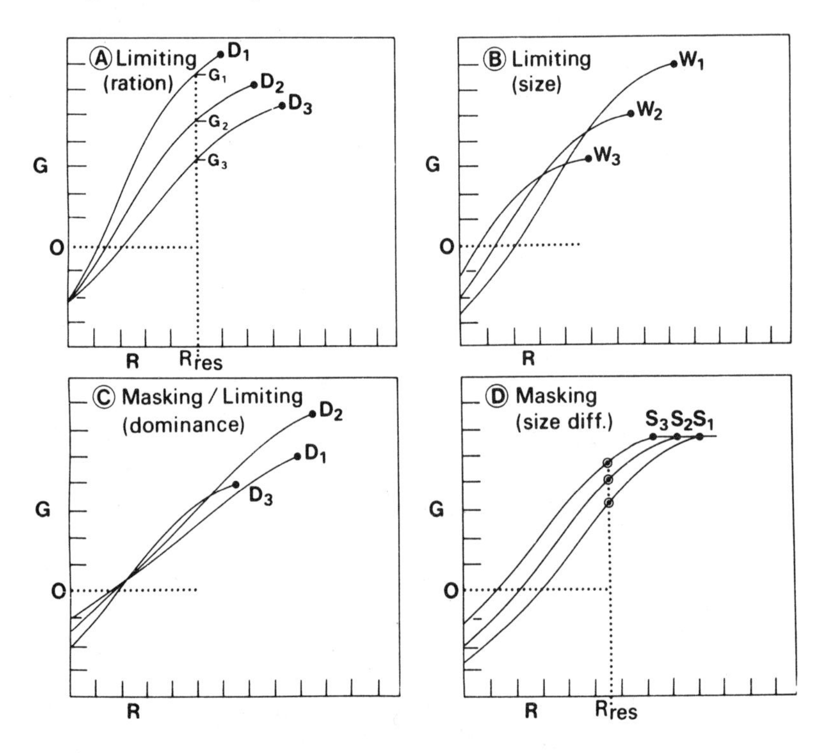

Source: Redrawn with simplifications from Brett (1979), with permission.

$$\text{and net efficiency } K_2 = \left(\frac{G}{R - R_{maint}}\right) \times 100 \qquad (4.7)$$

where G is the growth rate and R the ration size. Net efficiency will always be higher than the gross efficiency for a given combination of temperature and ration size. The gross efficiency at different ration

Figure 4.10: The Response of Gross Conversion Efficiency (*K*) to Changes in Ration Size and to Various Abiotic Factors (the Same Factors as in Figure 4.8). A. The Influence of Temperature, a Controlling Factor; B. The Influence of Salinity, a Masking Factor; C. The Influence of Light (in the Form of Static Photoperiods), a Directive Factor; D. The Influence of Oxygen, a Limiting Factor. The arrows marked 1, 2, 3 in each graph label the maximum efficiency under each set of conditions.

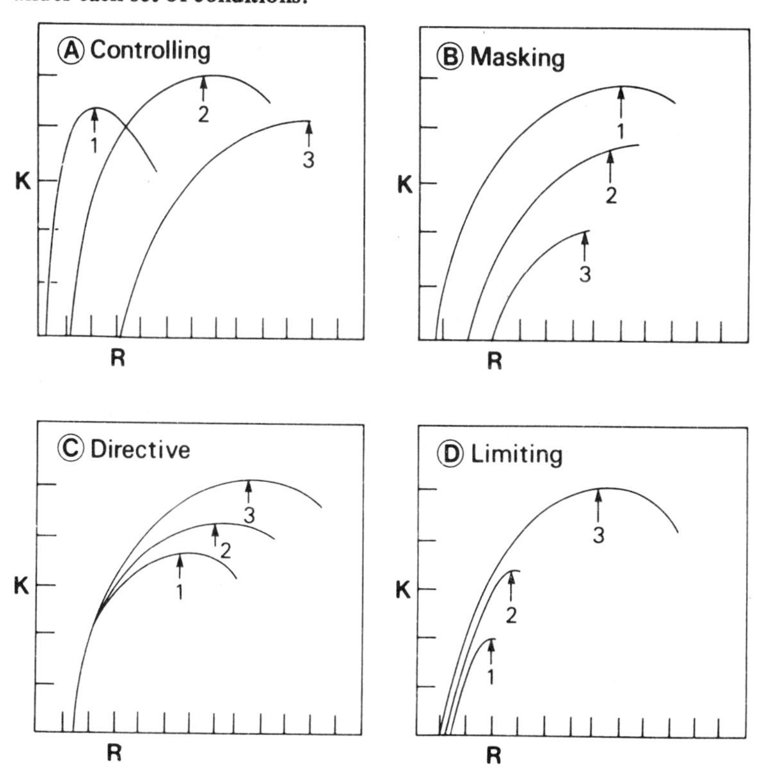

Source: Redrawn from Brett (1979), with permission.

sizes for sockeye salmon at 10°C is shown in Figure 4.7b.

Ration size is not the only factor to influence growth rate and gross efficiency. The biotic and abiotic environments influence fish activity and growth through metabolism (Brett, 1979). Fry (1947, 1971) used this approach to classify the effects of environmental variables. Four types were recognised. Controlling factors, such as temperature, govern the rate at which metabolic reactions occur.

Factors such as oxygen restrict the rate at which certain biochemical transformations can occur and are classified as Limiting Factors. A third type of factor prevents some other environmental feature from having its full effect; for example salinity imposes an extra load on the metabolic system so preventing it from responding fully to temperature changes. Such features are called Masking Factors. Finally there are the Directive Factors which cause the metabolic system to respond in an appropriate way. An example would be the way in which changing day length triggers the process of gonad maturation (see Chapter 11).

The influence of the four types of factor, both biotic and abiotic, on growth rate and efficiency, are considered by Brett (1979) and by Brett and Groves (1979). The principle findings are summarised in Figures 4.8, 4.9 and 4.10.

Growth rate and efficiency are not influenced only by each factor acting alone. Ration size is a strong influence on growth rate as shown in Figure 4.7a, but temperature moderates this influence and changes the response. The factors in Figures 4.8, 4.9 and 4.10 all interact in various ways, but we only have space here to mention the interaction between ration size and temperature. In 13 g sockeye salmon the optimum temperature for growth is 15°C on full rations of about 8 per cent body weight per day. Gross efficiency is greatest at 10°C and a ration of 5 per cent body weight per day. Other studies show that temperature, being a Controlling Factor, always acts in conjunction with Limiting Factors.

Strategies of Growth

Fish do not grow all the time. In tropical regions continuous growth may seem possible, but periodic events such as changes in light, nutrients, temperature, salinity and biological processes such as spawning or feeding can cause somatic growth checks. In temperature waters, in addition to these factors, the pattern of growth is clearly associated with the annual cycle in water temperature. For example LeCren (1951) suggested that perch growth in Windermere does not occur when the temperature is below 14°C, which means that no growth occurs through the winter.

Growth in size is not the only kind that a fish has to cope with. Gonads require energy input for their development and will compete with somatic growth for resources. Each kind of growth has costs and benefits which are intimately related to fitness. Bigger females have more eggs (Chapter 3) so there is sense in a fish delaying reproduction for a year or two, putting all available energy into somatic growth. This is a sensible strategy if the probability of the fish dying is low during the period of early somatic growth. The age at first maturity is then a balance between the advantages of being big and the increasing chance of dying as time goes by. Once reproduction has started somatic growth continues and it is of interest to see how the two types of growth are accommodated.

Iles (1974) has pointed out that in temperate marine fish such as the herring there is no sudden change in growth rate as the fish become mature and start to produce gonads. Annual somatic growth decreases each year, but the change is smooth and has no sudden step. Observations on many temperate species show that somatic growth usually occurs first in the season with the fish actively feeding on the rapidly growing zooplankton. The Atlanto-Scandian herring for example grows in the three months from May to July with both young and old fish growing at the same time. Growth during this period is fast to begin with, then settles down at a steady level until it tails off towards the end of the three months. In North Sea herring somatic growth takes place in April, May and June to be followed by gonadal growth in July, August and September. These seasonal cycles in mature fish are foreshadowed in the immature individuals (Iles, 1974). It is proposed that during the period of somatic growth an excess of materials is stored in the body to be later mobilised for gonadal growth. This strategy would make sense in species that only have access to abundant food for a limited period each year.

Some of the details of growth strategies have been supplied by Medford and MacKay (1978) and Diana and MacKay (1979) for somatic and gonadal growth in pike in relation to season, sex and age. In a standard 50 cm pike, gonadal growth in males and females starts in July in Lac Ste Anne, central Canada. By September the testes are fully grown but ovaries continue to grow throughout the winter until April. It is likely that males use endogenous energy to build up gonad tissue. During the growth of ovaries the females do not lose body weight, so it is likely that they keep feeding through the winter, taking more food than do males. By March both sexes have livers well stocked with protein, although the females have far greater reserves, with 163 per cent more protein in their livers than in those of the males. During final gonad maturation in March and April, female liver size decreases significantly. In males the decrease does not occur until the fish are spawning. Observations by Frost (1954) indicated that during spawning

Figure 4.11: Total and Somatic Production in Calories (1kcal = 4186J) by Female and Male Pike from Lac Ste Anne, Canada. Production is calculated between May of one year and March the next year.

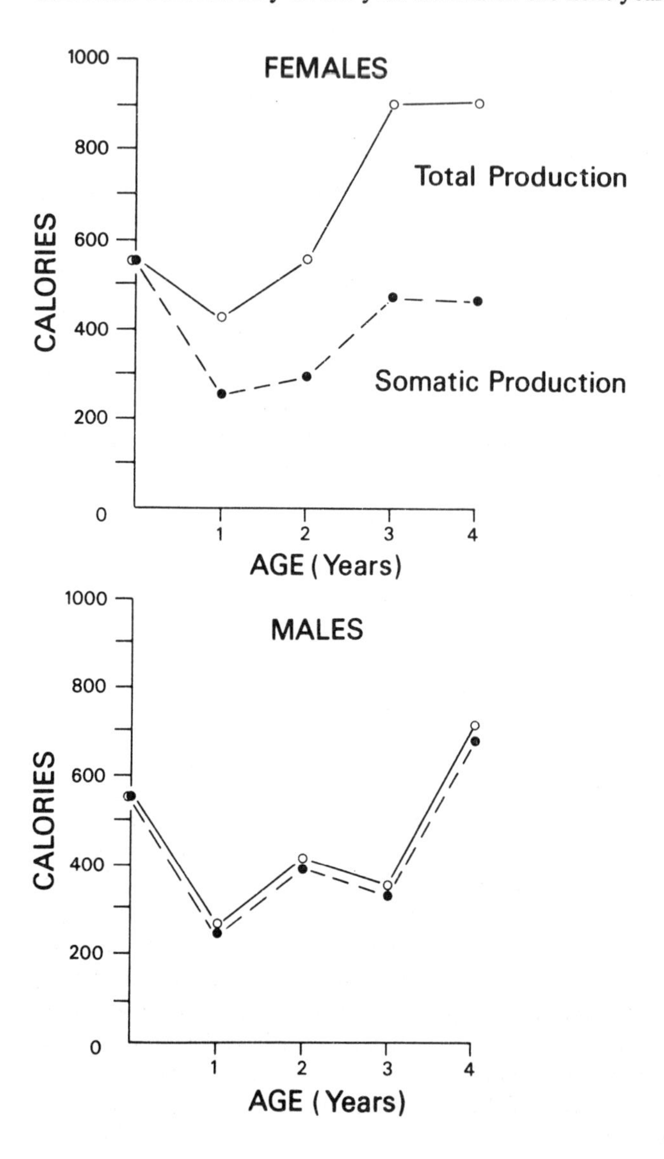

Source: Drawn from data in Diana and MacKay (1979).

pike stop feeding, so they must depend on endogenous energy. Somatic growth in females takes place only in the summer, while males continue somatic growth throughout the winter. The changes in production with age and sex are shown in Figure 4.11. From May to March female production is twice as high as male production, although expressed annually both sexes produce similar amounts of somatic tissue. Females have fifteen-fold higher gonadal production. Immature pike in their first year grow extremely fast, putting all energy into somatic growth, exemplifying the general growth/reproduction strategy described above.

Population Growth Parameters

No two fish grow at the same rate. Fishery managers need some measure of mean size at age and a method of modelling the average growth rate of a species (see Chapter 8). Although inequality of growth rates occurs, there are reasonably confined limits to the range of growth at each age in a particular habitat. The growth curve for roach from the River Nene in Northamptonshire, UK is shown in Figure 4.12a, giving an idea of the spread found in a natural population. The curve can be characterised by a list of the mean lengths at each age with a measure of the degree of variation. Means would be calculated from a sample of at least 20 fish per age group. Because of differential growth between the sexes (see below), growth data for the sexes must sometimes be obtained separately. Age is estimated using, in temperate species at least, some hard part such as scales, otoliths or vertebrae (Bagenal, 1978). Because tropical species are less likely to have cyclic interruptions to growth, their hard parts are not so useful for ageing. An alternative method of ageing is size-frequency analysis, reviewed by Macdonald and Pitcher (1979). For yields, it is better to use weight as a measure of size: some growth-in-weight curves are shown in Figure 4.12b. Weight is not as easy to measure as length, so that length is used more often.

Weight and length are related by a power relationship. Length is usually regarded as the independent variable, although it too shows variation. The relationship takes the form

$$w = a\,l^b \tag{4.8}$$

where $b \simeq 3$ and a is a constant. If weight and length are transformed

Figure 4.12: Growth in Length and Weight of Roach in the River Nene at Wollaston, Northants, UK. The solid circles show the mean length and the vertical bars show the 95 per cent confidence limits on the mean. The smooth continuous line is the von Bertalanffy fit to the data.

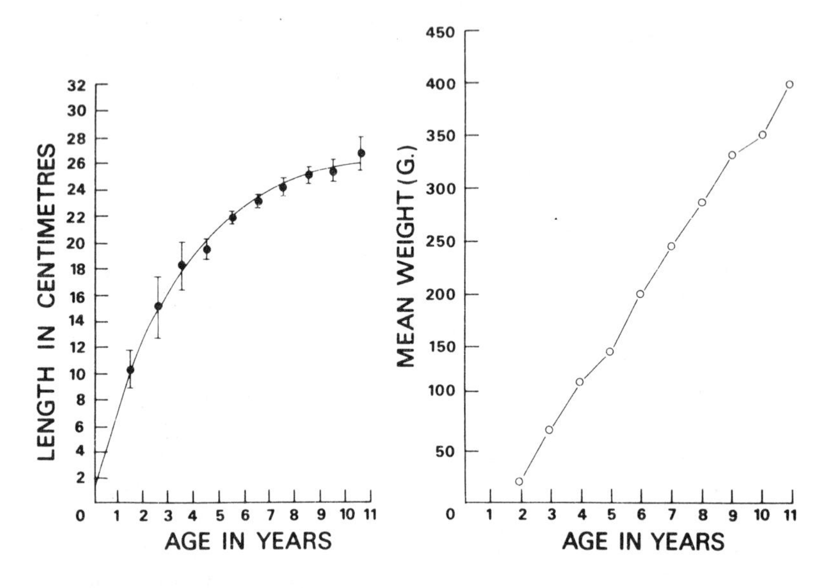

Source: Hart (1971).

to logarithms the above equation becomes

$$\ln(w) = \ln(a) + b \ln(l) \tag{4.9}$$

which is a straight line with slope b and intercept ln (a). This transformation makes it much easier to fit the relationship to data as a simple regression may be used. The exact statistical procedure that is appropriate is discussed by Ricker (1973).

The length-weight relationship has both applied and pure applications. Market sampling of fish of commercial importance will often only measure length, as fish are usually landed gutted (headed as well in Norway), and live weight cannot be measured. An estimate of it can be obtained using a predetermined length-weight regression. Care should be taken if both the coefficient of variation of mean length and the exponent b are large. Nielsen and Schoch (1980) have shown that under such circumstances the estimated mean weight can be 80 per cent different from the true mean.

In ecology the length-weight regression coefficient is often used as a measure of fish condition (e.g. LeCren, 1951). Condition is the volume of the fish relative to length and this is taken to be a measure of well-being. A cyclical change in condition occurs annually as the fish, particularly the female (see previous section), develops its gonads. While these grow the fish becomes relatively heavier, so that b increases and may be greater than 3. Directly after spawning the fish is thin and in poor condition so that b is less than 3. More traditionally it has been assumed that $b = 3$ at all times and that condition is defined by

$$k = \frac{w}{l^3} \qquad (4.10)$$

with k close to 1. When the fish is in good condition then w will increase so that k is greater than 1, and vice versa when w is low. Using b as a measure of condition is a generalisation of k, taking account of the changing relationship between length and weight.

Growth rate can be defined in a number of ways. The absolute growth rate is the increment of weight $w_2 - w_1$ over a known time interval, i.e.

$$\frac{w_2 - w_1}{t_2 - t_1} \qquad (4.11)$$

Often the growth rate is referred to the weight of the fish, in other words the relative growth rate is obtained. It is Equation 4.11 divided by the fish's weight (w_1) at the start of the time interval. When the time interval is made infinitely small, Equation 4.11 becomes a differential equation and we have dw/dt for the absolute rate and dw/dtw for the relative rate. So these become instantaneous rates. The instantaneous relative growth rate is shortened by Ricker (1979) to the instantaneous growth rate, and is often called the specific growth rate in the fisheries literature.

Growth in length can be modelled in a number of ways – see Jones (1976) and Ricker (1979) for reviews. A single model will not fit the growth curve from egg to old age, so a number of models are usually employed for different segments or stanzas of the growth curve. Because of our interest in modelling the fished part of the population, which is usually adult, we will only describe models applying to the final stanza of life. This can be taken as starting when the fish has recruited (Chapter 6).

The model most frequently used in fisheries work is due to von Bertalanffy (1957), although the original concepts underlying it derive from the work of Pütter (1920). Because the von Bertalanffy model will play an important part in Chapter 8, we will describe it here in detail. Any model must take account of the way in which the growth rate slows with time. The von Bertalanffy equation assumes that fish grow towards some theoretical maximum length or weight, and that the closer the length gets to the maximum the slower the rate of change of size. The equation is

$$l_t = L_\infty(1 - \exp(-k(t - t_0))) \text{ for length} \tag{4.12}$$

and $$w_t = W_\infty(1 - \exp(-k(t - t_0)))^3 \text{ for weight} \tag{4.13}$$

l_t and w_t being the length and weight at time t. The equation is most often used to describe changes in size from one year to the next. As a result t is usually the age of the fish, although this usage is not the only one possible. L_∞ and W_∞ are the length or weight of fish of infinite age. The growth parameter k is a measure of the speed at which the length approaches W_∞ or L_∞. The larger k, the faster W_∞ or L_∞ is reached. A survey of many species of temperate fish (Jones, 1976) has shown that L_∞ is usually inversely related to k, although the relationship is not a simple straight line. t_0 is a parameter which needs interpretation; its presence is a consequence of the mathematical structure of Equations 4.12 and 4.13. Gallucci and Quinn (1979) point out that the derivative of Equation 4.12 is

$$l' = k(L_\infty - l) \tag{4.14a}$$

which can be rearranged to

$$l' + kl = kL_\infty \tag{4.14b}$$

with the initial condition $l_0 = l(t_0)$. Equation 4.14b is, in mathematical terms, a first-order linear inhomogenous differential equation and as such can be solved using standard mathematical procedures. Any solution will be entirely general specifying an infinite set of solution curves, unless an initial condition is specified and values for k and L_∞ obtained. This is the role played by t_0 which is, in biological terms, the age at which $l_0 = 0$. When Equation 4.12 is fitted to data it can happen that t_0 is negative, which clearly has no biological meaning;

this phenomenon emphasises the fact that models of this sort are really only convenient ways of describing the data and do not say anything about the nature of the growth process itself. A more intuitive way of thinking about t_0 is as a scaling factor which moves the von Bertalanffy curve, specified in shape by k and L_∞, along the age axis until the curve fits the data points.

The model can be fitted using a least squares method (Allen, 1966). Alternatively, an often used classical technique is the Walford plot which can be applied easily without a computer (Ricker, 1975).

Another curve in which weight at age tends to an asymptote is the logistic equation (first introducted in Chapter 3),

$$W_t = \frac{W_\infty}{1 + \exp(-g\,(t - t_0))} \qquad (4.15)$$

where the new symbols are g, the instantaneous rate of growth as w_t tends to zero, and t_0, the inflection point of the curve. This is the point where the rate of change of weight equals zero (see Figure 3.3). Unrealistically, the logistic curve is symmetrical about the inflection point at a weight equal to $W_\infty/2$. As we will see in Chapter 8, the logistic curve is mainly used in fisheries to describe the change in the size of exploited populations. A logistic equation could also be developed to describe the change with age in length. It can be fitted using non-linear least squares methods (Somerton, 1980).

The final growth model we wish to describe is due to Gompertz (1825). The model has size tending to an asymptote with an inflection point asymmetrically placed, at an age less than half way up the curve, realistic for weight curves. The Gompertz equation for weight is

$$W_t = W_\infty \exp(-g\,(t - t_0)) \qquad (4.16)$$

where g is the instantaneous rate of growth when $t = t_0$, whilst t_0 is the inflection point of the curve at $W_\infty/\exp(1)$. Methods for fitting this model are given by Ricker (1975, 1979), where details of other models can also be found (see also Table 4.3).

Further refinements of the von Bertalanffy equation have taken account of the seasonal nature of growth in temperate species, for example, Pitcher and Macdonald (1973a). No one equation is intrinsically better than another; each must be fitted to the data and the best one chosen. This implies that the parameters of the models have no absolute meaning. Comparisons between parameter values can be useful when the same model is fitted to data from different stocks of the same species, the different values for k and L_∞ in Table 3.1, for example.

Table 4.3A. The Different Models of Fish Growth and Mortality Used by Allen (1971).

Name of Function	Mathematical Form
1. *Mortality functions*	
A. Simple exponential	$N_t = N_0 e^{-Zt}$
B. Multiple exponential	$N_t = N_0 \exp(-[(Z_n(t - T_{n-1}) + \sum_{i=1}^{n-1} Z_i (T_i - T_{i-1})])$
C. Linear models	$N_t = N_0 (1 - t/T)$
D. Fixed lifespan	$(N_{t+1} = N_t, t < T)$ $(N_{t+1} = 0\ , t > T)$
2. *Growth functions*	
A. Exponential growth	$w_t = w_0 e^{Gt}$
B. Simple asymptotic	$w_t = W_\infty (1 - \exp(-kt))$
C. von Bertalanffy	$w_t = W_\infty (1 - \exp(-kt))^3$
D. Linear growth in length	$w_t = w_T t^3/T^3$
E. Linear growth in weight	$w_t = w_T t/T$
F. Growth proportional to t^m	$w_t = w_T (t/T)^m$

Table 4.3B: The P:B Ratios that Result from All the Combinations of These Models.

Mortality Function	P:B Ratio for Growth Functions A	B	C	D	E	F	Mean Age	Mean Lifespan
A	G	Z	Z	Z	Z	Z	$1/Z$	$1/Z$
B	G	$f(k,Z)$	$f(k,Z)$	$f(Z)$	$f(Z)$	$f(Z)$	c.f.[a]	$\Sigma^n D_n/Z_n$
C	G	$f(k,T)$	$f(k,T)$	$5/T$	$3/T$	$m+2/T$	$T/3$	$T/2$
D	G	$f(k,T)$	$f(k,T)$	$4/T$	$2/T$	$m+1/T$	$T/2$	T

Note: a. Complex function of Z_n, D_n, T_n and S_n. D_n – proportion of original population dying during the nth period; S_n – proportion surviving at end of nth period.

Fish Production

Production is an increase in biomass over a given period of time. The increase can come from growth of new tissue or by the production of new offspring. During the period of assessment, biomass will also be lost through deaths and emigration. Net production is therefore a

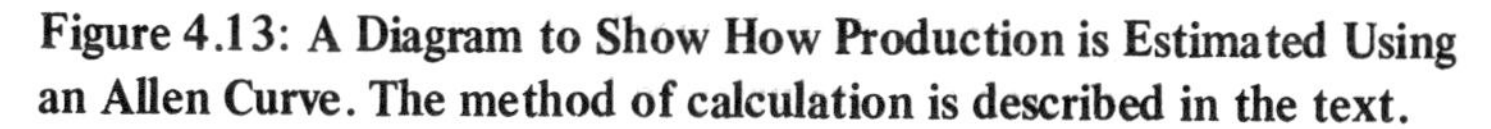

Figure 4.13: A Diagram to Show How Production is Estimated Using an Allen Curve. The method of calculation is described in the text.

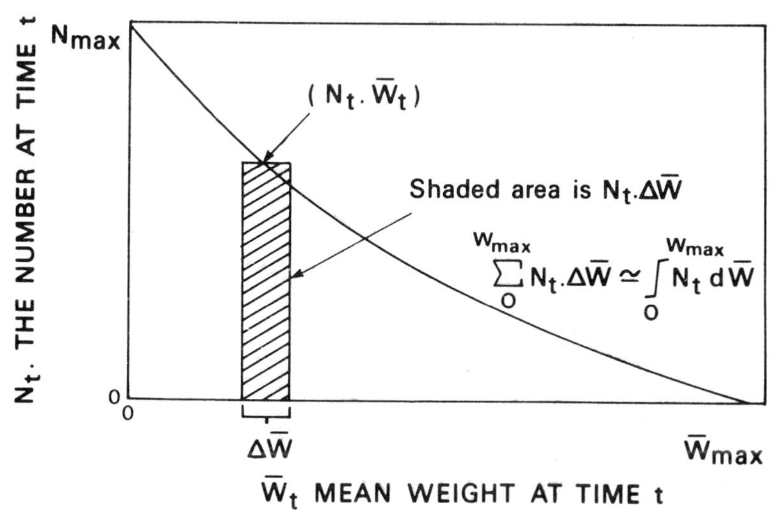

balance between gains from growth and reproduction, and losses from deaths and emigration, so these items must be measured to calculate production. Production can be estimated graphically using an Allen curve or calculated directly by making assumptions about the form of growth and mortality over time. We will first discuss the use of the Allen curve.

Imagine a pond containing one year class of fish which have just hatched. Once a month over the next twelve months the numbers surviving, and their mean weight, are estimated and plotted on a graph. This is shown in Figure 4.13 where it is assumed that there were 1,000 fish alive (N_1) at the start of the year. After mortality had reduced N_1 to N_2, the individual fish had grown from w_1 to w_2 grams. The production P_T for this period is $\bar{N}(\bar{w}_2 - \bar{w}_1)$ grams. The whole time period of the study could be split up into similar time (one month) segments, the sum of all the small rectangles providing on estimate of the production for the year. We can write

$$P_T = \sum_{i=0}^{12} P_i = \sum_{i=0}^{12} \bar{N}_i \, (\bar{w}_{i+1} - \bar{w}_i) = \sum_{i=0}^{12} \bar{N}_i \Delta \bar{w}_i \qquad (4.17)$$

This is estimating the area beneath the curve in Figure 4.13. The weight intervals could be made smaller and smaller giving increasingly accurate estimates of the area. The limit of this procedure is the exact area and can be represented by the integral

$$P_T = \int_{w_0}^{w_{12}} N_i \, \mathrm{d}w \tag{4.18}$$

The useful aspect of this method is that it makes no assumptions about the way in which numbers decrease with time, or the way the fish grow. Confidence limits for w and N can be used directly to give confidence limits for P (Pitcher, 1971). The area under the curve in Figure 4.13 is given by simply adding up the small rectangles, one of which is illustrated. In our figure the equal division of the horizontal axis is arbitrary. In reality it may take longer for the fish to put on an equal weight of new flesh at the end of the twelve-month period than at the beginning.

It is not always possible to follow a cohort over a period of time. Ricker (1946) devised a way in which production could be calculated given that one already had data on the instantaneous rate of growth G, and the instantaneous mortality rate Z. If G and Z remain constant over a year then

$$\frac{\mathrm{d}B}{\mathrm{d}t} = (G - Z)B \tag{4.19}$$

This formulation assumes that numbers decrease and weight increases exponentially. Integrating Equation 4.19 gives

$$B_t = B_0 \exp((G - Z)t) \tag{4.20}$$

where B_0 = biomass at time $t = 0$. The average biomass over the year is then

$$\bar{B} = \int_0^1 B_0 \exp((G - Z)t)\mathrm{d}t = B_0 \frac{(\exp(G - Z) - 1)}{G - Z} \tag{4.21}$$

Production can now be calculated from

$$P = G\bar{B} = GB_0 \frac{(\exp(G - Z) - 1)}{G - Z} \tag{4.22}$$

This only holds as an annual estimate when growth and mortality are exponential, but the equation can be used over shorter periods when this assumption is broken.

It is often useful to be able to use a simple multiplier to scale up the data on biomass to give an estimate of production. This assumes that P/B is constant. Where $P = G\bar{B}$ can be used (as above), $P/\bar{B}$ will equal

G, so production is easily obtained. This simple method only applies when a single cohort is considered at a time. Allen (1971) examines what happens when more than one cohort exists in the population at any time and when growth and mortality are not necessarily exponential. If more than one cohort is present then production is

$$P = \sum_{n=1}^{N} P_n \text{ for } N \text{ cohorts,} \tag{4.23}$$

and mean biomass is

$$\bar{B} = 1/T \sum_{n=1}^{N} B_n^* \tag{4.24}$$

where T is the period of time being considered, and B_n^* is the total biomass present during T belonging to cohort n. B^* is calculated from

$$B^* = \int_0^1 N_t w_t \, dt \tag{4.25}$$

The ratio of production to biomass

$$P{:}B = \frac{1/T \sum_{n=1}^{N} P_n}{1/T \sum_{n=1}^{N} B_n^*} = \frac{\sum_{n=1}^{N} R_n B_n^*}{\sum_{n=1}^{N} B_n^*} \tag{4.26}$$

when the cohorts are identical in their growth and mortality rates the ratio for the whole population is the same as the ratio for the individual cohorts.

When mortality and growth are described by functions other than the exponential the value of $P{:}B$ is described by different equations. The combinations of mortality and growth functions examined by Allen (1971) are given in Table 4.3a. The $P{:}B$ ratios are constant at G when growth is exponential, whatever the model used to describe mortality. Likewise, a simple relationship exists for the ratio when mortality is described by the simple exponential model. In this case the value of $P{:}B$ is the same as the total mortality rate Z. The table shows that before estimating production from biomass data it is desirable to have as much information as possible about the form of growth and mortality.

If production and biomass data are for a year then the reciprocal of the $P{:}B$ ratio is a measure of the time it takes for the population to

Figure 4.14: Production Parameters for a Population of Minnows in the Seacourt Stream, Berks. Cohort production, biomass and the P/B ratios were calculated on a four-weekly basis. Population biomass and production include contributions from each of the three cohorts present at one time. Values are based on an average cohort.

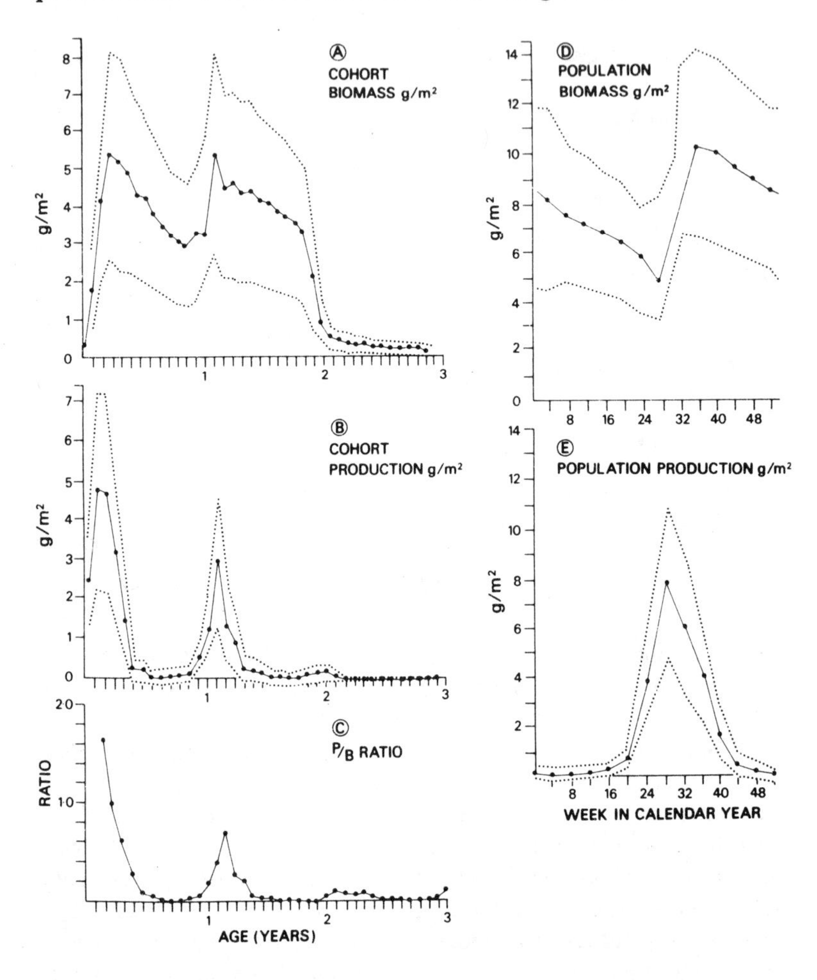

Source: Pitcher (1971).

reproduce its own weight; it gives a turnover time. When mortality is described by the single exponential function, mean age and lifespan are also estimated by the reciprocal of the *P:B* ratio. In general the short-lived organisms, such as copepods in the plankton, have high rates of turnover and low mean lifespans. Fish, as large organisms, have slower turnover rates and higher mean lifespans. Some examples were provided for different ecosystems in Chapter 1 and a specific example is given in Figure 4.14 which shows the cohort production, biomass and *P:B* ratio for minnows from the Seacourt stream in Berkshire, UK (Pitcher, 1971). Also shown is the population production and biomass for a period of one year. Both sets of data illustrate how biomass and production vary seasonally, mirroring the upsurge of growth during the spring and summer, followed by the effects of mortality as winter approaches and growth stops. These small fish also suffer heavy summer spawning mortality, curtailing summer production in the cohort after maturity at two years old. Small minnows are tending towards a semelparous (Chapter 5) reproductive strategy in contrast to most other UK cyprinids which grow large and are strongly iteroparous.

This chapter has considered the processes influencing growth, one of the main driving forces in the dynamics of a fish population. The next chapter will consider the evolutionary forces that have shaped the characteristics of fish populations through selective mortality. We will see that selection can occur through the activities of man as well as by natural processes.

5 EVOLUTIONARY EFFECTS OF MORTALITY

Tracking the Environment

All animals are genetically primed to stay adapted to their environment. As this is constantly changing perfect adaptation is never achieved, but individuals manage to keep close enough to the ideal to survive to reproduce with some degree of success. The environment of animals is multidimensional; change in one dimension does not mean that the other dimensions will change too. In tracking changes, a species cannot hope to adjust optimally to each dimension but must achieve the compromise that will yield the greatest fitness. Response times vary from seconds to hundreds of years depending on the mechanism; biochemical and physiological responses are fast, behavioural somewhat slower and genetically determined morphological changes slower still. The classification of responses as shown in Figure 5.1 (from Wilson, 1975) vividly summarises the levels of response of environmental tracking and their time scales.

Failure to respond sufficiently well to an environmental change has the final penalty of death. This can be brought about slowly by disease or starvation, or in a brutally decisive way by predation. Over evolutionary time individuals evolve which are able to respond to these pressures; in species which have survived there must by definition be an acceptable probability of survival. Other things being equal, no natural disease organism or predator is so efficient that its prey is obliterated without having a chance to respond genetically to the new challenge. In the past fifty years the fisherman, as a predator, is possibly the one exception to this generalisation. Before large steam-powered trawlers were built, the rigours of the marine environment limited the efficiency of fishing fleets, so that fishermen and fished stock could co-exist; this has been true for thousands of years (see Chapter 8). It is even possible to speculate that long-standing fisheries such as that for spawning cod off Lofoten, Norway, have been the cause of microevolutionary changes in the species. Nowadays most fisheries are so efficient that they must be the major component of mortality (Moav *et al.*, 1978). This is an unfavourable state to be in, with many unknown consequences. The short-term responses of a fish population to fishing are easier to predict and examine and are described in detail in Chapters 6, 7 and 8. Such short-term responses have important consequences for management, but in this chapter the ability of the stock to respond on a longer,

Figure 5.1: The Hierarchy of Biological Responses to Environmental Change. The response curves shown are imaginary.

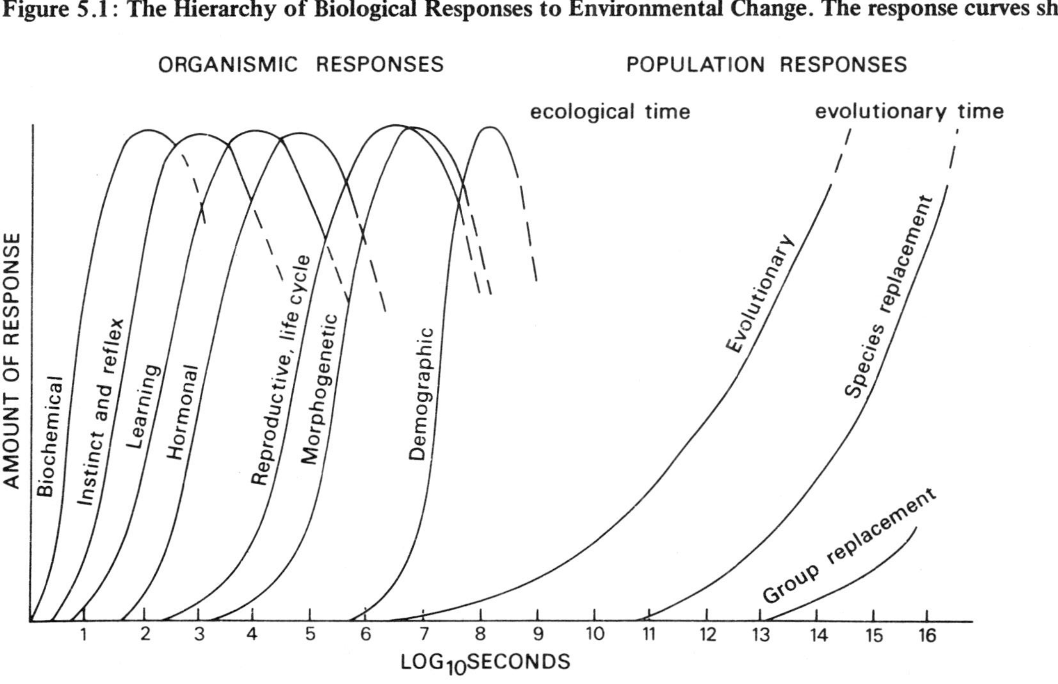

Source: Redrawn from Wilson (1975), with permission.

evolutionary, time scale will be examined. These long-term responses are relevant to the state of managed stocks but at present few precise details are known. In this book we will attempt to survey the potential that fish species have for response to long-term changes and identify the ways in which they could affect management.

The type of response we are concerned with is that mediated through the fish's genetic system. All diploid sexually-reproducing organisms recombine their genetic material in ways which produce phenotypic variations. These will generally be slight, but they do allow the animal to go part of the way towards adapting to slowly changing environmental conditions. This is true so long as a significant proportion of the variation in a character is of genetic origin. It is the phenotype which faces the inquisition of selection and unless it strongly reflects the character of the genotype the selection will have no lasting effects. A coarse measure of the closeness of the link between genotype and phenotype is heritability, which is a concept vital to our attempt to assess the evolutionary responsiveness of fish.

To understand heritability it is first necessary to make some observations about the variance of a certain class of characters. These are those features of an organism that can take any value between two extremes. As an example let us choose the length of North Sea cod on their third birthday. A sample from the population would show that this length varies, taking any value between an upper and lower limit. The main feature of this class of characters is that a frequency distribution of the different values attainable by the character can be approximated by a normal distribution. Genetic studies have shown (see for example Cavalli-Svorza and Bodmer, 1970) that the variation of this type of character can be explained by assuming that a number of genes contribute to the size or magnitude of the character. In the simplest case, where two genes with two alleles each are invoked, the difference between the mean values for the two is a consequence of the different genes and is called the additive genetic variance (V_A). Sometimes the heterozygote shows characters exactly midway between the mean values shown by the homozygotes. This is especially true for enzymes determined by relatively short lengths of DNA called cistrons. In these cases no one gene is dominant. For other characters the heterozygote may have a mean value nearer to one or the other of the means of the homozygote. In this case one gene is said to be dominant over the other. Dominance occurs typically in 'Mendelian' genes made up of lots of cistrons affecting major morphological characters. This element of the variance is called the dominance variance (V_D). Any variance in homozygotes must be a result of environment and this variance is called environmental variance (V_E). Total phenotypic variance is assumed to be the sum of the three types:

$$V_P = V_A + V_D + V_E \tag{5.1}$$

We can now define precisely what is meant by heritability, h^2. It is the proportion of the phenotypic variation caused by additive genetic variance or

$$h^2 = \frac{V_A}{V_P} \tag{5.2}$$

The important point for evolutionary studies is that the degree to which the organism can respond to selection pressures is a function of the amount of the phenotypic variance which is itself a result of additive genetic variance. This is the only part of the phenotypic variance determined by different genes and which can be driven by selection pressure to new frequencies in the population.

Each species has a store of variation, only part of which is expressed at any time. This characteristic was made significant by Fisher (1958) when he derived his Fundamental Theorem of Natural Selection which states that the rate of change of fitness (i.e. evolution) is proportional to the amount of variability in fitness in the population; greater variability means a superior capacity to respond to environmental change. This makes intuitive sense because a greater number of alternative responses available to a species makes it more likely that at least one will be appropriate.

Although the gene pool of each species can contain a wide range of alternative alleles (Lewontin, 1974), this is finite. After a certain time, recombination alone will not save the animal from the depredations of environmental change (including competition with other organisms). Williams (1966) neatly encapsulated the problem with his concept of 'running out of niche'; there is no longer a place in environmental space in which any genotype thrown up by the species is suitable. The only way that individuals in the species can respond is by introducing new variation, the single source of which is mutation. On a more immediate time scale the migration of animals causing gene flow can bring new genetic elements into a local population, but in the long term mutation is the sole innovator.

The ability of the gene pool to respond to change is then a function of heritability and the store of variation. The rate at which a particular allele increases in the gene pool depends upon its contribution to the fitness of the individuals that possess it. The important feature about this system is that the fitness of a particular genotype is measured relative to the average fitness in the population. This concept can be related to models of population growth in the following way (Wilson, 1975). Let n_x be the number of an allele x in a population which has a total of N genes. The frequency of gene x is $p_x = n_x/N$. Using calculus the rate of change of p_x can be written as

$$\frac{\mathrm{d}p_x}{\mathrm{d}t} = \frac{\mathrm{d}}{\mathrm{d}t}\left(\frac{n_x}{N}\right) = \frac{N(\mathrm{d}n_x/\mathrm{d}t) - n_x(\mathrm{d}N/\mathrm{d}t)}{N^2}$$

$$= \frac{n_x}{N}\left(\frac{\mathrm{d}n_x/\mathrm{d}t}{n_x}\right) - \frac{n_x}{N}\left(\frac{\mathrm{d}N/\mathrm{d}t}{N}\right) = p_x(r_x - \bar{r}) \qquad (5.3)$$

The *r*'s are equivalent to the intrinsic rate of population increase in the logistic model of population growth (see Chapter 3); r_x refers to the rate of growth in numbers of individuals possessing the x allele, whilst $\bar{r}$ is the mean population growth rate including all individuals without x. The important feature to be noted about Equation 5.3 is that the rate of change of p_x is equal to the existing value of p_x times the difference between the growth rate in numbers of the individuals with the x allele and those without it.

The rest of this chapter will examine heritability in fish in order that we may assess their potential for genetic response to selection. We will also outline some of the selective forces active in fish populations. Fishing and pollution will be evaluated as potential agents of selection.

Heritability in Fish

Heritability in fish has mostly been studied in characteristics which could be of importance to fish farming. Growth, resistance to disease and the proportions of flesh produced on a carcass are of particular importance. As we saw in Chapter 4, growth in fish is very plastic, as clearly demonstrated by Purdom (1974b), who examined growth variability in plaice, sole and plaice/flounder hybrids kept individually and at varying densities. The coefficient of variation for length of fish kept alone fell as the fish grew, but when several fish were kept together the variation increased as time passed. In Purdom's experiments there was a positive correlation between the size of the coefficient of variation and the number of fish in the group. Purdom noted that the flatfish in groups developed size hierarchy behaviour and this meant that bigger fish grew faster. This idea is invoked by Purdom to explain the increasing length variation he found. Other experiments showed that if plaice were grown at different temperatures then the coefficient of variation for length increased with time. These data are taken to imply that the genetic element in fish growth is quite low. If it were high then it could be expected that growth would not vary much from one set of environmental conditions to another.

Purdom (1974b) also calculated the repeatability of growth characteristics as an upper estimate of heritability. Repeatability is the ratio of genetic and environmental variance to total variance of a character. It is calculated using measurements of a character made within and between individuals over a period of time and is therefore easier to measure than is heritability in fish that breed only annually. In domestic animals this statistic can vary between 0.4 and 1.0, but for the plaice growth rate it was 0.10 and for the plaice/flounder hybrid it varied between 0.08 and 0.12 depending on population size. These data confirm that heritability for growth rate in teleost fish is low.

In salmonid species the behavioural tendency to form hierarchies, which is itself possibly determined by differing neurological structures, may be a more useful criterion for the breeder to examine than growth rate *per se* (Purdom, 1979). However, in the wild, selection for final size in a particular environment may be more important than growth rate. This would be especially true for species living in the more confined spaces of lakes and rivers and this factor is particularly important to sport fishery management.

Heritability has been measured in a few fish species and some examples are given in Table 5.1 together with data for domestic animals. The values for fish, mostly relating to growth, are at the low end of the range but are not enormously different from heritabilities in cows, sheep, pigs and chickens. The importance to the breeder is that the response to selection is proportional to the amount of selection applied; but it is instructive to remember that this is also true of the animals in the wild. The extremely high fecundity of fish means that they can tolerate higher selection pressures and this can be thought of as compensating for the lower heritabilities. Consequently, as we will see at the end of this chapter and in Chapter 10, despite the difficulties to the fish breeder, natural selection for changed growth rate is possible in fish.

This section ends with a word of warning. Measures of heritability are not absolute but depend on the environment in which they are

Table 5.1: Examples of Heritability in Domesticated Animals and Fish.

Species	Character	Heritability	Authority
Cattle	amount of white spotting in Friesians	0.95	Falconer (1960)
	butterfat %	0.6	"
	milk yield	0.3	"
Pigs	body length	0.5	"
	weight at 180 days	0.3	"
	litter size	0.15	"
Sheep	body weight	0.35	"
Poultry	body weight	0.2	"
	egg production (annual)	0.2	"
Salmon (Atlantic)	size of fingerlings	0.1 –0.2	Gall & Gross (1978)
	growth rate	0.1 –0.2	Gjedrem (1975)
	resistance of fingerlings to vibrio disease	0.07–0.1	"
Rainbow trout	weight at 150 days	0.26–0.29	Gall & Gross (1978)
	post-spawning weight at 2 years old	0.21	"
Carp	growth rate (selection for slow growth)	0.2 –0.3	Gjedrem (1975)

measured. This is because the denominator in the equation for the heritability coefficient includes the influence of environmental factors and these are not fixed. For example, despite the high heritability of the height of humans there has been a steady increase in mean height over the last 50 years which can only be due to environmental influences (Cavalli-Svorza and Bodmer, 1971). So long as this proviso is taken into consideration, heritability is a useful indication of the way a species will respond to selection pressures.

The Principal Selective Forces in Fish

In Equation 5.3 the rate of change of an allele is proportional to the difference between the intrinsic rate of increase of the genotype possessing the allele and the mean intrinsic rate of increase for the population, or

$$\frac{dp_x}{dt} = p_x(r_x - \bar{r}) \qquad (5.4)$$

It pays to remember that $r = b - d$, where b is the birth rate and d the death rate. This means that a genotype could have a faster growth rate than the population mean in two different ways, either because births are greater than average or because deaths are lower. The birth rate in fish will depend on inherent differences in specific fecundity and in size at maturity, and in egg viability, thus varying fertility. The average birth rate in a population will depend on these individual factors plus the age distribution of females (Chapter 3). So a genotype which did better than average could result from a number of changes. Death can occur at all times in the life of the fish and when it strikes it signifies failure in the struggle to survive. In the next section of the chapter we will discuss some causes of mortality in fish and their possible selective effects.

The Contribution to Mortality of Non-biological Factors

All species have adaptations that allow them to cope with the temperatures, salinities, currents and pressures experienced in their particular habitat. These physical and chemical factors are often easier to measure than are biological features, and this has led to large collections of physical data. When a biologist wants to try and explain the rise and fall of population abundance he will often first turn to these physical data. This expedient is done with some justification. Alderdice and Forrester (1971) found that the eggs of the Pacific cod have an optimum temperature range for development which is between 3 and 5°C. When the temperature is between these optima more of the eggs survive to hatch and when hatched the larvae are larger than those coming from eggs that have developed at suboptimal temperatures. These and similar findings provide the background to many studies on a wide range of species in the sea and in fresh water which have correlated year-class strength with temperature, making the assumption that low or high temperatures cause changes in mortality.

Between 1965 and 1973 the population of Atlantic cod in the North Sea increased greatly (see also Chapter 8). Dickson *et al.* (1974) correlated year-class strength with temperature in an area in the central North Sea and found a negative correlation between abundance and temperature in the early part of the year. Further support for this correlation comes from the fact that in the southern North Sea the cod are at their southern-most limit and might therefore be expected to be more sensitive to changes in the environment's physical characteristics.

It is unlikely that only one factor at a time will be exerting a selective force on a population through mortality. Abundance fluctuations in the plaice population in the southern North Sea provide one of the best examples of influence by more than one factor. Biological events are also involved, but these will be discussed in the next section. The mature adult plaice migrate in November and December to the spawning grounds in the Flemish Bight. Spawning is in December and January, after which the adult fish move northwards to disperse over a wide area of the North Sea for the summer feeding period. Despite the benthic habit of the adults, the eggs and post-larvae are planktonic. Once in the plankton the eggs, then larvae, drift north-eastwards, growing as they go. When they are in the region of the Friesian Islands, the larvae begin to metamorphose and move deeper and get carried by onshore currents into the Waddensee. This is their nursery ground where they become truly benthic and spend the first year or so of their life close to the tide mark. Once in the plankton the eggs and larvae are affected by at least two physical factors, temperature and speed of the north-east-flowing residual current (Cushing, 1974; Talbot, 1978; Harding *et al.*, 1978). Temperature has two possible effects; it controls growth rate via metabolic rate (Chapter 4) and also influences the amount of planktonic food available to the larvae. This is so because the meteorological events that cause cold temperatures create a stable water column which increases the rate at which the spring bloom of phytoplankton develops. The influence on growth rate also determines how long larvae are exposed to predation (see Chapter 6; Cushing, 1974). These effects mean that temperature itself is not necessarily the selective force; changes in temperature recast slightly the niche to which the species is trying to remain adapted. Current speed plays a similar role. Should the north-east-flowing current through the Straits of Dover be stronger than usual, there are two possible outcomes depending on the temperature. A high temperature could mean fast growth so that the larvae will arrive at the entrance to the Waddensee at the correct time in their life history. They will have the appropriate behavioural repertoire ensuring that they move deeper in the water column, so being carried into the nursery area by currents moving onshore. If the greater than average north-east flow is accompanied by low temperature, then the larvae will grow more slowly and as a consequence will be at too early a developmental stage when the entrance of the Waddensee is reached. As a result many will be swept further north-east ending up in unsuitable areas where mortality is higher.

The essence of the variation found in these physical factors is its unpredictability. It is therefore unlikely that fish such as the plaice will closely track rapid year-to-year changes in temperature, current speed or salinity. Perhaps slow, large-scale oceanographic trends over long periods, such as those observed in the western English Channel in the past 60 years (Southward, 1980) would induce evolutionary changes, although their effects on fish community composition are better documented than their effects on individual characters. It is possible that plaice fecundity, which differs by a small but significant amount between stocks (Bagenal, 1971a), has been selected in this way. How then have fish responded to the unpredictability of non-biological selective factors?

The most important point to understand about the life histories of fish such as cod, plaice and herring is that the eggs and larvae suffer most of the mortality. Mortality of the adults varies from 5 to 10 per cent per year (Woodhead, 1979) although there can be an increase with age. This can be compared with larval mortality of 2 to 10 per cent per day. There are two possible evolutionary responses when mortality mostly affects the young pre-reproductive stages: increased parental care with a corresponding reduction in the number of offspring or a large increase in fecundity. The process can be likened to a person trying to score a bullseye on a dart board. He could decide to take just one chance but build an extremely accurate crossbow device which would launch the dart with a high probability that it would hit the target. Alternatively the competitor could choose to have 20 throws and launch the darts by hand. The chance of any one dart scoring the bull might be low but the probability that at least one would go in could be as high or higher than the probability of success with the crossbow. The nature of the open sea environment makes increased parental care unlikely, although a few species, such as the redfish and some elasmobranchs, have solved the problem by developing viviparity. Parental care requires a place where the adult can build some form of nest to guard the young; it must be possible for the adult to return to the place after an absence and to defend it against intruders. Cod and plaice live and spawn over deep water with strong currents and there is an ever-changing relationship between features of the environment. In terms of the dart-board analogy, saturating the target has been the most common solution in teleosts.

Fish also have had to adapt to the fact that the intensity of the mortality factors varies stochastically. The dart player now has to contend with an additional problem. At random intervals someone deploys a strong magnet in the path of the dart as it flies towards the board. As a result the dart is deflected from its path and is guaranteed to miss the target. Similarly with the spawning of fish, at unpredictable intervals some environmental event causes more larval mortality than normal so that the spawning success of an individual is reduced to near zero for the year. It is this process that was not taken into account in the earlier life-history models (e.g. Gadgil and Bossert, 1970; see Chapter 3). Fish appear to have coped with the problem not only by having high fecundity, but also by extending the number of seasons in which spawning occurs. In north-east Atlantic herring populations it has been found that those living in shallower areas, where the production cycle is relatively more predictable than in deeper water, have shorter life spans and greater reproductive effort per occasion than the Atlanto-Scandian stock which lives in deeper water (Mann and Mills, 1979). This strategy has been called 'bet-hedging' by Stearns (1976) and is a direct consequence of high pre-reproductive mortality.

Contribution to Mortality by Biological Factors

In fisheries biology the influence of food availability has played a particularly prominant part, mainly as it influences larval survival. In teleosts, pelagic fish eggs have a supply of yolk which provides for the developing embryo, and the remains of the yolk sac keep the larva going for a few days immediately after hatching. Newly-hatched larvae are apparently inefficient feeders and take time to perfect the skills necessary for food capture (May, 1974). The Norwegian fishery biologist, Johan Hjort (1914), suggested that this period of change, from the internal to the external source of energy, was a delicate one for larvae and could be called a 'critical period'. This concept has stuck in the literature and has been invoked many times to explain the vast differences in the success of different year classes. The concept that food availability is critical to year-class success has also been used by Jones and Hall (1974; see Chapter 6) and Jones (1978), but not in the narrow sense used by Hjort. A further extension of the idea has been developed by Cushing (1974), but food availability now becomes a secondary cause of mortality, predation becoming the principle cause. These ideas will be discussed and related to the reproductive strategies adopted by fish species.

The critical period idea was thoroughly reviewed by May (1974). There are three lines of evidence critical to the concept: the shape of the curve describing egg and larval mortality, the occurrence of starved larvae at sea and the degree to which larvae are sensitive to a shortage of food.

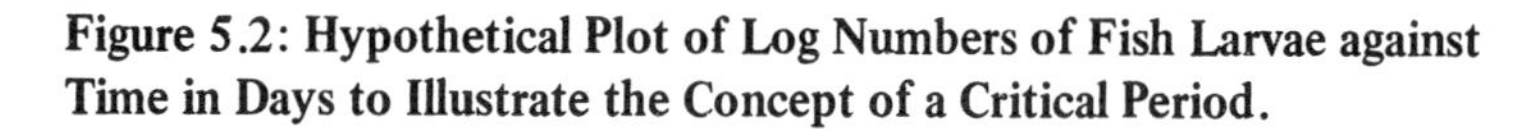

Figure 5.2: Hypothetical Plot of Log Numbers of Fish Larvae against Time in Days to Illustrate the Concept of a Critical Period.

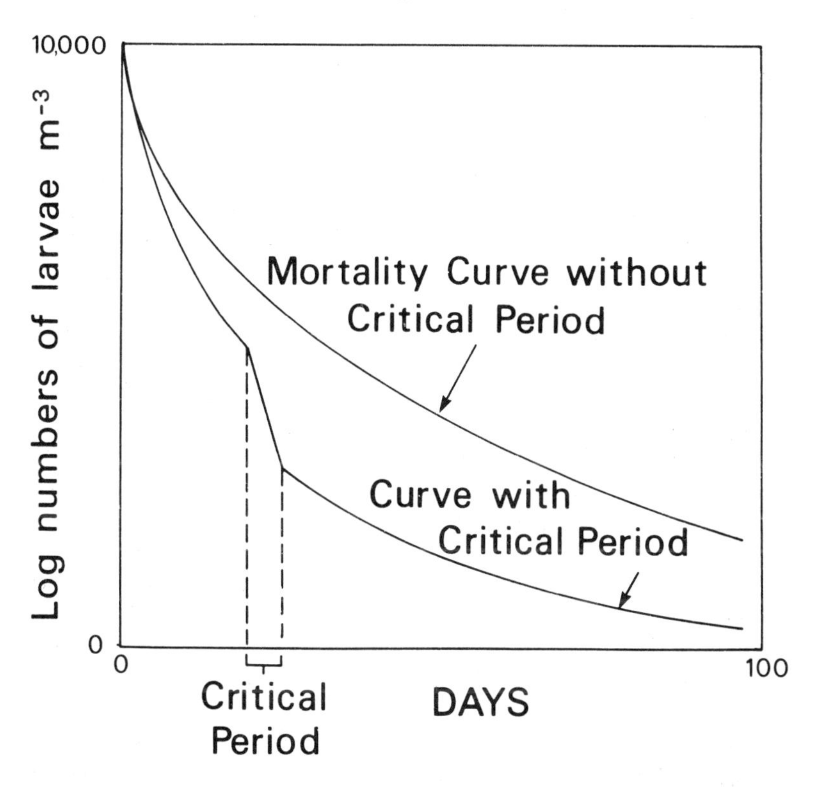

An exponentially-decreasing curve is a rough description of the reduction in numbers through time of eggs and larvae (Figure 5.2). This is labelled a Type III mortality curve in the standard ecology texts (Krebs, 1979; Pianka, 1978). If a critical period occurs then the stepped curve in Figure 5.2 would be expected, caused by a much higher rate of mortality over a relatively short segment of the time during which the fish are larvae. To obtain a graph like the one in the figure, we would need accurate estimates of the numbers of a cohort of eggs or larvae through time and these are hard to obtain. Small larvae pass through the net and the larger ones can escape it. A further problem is that plankton is patchily distributed which makes it difficult to sample each age class with equal intensity. This presupposes that one can age the larvae accurately, but this is not always so. Taking these problems into account May (1974) considered data on mortality in eleven species, all but one of which are pelagic, and concluded that the majority do not appear to show convincingly a critical period. It seems more likely that presence or absence of a critical period will be dependent on species and season. In other words it will be a much more variable phenomenon than was assumed in earlier works (see Chapter 6).

Evidence for the occurrence of starved larvae at sea comes from two sources – the proportion of larvae in a sample that have food in their stomachs, and the association between the condition of the larvae and the quality and abundance of the plankton. Again there are sampling problems which distort the proportion of full stomachs. Diurnal feeding patterns make the timing of the sampling critical whilst the trauma of capture can often lead to regurgitation or defaecation. Nevertheless, positive correlations between the proportion of larvae with full stomachs and density of plankton have been found. For example Berner (1959) found that the larvae of the anchovy only had food in their guts when they were taken from areas with high food concentration. Theoretical calculations by Zaika and Ostrovskaya (1972) showed that a low incidence of full stomachs in larvae may well be a result of the plankton densities found as a matter of course in natural waters. Further details are given in May (1974).

Evidence is scarce for a high incidence of weak larvae in poor plankton patches. Soleim (1942) found large numbers of dead herring larvae in sea areas with very low numbers of copepod nauplii, which form the main food of the larval fish. On the authority of Marr (1956), May (1974) dismisses this report as being unreliable and probably the result of Soleim not washing his nets well enough between hauls. Recently Wiborg (1976) has supported Soleim, pointing out that he was a meticulous worker. Shelbourne (1957) assessed the condition of plaice larvae from good and bad plankton patches. When plankton was sparse, plaice larvae at the end of the yolk sac stage were in poor condition, but when plankton was dense, larvae at a similar stage of development were in good condition. Experiments on survival of larvae in relation to food supply are discussed in detail in Chapter 6.

The evidence for a discrete critical period is equivocal, but food is clearly vital to larval survival. This is emphasised by the close links that exist between spawning time in temperate marine fish and the time of greatest abundance of planktonic food (Cushing, 1977). The herring stocks around the British Isles are an example. There are three groups of three stocks, each spawning at different seasons as shown in Table 5.2.

The forms of the production cycles from Colebrook and Robinson

Table 5.2: The Different Stocks of Herring Around the British Isles, Their Spawning Times and the Type of Production Cycle They Are Associated With.

	Season		
	Winter	Spring	Autumn
Stock	Downs (Nov.-Dec.) Plymouth (Dec.-Jan.) Dunmore (?)	Hebrides (Feb.-Mar.) Norwegian (Feb.-Mar.) N. North Sea (Mar.-Apr.)	Dogger (Sept.-Oct.) Buchan (Aug.-Sept.) Icelandic (July-Aug.)
Type of production cycle	shelf and intermediate Atlantic	oceanic	bank

Source: Cushing (1967).

(1965) are illustrated in Figure 5.3, which shows that there is a broad association between the production of herring eggs and the peak of production. A finer analysis indicates that this association is consistently closer than one would expect than if chance alone was operating. A detailed analysis for the spring-spawning Hebrides stock is shown in Figure 5.4 (Cushing, 1967) which shows that eggs are produced in February and March, which is sufficiently close to the time of increased phytoplankton production to ensure that there is plenty of food available for the larvae when they hatch. There is also an association between the egg sizes of herring and the time of spawning (Bagenal, 1971b). The lag between egg production and good food abundance is greatest for winter and spring spawners; fish from these stocks have larger eggs than do those spawning in the autumn. The larger eggs have more yolk, so making it possible for the longer lag to be accommodated. There is evidence for a similar relationship between time of spawning and time of plankton production in plaice (Bagenal, 1971a), blue whiting (Coombs and Pipe, 1978), and several other species (see Mann and Mills, 1979).

In temperate fresh waters there is no close link between egg size and production cycle (Bagenal, 1971b). As allochthonous organic matter is much more important, particularly in rivers (Chapter 1), the onset of spring is not such a vital event to fish. Production and reproduction in river fish are linked, as shown by bullheads in a chalk stream in the south and a moorland stream in the north of England (Fox, 1978). In the more productive chalk stream, female bullheads spawn three or four times a summer, whilst in the less productive

Figure 5.3: The Different Production Cycles Found Around the British Isles. Production is measured in terms of greenness on the filtering silk of the continuous plankton recorder. The greenness is judged by eye to be 'very pale green', 'pale green' and 'green'. These records have been quantified by comparing acetone extracts with standard solutions and dividing the range arbitrarily into three segments with means of 1.0 for 'very pale green', 2.0 for 'pale green' and 6.5 for 'green'. The vertical scale in the figure shows the standardised mean seasonal variation in abundance. The various production cycles are associated with herring spawning times in Table 5.2.

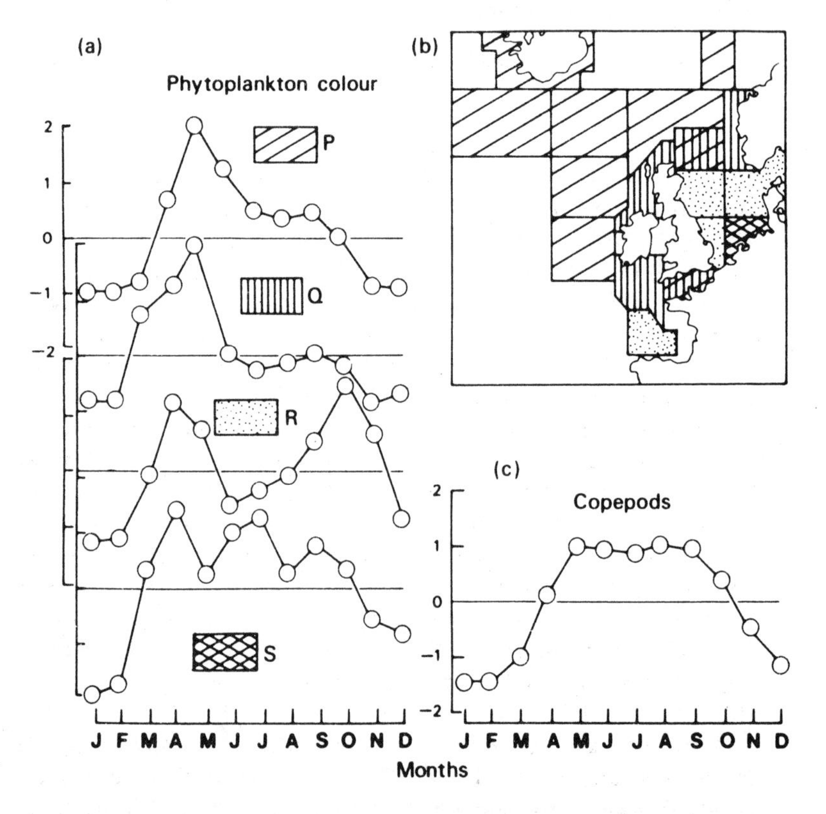

Source: Redrawn from Colebrook and Robinson (1965), with permission.

Note: P, Q, R and S in (a) refer to the various shaded areas in (b).

Figure 5.4: The Seasonal Cycle of Primary Production as Related to the Spawning Time of Herring to the West of the Outer Hebrides. Primary production is shown by the solid line and the dotted line labelled *Iqb* and *kq* respectively. The labelling refers to two statistical rectangles used for data analysis. The vertical scale is in arbitrary units of greenness in the sample (see legend to Figure 5.3). The dotted horizontal bar shows the period over which spawning extends while the solid arrows show the period of larval survival, from fertilisation to the time at which only 50 per cent of the larvae remain.

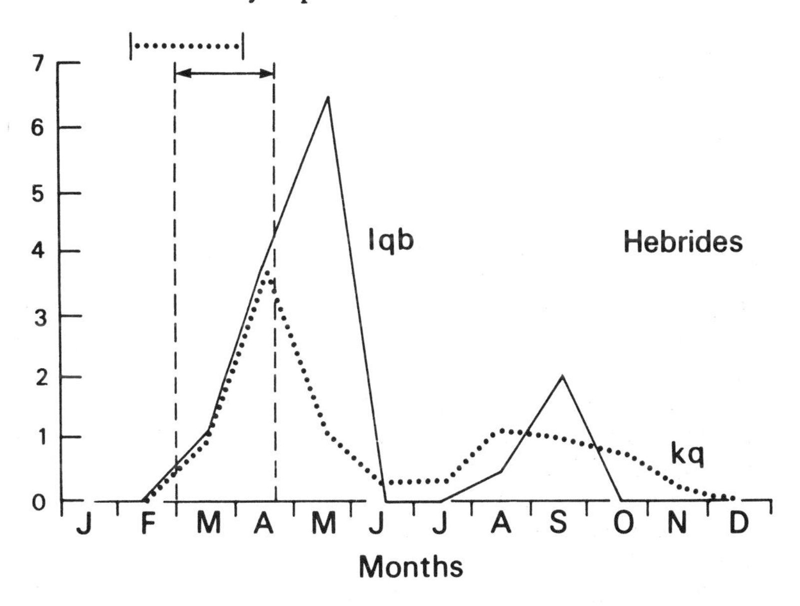

Source: Redrawn from Cushing (1967), with permission.

moorland stream spawning takes place only once.

Predation can be an important cause of mortality. Cushing and Harris (1973) proposed that as fish larvae grow older and bigger they become available to larger predators. The predators eating the smallest larvae are the most numerous, for example carnivorous copepods, pleurobrachians and chaetognaths. Poorly-fed fish larvae grow slowly and as a result remain subject for longer periods to predation by numerous predators. Alternatively, well-fed larvae grow fast and are exposed to intense predation for a shorter time. The elements of this idea are true but whether the system works to regulate numbers is as yet unknown. Jones and Hall (1974) argue convincingly that

Figure 5.5: The Proposed Chain of Events that Determines the Size of a Perch Year Class in Windermere, Cumbria, UK.

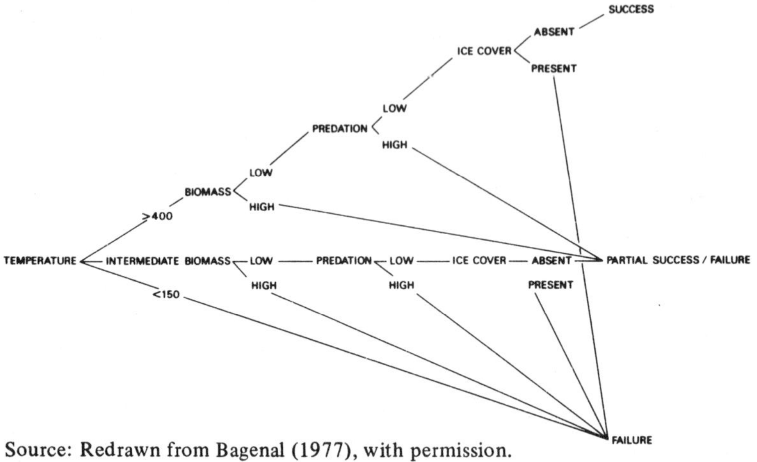

Source: Redrawn from Bagenal (1977), with permission.

predation cannot account for the larval mortality rate of 2–10 per cent per day which is commonly found in species living in the temperate regions.

Once fish have got past the larval stage, predation is more likely to become the primary cause of mortality. This is illustrated by perch populations of Windermere. The abundance of this fish has been monitored since trapping began in 1939 and the abundance of pike, the major predator on perch, has been observed since the middle of the 1940s. The most abundant year class of perch has been 400 times greater than the least abundant and LeCren *et al.* (1977) analysed the data to determine possible causes of year-class success. Figure 5.5 from Bagenal (1977) shows a scheme which might explain how good and bad year classes in perch come about. Predation by pike plays an important part. The netting programme to remove pike took out fish bigger than 55 cm. The numbers of smaller pike increased, thus effectively increasing the predatory capacity of the pike population; in the first two years of its life a pike is thought to consume 60 per cent of the food it is ever going to eat (Frost and Kipling, 1967). Before the netting programme for pike started three-year-old perch greater than 15 cm were rare. Ten years later few three-year-old perch were less than 15 cm. The conclusion is that the increased pike predation, by the more numerous small pike, reduces the numbers of young perch, so allowing those remaining to grow faster. Mann and Mills

(1979) consider that this increased predation has increased the variability of year-class success. A similar hypothesis was put forward by Hart and Pitcher (1973) to explain differences in growth of roach at separate sites on the River Nene in Northamptonshire, England.

We turn now to the evolutionary effect predation has on fish structure and behaviour. Both the predator and the prey species are changed by the continual interplay. Huntsman (1979) discusses the role of predation in structuring fish communities on coral reefs and gives examples of adaptations by predator and prey to the exigencies of their relationship. Further examples are also given by Hobson (1979).

There are numerous ways in which prey attempt to avoid predation. Some hide in holes during the day, only emerging at night when the risks to life are less. Others have used various invertebrates as hiding places. For example, cardinal fishes hide amongst the spines of living sea urchins whilst other groups use sea anemones or the coral heads themselves. There are also species that venture out into the open water but dive for cover as soon as any danger approaches. Prey species have also developed a wide range of morphological adaptations helping to avoid death by predation. Some, like the puffer fish and trunk fish, have odd shapes which are difficult to grasp; they are also equipped with poisonous skins. Others, such as the scorpion fishes, have poisonous spines. Spines have also been developed to make it harder for a predator to swallow a small fish, for example in the stickleback of British waters (Hoogland *et al*., 1956). Like many insects fish have used colour and pattern to make themselves less conspicuous or more confusing. Many have what appears to be an eye at the tail end, which is in fact a dark spot of colour. Other species have special colouration to disguise the eye; for example the European minnow has a dark stripe along the length of the body going through the eye. This makes it hard to distinguish which is the head and which is the tail and it is possible that this makes it harder for a predator to predict which way a resting minnow is going to dart. Like foul-tasting insects, some fish with poisonous spines have bright colours, perhaps to advertise this fact so warning potential predators who presumably learn to associate the colour with the unpleasant sensations. Finally mention should be made of a species in the Indo-Australian region, *Calloplesiops altivelis*, which has an ocellus on its rear end such that when the fish is head first in a crevice with its tail poking out it looks like the head of a moray eel.

Schooling is one of the most dramatic behaviours shown by fish and this too could be a defence against predation (Pitcher, 1980). It is also possible that schooling is a device for saving energy whilst swimming, but the way this operates is not yet clear (see Pitcher *et al.*, 1979).

The Evolutionary Response of Fish to Human Activities

At the start of this chapter we mentioned that fishing must now be a major component of mortality in fish populations and as such it could be expected to have significant selective effects. Man is also changing the environment of the fish with the effluents that are daily dumped into the sea both directly and via lakes and rivers. Pollution can change the environment in subtle ways, the selective effects of which will be hard to predict. For example, it is possible that fish communicate in part by means of pheromones released in small quantities in the water. Pollutants could interfere with these and this would lead to a breakdown of communication, perhaps at the breeding season, resulting in a spawning failure. In this section we shall examine the possible selective effects of fishing and pollution. Neither topic has been investigated seriously so we cannot hope to do more than give an outline of the problem.

The Effects of Fishing

Most fishing gear selects certain sizes of fish. Big hooks cannot catch small fish and wide meshes allow small individuals to wriggle through. As we saw in Chapter 4, there is considerable variation in growth rate in fish of the same age. As a result it is reasonable to conjecture that fishing might select out the fast growers in an age class. If growth rate is partly determined by genetic factors, it is likely that prolonged selection against fast growth will lead to the evolution of slow growth.

The critical assumption in this argument is that genetic factors significantly influence growth rates. At the beginning of the chapter we saw that the heritability of growth is 0.2–0.3 or lower in those species that have been investigated. We now look at a few more examples which bear more directly on the genetic influence of fishing.

In a programme designed to improve the properties of a stock of chinook salmon, Donaldson and Menasveta (1961) found that growth was improved by selecting the fast growers. Because there was no control line it was not possible to disentangle the effects of improved husbandry from genetic changes. A more careful study was done by Kincaid *et al.* (1977) using rainbow trout. They found that the mean weight of progeny at 147 days old, produced by two-year-old parents, was increased by 161 per cent by the fourth generation; 30.1 per cent of the increase was attributed to genetic factors. Purdom (1979) criticises the feeding regime used in this study. There is also evidence for the genetic determination of growth rate in carp (Moav *et al.*, 1978). The Chinese big-belly carp, a strain of the common carp, grows fast in a pond environment for the first 90 days of its life after which the growth rate falls away. A second strain, the European domestic carp, grows slowly at first but at ages greater than 120 days it grows faster than the big-belly variety. The F_1 generation from a cross between male big-belly and a female European carp grows fast up to 90 days and then continues with rapid growth; this pattern is a mixture of the two parental lines, indicating a genetic influence, although this may be a result of epistatic effects only. These examples establish that some small and variable part of fish growth rate is genetically determined.

Evidence for the effects of fishing on the genetic structure of fish populations is growing. Moav *et al.* (1978) think that the state of some stocks is so serious that a programme is necessary to introduce special strains of fish to improve genetically the wild stock. A possibility of change in genetic structure as a result of fishing has been suggested for various percid populations (Spangler *et al.*, 1977). The theory of life histories also has something to say about the effects of fishing. It is predicted that in the Atlantic salmon the removal of the larger individuals by the sea fishery will result in an earlier age of return to the parent stream for spawning (Schaffer, 1979). This would be a genetically-mediated effect. The prediction is in general agreement with data.

Silliman (1975) experimenting with *Sarotherodon mossambica* showed that short-term selective effects could be brought about by exploitation. Two stocks of the cichlid were raised under identical conditions. Once the abundance of the populations had stabilised, the two groups were subjected to different types of exploitation. In the first group all sizes above fry stage were unselectively exploited at a rate of 10 per cent, rising to 20 per cent per two months. The second group was selectively exploited in that all fish that could not pass through a 25 mm (later 22 mm) slit were removed. The process was continued for 36 months, which included three or four generations. At the end of this period Silliman determined whether the selective exploitation had had a genetic result. Forty-six mature fish were taken from the control group and the selectively-fished group respectively, samples being as similar as possible in length composition and sex. The fish were then fed a standard diet for 150 days during which the fish weight in each group was determined three times. The weights of individual fish were measured at the end of the growth period. Female growth was the same in the control (unselectively-fished) and the selectively-fished group, but males from the control group grew faster. Silliman took these results to indicate that the selective fishing regime had definitely caused genetically determined slower growth in males. The lack of response in the females was considered to be a result of sex-linked differences in the capacity for growth. Under natural circumstances tilapia males tend to be bigger than females.

Favro *et al.* (1979) used a simulation model to examine the effects of fishing on growth rate in brown trout, assuming that growth rate is determined by two unlinked loci with two alleles at each locus. The model was run on a computer for simulated time periods of up to 30 years. It was found that the numbers of large fish decreased as the fishing pressure increased. A further result was that the decrease in abundance of large fish was inversely related to the size limit in the fishery. A small size limit meant a large decrease in abundance of big fish. These results were consistent with data from brown trout populations in the Au Sable River in Michigan, USA, which had been fished mostly with a size limit of 25 cm between 1959 and 1977. This agreement does not verify the model, but the results show that it is possible to base an explanation for the observed decrease in abundance of large fish on a process determined by heritable factors.

It is certain that fisheries have some selective effect on populations. Favro's simulation study shows that if growth is a heritable factor, then an explanation based on genetic processes can explain changes in size in a fished population. What is not clear yet is the extent to which growth rate is genetically controlled. It seems reasonable to believe that the results so far indicating low heritability of growth rate are near the truth, so that only heavy fishing could be a potent selective force on growth rate. However, as many stocks are in fact heavily fished, this has serious implications for the long-term management of fish stocks. Silliman (1975) proposed that the best management strategy, considering the outcome of his experiments, was to try to exploit as wide a range of sizes as possible.

We have only considered fishing as it affects fish size. There are other behavioural, physiological, and morphological traits that could be similarly affected. Some of these might be size at first maturity, number of eggs per unit weight, the ability to escape nets and body proportions. Commenting on the reduction in age of maturity and increased pregnancy rates in heavily-exploited whale and seal stocks, Estes (1979) considers that exploitation artificially imposes r-selection on animals which themselves evolved through K-selection. In contrast to fishing r-strategists like teleosts, disruption of the traits naturally selected to maximise fitness in resource-limited K-strategists can exacerbate decline of stocks.

The Effects of Pollution

The selective effects of pollutants are unlikely to be as obvious as the effects of fishing. Heavy doses of some polluting materials are likely to kill fish outright leaving no change for an adaptive response. Pollution that causes a gradual change in the habitat of the fish opens the way for biological processes such as competition or food depletion which slowly oust the indigenous species. A particular example of this is eutrophication in freshwater lakes. In Lake Constance, on the borders of Switzerland, West Germany and Austria, a gradual enrichment with nutrients has changed the habitat to such an extent that the original fish fauna of coregonids, salmonids and burbot has given way to a fauna dominated by pike, perch, pike-perch and cyprinids (Numann, 1972). These species are characteristic of eutrophic conditions (see Chapter 1) which includes higher productivity, high turbidity and a non-existent or low level of oxygen in the hypolimnion. The original fauna was, one presumes, put at a disadvantage by the changed conditions; as a result cyprinids and their predators were able to gain a hold and increase in numbers. This effect was no doubt aided by the heavy fishing for coregonids. The 60 years it took for these limnological changes to occur has been too short a period for an adaptive response by the coregonids and salmonids.

Because of their immediate toxicity, pollution by heavy metals, hydrocarbons and pesticides is more likely to lead to an adaptive response. A commonly used test for the toxic effect of a poison is the LD50 (LD = lethal dose). This is the concentration at which 50 per cent of the test population die. The statistical character of the response indicates that some individuals are more able to cope with the poison than are others. Further work has shown that some fish species are able to put forward individuals that are resistant to certain toxins. A particular example is the mosquito fish, from Texas, USA, which has become resistant to an insecticide (Dzuik and Plapp, 1973). In the long term such an adaptation should mean that a fish species can survive successfully despite the presence of a poison. Such a conclusion is, however, too simple and assumes that the properties of the resistant strain are identical to those of the parent strain. It also assumes that the rest of the environment remains unchanged.

The likelihood of environmental constancy during pollution is low. This is illustrated by the work of Södergren (1976) who monitored the salmon population in the Rickleå, North Sweden, from 1961 to the present, a time when heavy metal pollution increased. Between 1961 and 1966 the abundance of the salmon population declined. Organic pollution and spawning failure were ruled out as possible causes of this. In parallel with the decline in salmon has been a decreased abundance of many benthic species which serve as food for the young salmon. In 1963 a diamond-processing plant began to discharge an effluent con-training nickel (up until 1967) and cobalt. Experimental work showed that an important invertebrate species, the mayfly *Ephemerella ignita*, was sensitive to sub-acute concentrations of cobalt nitrate. It was concluded that the heavy metals had caused the decline in invertebrate abundance which in turn caused the decline in the salmon population. This example shows how a fish population can be affected by pollution even though it could itself tolerate the poison or adapt to it. A similar conclusion must be drawn from the work of O'Connors *et al.* (1978). A phytoplankton population was kept with polychlorinated biphenyls (PCBs) at 1–10 mg l^{-1} in dialysis bags *in situ* in an estuary. The phytoplankton biomass and cell size was reduced by the toxin. Smaller particle size would favour species like jellyfish rather than copepods, thus diverting energy away from commercial species of fish.

It is hard to assess the truth of the assumption that resistant strains are the same as the parent stock in all other respects. Sub-lethal levels of the toxin can have a debilitating effect on an organism, which greatly reduces its Darwinian fitness. McFarlane and Franzin (1978) compared population parameters of the white sucker in Hamell Lake, Saskatchewan, Canada, which is contaminated by zinc, copper and cadmium, with Thompson Lake, Manitoba, which has much lower levels of these metals. The Hamell Lake population had an increased rate of growth, higher fecundity and an earlier age of first maturity than the population in Thompson Lake, but spawning success and egg and larval survival were reduced as was egg size and adult longevity. Larsson *et al.* (1976) studied the effects of sub-acute levels of cadmium on the physiology and biochemistry of the flounder. Low levels of the metal caused blood anaemia, disturbance in the balance of divalent ions and altered carbohydrate metabolism. Fish so affected survived for long periods but their ability to function normally in all respects must have been impaired. This type of response would suggest that a poison does not necessarily act on one character alone, so that adaptation to a toxin at low concentrations is likely to be a slow affair involving several characters at once and considerable interactions with a changing environment. Consequently an ecotype adapted to a toxic substance is unlikely to be the same in all other respects to the parent stock.

Because individual fish are differentially debilitated or killed by pollutants the latter must be an important selective force. How this force is responded to must be the subject of future research.

In this chapter we have discussed responses to selective pressures over an evolutionary time scale. On a shorter time scale, Chapter 6 will examine the process of recruitment to the fishery from year to year, how recruitment may be measured and modelled, and the detailed evidence concerning larval survival. We will return to the wider questions of adaptation and life history characteristics.

6 RECRUITMENT

In this chapter we will be concerned with the recruitment of new fish to the exploitable population. Fisheries generally consist of an organised attempt to harvest at a particular place and time. For example, the autumn mackerel fishery off the west of Scotland exploits the annual migration of these fish with warmer water and plankton to the north. Only fish above a certain size take part in such migrations, juveniles remaining behind in shallower 'nursery' areas. When the juveniles grow large enough to stay with the main body of adults they are said to have been recruited to the fishery. This natural renewal of harvestable fish is of crucial importance since it ensures the continuity of the fishery. However, the recruitment process is complex since it is the result of a whole chain of natural stages in the life cycle through spawning, egg-laying, hatching, larval growth, metamorphosis, growth and survival in the nursery areas, and finally migration to the adult feeding grounds. In short, each spawning produces a new cohort (see Chapter 3), which has to survive a succession of hazards until it is recruited to the adult stock. Over evolutionary time each species has adjusted its fertility and the way it responds to increased density, so that it persists without depleting its resources of food and space, but we should remember that this evolution took place in the absence of fishing. Man can fish only on sufference by removing part of the species' natural buffer against extinction. It follows that any cropped species becomes more vulnerable to hazards. The point during the life cycle at which the fish enter the fishery affects that vulnerability.

We need to know about recruitment in order to manage a fishery, but the young stages of the life history of fish are inherently the most difficult to study, especially in the majority of marine and lacustrine teleost species with pelagic planktonic larvae. How much of the ecology of these early stages do we need to know about? Until quite recently, all that has been available to the fishery manager have been poor estimates of the numbers of recruits, usually calculated from the numbers of the youngest age group in the stock, and an equally poor estimate of the numbers of spawning adults which gave rise to them one or more years previously. The classical fishery recruitment models, which we will describe later in this chapter, all represent attempts to relate these two sets of numbers with a simple predictive law. None was particularly successful in its original form and, from the 1960s on, recruitment failures in a number of important stocks (see Chapters 7 and 8) pinpointed the need for greater insight into the ecology of the pre-recruitment life stages. Ten years ago, reviews were stating, rather hopefully, that 'recruitment is the last unsolved problem of fish population dynamics'. As we shall see, a deeper understanding of the nature of the processes at work in the young stages of fish life cycles has emerged since then, but it remains true that there is no simple unified solution to the 'problem'. It seems increasingly likely that the reason for this is that recruitment in each species has its own special characteristics which have to be researched and analysed if the classical models are to be improved upon. In recruitment we also encounter the quandary, familiar to all scientists, of balancing what actually needs to be understood in order to solve today's problems, with what could potentially be discovered given unlimited research effort.

Recruitment is an artificial discontinuity in the life of the fish caused by the way in which man operates his fishery. However, this need not necessarily be so, because becoming vulnerable to fishing gear often coincides with a natural change of habitat, nutrition and behaviour of the growing fish. Such natural changes vary from abrupt to gradual in different species. Recruitment depends on where fish of a catchable size are located, and to some extent this is going to depend upon the kind of gear and fishing boats used. Where the boats decide to go to obtain marketable catches is going to be the deciding factor, however, and since this is not likely to alter very much over the medium term, the age, size and location at which fish recruit to a given fishery can be considered fixed. This means that we can make a vital distinction between two points in the fish's life: the age (and size) at recruitment to the stock, and the age (and size) at which we decide to capture the fish with our fishing gear. For most fisheries this latter is changed quite simply by altering the size of the mesh in the nets. Choosing the optimum size/age of first capture is an essential part of fishery management, and is dealt with extensively in Chapters 7 and 8, so that in this chapter we will be concerned only with recruitment to the potentially fishable stock.

The reason for the distinctness of the juvenile stages of teleosts is fundamentally one of feeding; the diet of any small fish is restricted by mouth size and swimming speed. As we shall see, food type and availability is the central factor in understanding the problems of recruitment, although shelter and protection from predators are also important. Growth and mortality rates are high during this phase: for example, haddock increase their weight by a factor of 10,000 during the first year of life, but only 1 in 10,000 will survive to do this. Such fish have evolved specialised larvae to exploit the abundant planktonic food supply. Although the plankton contains voracious invertebrate predators such as arrow worms, the active young fish larvae soon grow too big for these to tackle; most of the fish's own food is passive or relatively immobile in comparison, for example copepods and the larvae of sessile invertebrates. The body design of the larva at this stage is specially adapted to remaining in the plankton (e.g. the flattened and transparent leptocephalus larva of eels). At the end of this larval stage, metamorphosis to the adult design occurs, and the fish then swim to areas where food of the adult type is available. In the case of small fish like the anchovy they recruit to the fishery soon afterwards, such an abrupt recruitment of one age and size being termed 'knife-edge' recruitment. In other cases recruitment is a more gradual process, the young fish progressively migrating into deeper water as they grow larger, gain larger mouths and can swim faster. For example, whiting in the Irish Sea move off the Down Banks south of Ulster as they grow and join the main fishery off the Isle of Man and in Liverpool Bay. An intermediate situation exists with the Arcto-Norwegian cod (Chapter 2), where young codling leave the surface waters at about one year old at the end of their northerly drift in the plankton. Some cod are caught at two years old, but most do not recruit to the main fishery until four or five years old.

Most flatfish have pelagic larvae which live in the plankton, metamorphosing to the characteristic body shape after about ten weeks as they drift in the currents in temperate water. Few flatfish are recruited to a fishery at this age, but live in nursery areas in sheltered waters close inshore. This probably avoids predation. Different species of flatfish exploit characteristic habitats at this nursery stage. For example, in the British Isles, baby plaice in their first year live in sheltered areas of fine sand in depths less than 4 m. Young sole, whose parents are atypical inshore spawners, are found in deeper water with a muddy substrate. Flounder juveniles move into waters of low salinity and begin to adopt an intertidal feeding rhythm like their elders. In contrast, turbot in their first year prefer the rigours of very shallow water on exposed sandy beaches where there is heavy wave action. In their second year all species begin to move into deeper water but still retain their substrate preferences. The nursery areas for the older fish may contain fish of marketable size, and gregarious species like plaice are easily caught in small beam trawls.

In some cases fisheries exist specifically for juveniles which are pre-recruits to the main stock. Very small plaice are caught for fishmeal in the shrimp trawl fishery off the Flemish coast in the North Sea. Young-of-the-year herring are caught in the Clyde and Thames estuaries and sold as whitebait. The summer mackerel fishery off Cornwall largely consists of one-year juveniles which have not migrated north with the main stock and are immature. Fisheries for such juvenile stock need to be very carefully controlled indeed. The Cornish fishery may in the future be banned, and the Thames herring fishery is (1980) under the overall herring fishing ban. The quantitative reasons for this caution are explained in detail at the end of Chapter 8. In general, sufficient spawners must be maintained to provide the reproductive buffer against hazards which we have discussed earlier (see Chapter 5).

Low natural mortality in adult fish, typically spread over many age classes (Chapter 3), implies that the density-dependent factors (Southwood, 1978) which regulate the populations operate before recruitment. Egg mortality is quite low, of the order of 2–5 per cent in most species. Although a number of environmental factors such as oxygen concentration, turbidity and fungus can cause deaths of eggs, they do not usually operate in a density-dependent fashion. An exception is the mortality of salmon eggs in gravel redds which at high stock densities may be dug up by later-spawning females seeking to lay their own eggs.

It is worth emphasising that although many standard fishery texts state that there is 'no relationship between stock and recruitment', and this is usually made in reference to a scatter plot such as Figure 6.1d (plaice), the very absence of a trend of recruitment proportional to stock numbers allows us to infer that a density-dependent mechanism must be operating! However, since recruitment subsumes so many natural processes which will vary in nature from species to species and also from year to year, it is likely that any relationship of recruits to stock will exhibit a great deal of scatter, perhaps even obscuring any clear smooth trend. Paradoxically, as well as including density-dependent regulating factors, recruitment may also be the key factor in that it is the major factor determining subsequent adult numbers in the cohort. Although this is usually considered to be of greatest significance for short-lived exploited species with few age classes in the stock (e.g. sandeels), the same effect can be important even in long-lived multi-age-class species. This is not purely a consequence of fishing reducing the spread of age classes, but can be the result of sporadically exceptional

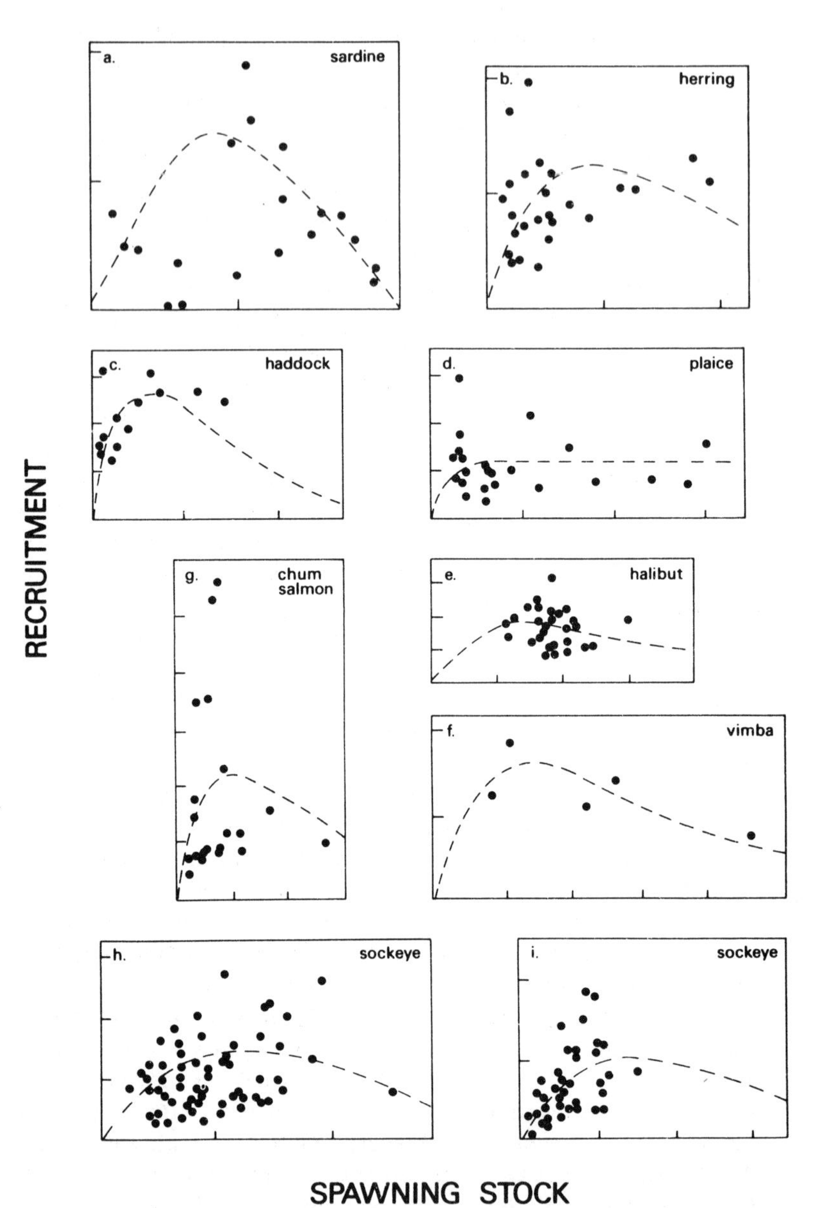

Source: (a) Radovich (1962); (b), (c), (e) and (i) Cushing (1977); (d) Beverton (1962); (f) Backiel and LeCren (1978); (g) Ricker (1975); (h) Rounsefell (1958).

Figure 6.1: Recruiting Fish (*R*) Plotted Against Spawning Adults (*S*) Which Gave Rise to Them for a Number of Fish Stocks to Illustrate the Form of the Relationship. Note the large amount of scatter. The lines drawn on the plots have been fitted by a variety of methods (see below), but usually do little more than indicate the general trends of the relationships, and in some cases are not particularly convincing. Note that the scales of the axes vary considerably between plots, but the details (given below) are not important for this illustration. (a) and (b) are clupeids, showing a great deal of scatter about a hump and some evidence of reduced recruitment at high stock densities; points well below the peak of (a) were produced just prior to the collapse of the stock. (c) is a gadoid with a humpy curve similar to the cod stock in Figure 6.6. (d) and (e) are flatfish with no evidence of reduced recruitment at high stock densities whereas (f) is a freshwater cyprinid where such reduction clearly occurs. The remainder are all Pacific salmon stocks exhibiting a variable amount of hump and very wide scatter. (h) and (i) are from different stocks of sockeye salmon to illustrate that differences can occur even between stocks of the same species. *Details*: (a) Californian sardine. Axes are in numbers, line was fitted by eye. (b) North Sea herring – Buchan stock. Axes in numbers, line is a Ricker curve fitted by least squares. (c) North Sea haddock. Axes are in numbers, line is a Ricker curve fitted by least squares. (d) North Sea plaice. Axes are stock weight indices, line is a Beverton and Holt curve fitted by eye. (e) North Pacific halibut. Axes are numbers, line is a Ricker curve fitted by least squares. (f) Vistula River vimba (a bream). Axes are indices of numbers, line was fitted by eye. (g) Chum salmon, Tillamook Bay stock. Axes are weights of fish, line is a Ricker curve fitted by regression methods. (h) Sockeye salmon, Karluk River stock. Axes are numbers, line fitted by eye. (i) Sockeye salmon, Skeena River stock. Axes are numbers, line is Ricker curve fitted by least squares.

year classes occurring rather less often than the number of years represented by the maximum age in the stock (e.g. cod, plaice, herring). In such cases the fishery can be dominated by one of these exceptional year classes for some time (see haddock) in Chapter 3). When density dependence is embedded within such a complex phenomenon it may be quite difficult to detect (Southwood, 1978). In the fished population we need only sufficient recruits to replace the adults before they die. It is worth remembering that at one extreme of the 'r' and 'K' spectrum

(Chapter 3) this might mean that only one year class in ten needs to be successful, and here one would need a great amount of data to detect density dependence in operation. It is therefore probably a vain hope to expect to make accurate predictions from generalised stock-recruitment models, although as we shall see there are ways in which such models can be put to good use.

Stock recruitment scatter plots for a range of species are reproduced in Figure 6.1. Such published graphs invariably omit the confidence limits on the estimated stock sizes, which can be large. The general feature is that in all species there is a great deal of scatter, but the trends of the relationship fall into two broad categories. Plaice, for instance, exhibit an asymptotic trend in recruitment with a wide scatter over a wide range of stock sizes, whereas cod have a low recruitment from high stock sizes. In fish like the cod and salmon there is evidence for a hump-shaped stock-recruitment curve, often termed 'domed'. The upper limit on recruitment for plaice is probably imposed by a limitation of suitably-sized food when the fish first take to the bottom after metamorphosis. For cod, cannibalism by the adults can produce a mortality that is dependent both on the density of the young and upon the density of the stock. The average rate of cannibalism in the Arcto-Norwegian stock is thought to be half a recruit per adult per year (Cushing, 1975a). In haddock, and in freshwater coregonids (whitefish and ciscoes), the food supply for the planktonic larvae is the most important.

In attempting to predict recruitment, details of the biology of the species are clearly going to be crucial. Food size available for planktonic larvae may be just as critical as overall density; the food will be no use to those with mouths too small to engulf it! Overlap in the sizes of food eaten by three species of larva in the Californian current is illustrated in Figure 6.2. Young jack mackerel larvae can take a wide range of food sizes, but the sardine has the narrowest range of the three. It is instructive to note that the sardine stock suffered a recruitment collapse under heavy fishing in the 1950s, and has since been virtually replaced by the anchovy.

Although the marine species of teleost with planktonic larvae appear superficially to be adopting the same larval feeding niche, in those few cases where detailed information about the food web is available each species seems to specialise in a different range of planktonic prey. This represents an evolutionary avoidance of food competition through selection of mechanisms which allow more efficient capture of certain types of prey. Last (1980) has published detailed quantitative data on

Figure 6.2: The Size Range of Food Organisms Eaten by Different Sizes of Pacific Sardine, Northern Anchovy and Jack Mackerel.

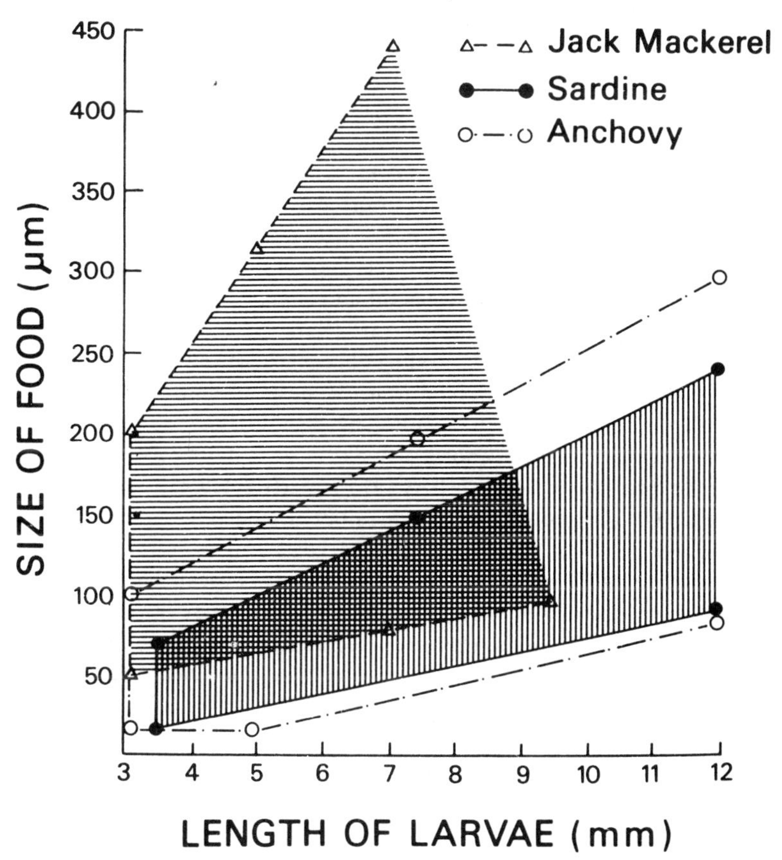

Source: Redrawn from Arthur (1976).

the food of 20 species of fish larvae in the North Sea, of which 12 are commercially important. The data were obtained from a towed high-speed plankton sampler which dived from the surface to the bottom and returned to the surface again in about ten minutes. The spectrum of food organisms consumed by the fish larvae was quite similar, but there were important quantitative differences. Individual fish contained 2–8 food items and several hundred individuals of each species were examined, so the diet figures are a true population measure.

Three species of flatfish specialised in eating appendicularia, but plaice took more lamellibranch larvae whereas lemon sole took more copepodites of *Oithona*. Witch specialised even more in appendicularia than the other two species. The two other flatfish larvae present had a different diet; dab took copepodites of the calanoid *Temora* and more nauplii than the others. Larvae of the solenette (a small rock-clinging flatfish of no commercial importance) did not compete as they specialised in polychaete larvae, unlike any other fish. Nearly all the larvae ate copepod nauplii, but whiting had over 50 per cent in their diet whereas the cod, the other gadoid studied, specialised in tintinnids (62 per cent). Mackerel larvae were the only species that concentrated on cladocerans (34 per cent of diet). The two clupeid species in the samples, herring and sprat, ate similar diets of nauplii and copepodites, but the herring took the larger items and species. Last's data also revealed that the fish larvae were visual predators, feeding almost exclusively in the daytime and building up to an evening peak in most species, except for the flatfish and mackerel which exhibited no clear daytime rhythm.

How much detailed knowledge like this is necessary in making accurate predictions for a fishery? How much detail can be sacrificed to generality in building effective recruitment models?

Actual numbers of recruits in past years can be estimated from fishery data by the back calculation method of cohort analysis described in Chapter 8 and Appendix 1, although this cannot be used in the development of a virgin resource. Recruits in years to come can be estimated by continuously monitoring the pre-recruit stages with special tow nets, surface nets, and other special gear. In freshwater nursery areas for cyprinid fry, there is even a trap net which rises to the surface when a 'Polo' mint dissolves, freeing a link to a weight (Bagenal, 1974). In addition to providing abundance in each year, to be used in predictive management models (Chapter 8), such larval and pre-recruit surveys can give valuable information on the location and movement patterns of the young fish. The results of five years of such surveys on juvenile North Sea gadoids are summarised in the maps in Figure 6.3 where the average densities of the first two age classes of cod, whiting and haddock are drawn as smoothed contours. Blacker (1980) described striking changes in the distributions between species and ages. First-year cod are concentrated in three main regions near to the spawning areas, whereas older cod are less dense and, in the southern North Sea, more coastal. Juvenile haddock are restricted to one large zone east of the Shetlands. The one-group haddock are not dispersed much more than this, whereas the whiting concentration south-west of Shetland moves southwards and inshore as these fish

Figure 6.3: Results of Some Pre-recruit Surveys for Three Species of Gadoids Carried Out in the North Sea in 1978. For each species two age classes were surveyed. The maps show the average density of pre-recruits.

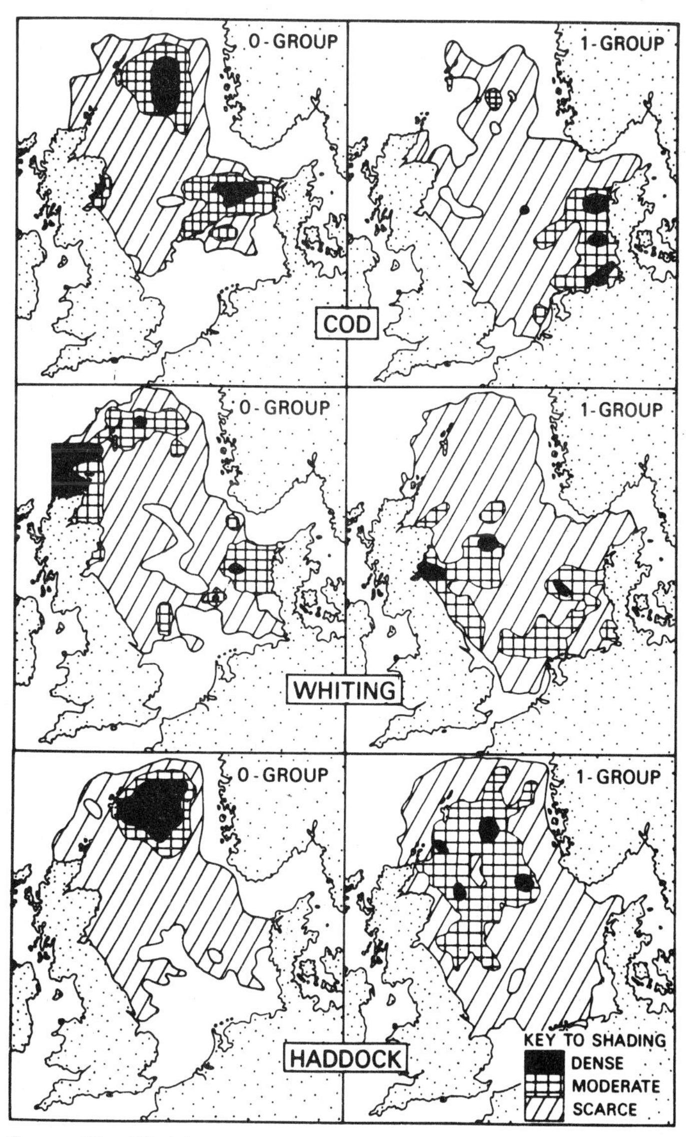

Source: Simplified from maps in Blacker (1980).

become older. Surveys of pre-metamorphosed larvae and eggs are also valuable but more difficult to interpret the greater their difference in age from that of the stock, as for example the survey of the potential of thread herring off Florida by Houde in 1977.

For a particular fish stock, such empirical data are expensive but effective. A cheaper method would be to predict likely recruitment from a model of some type. Considering the discussion above, it is not surprising that most attempts to date have not met with much success (with the notable exception of the management of Pacific salmon in Canada discussed below). In the rest of this chapter we will examine general stock-recruitment models which have been devised, and consider their success. We will then look in some detail at the problems of larval survival and growth in relation to food supply, expanding on the discussion of the critical period which we met in Chapter 5. This is followed by an account of two reasonably successful recruitment models for a haddock and a flounder fishery. We conclude this chapter with a discussion of two recent ambitious attempts to link recruitment, growth and production to the life history strategies of fish.

Stock/Recruitment Models

Two types of curve have commonly been used to describe stock and recruitment relationships, those of Ricker (1954) and of Beverton and Holt (1957). Although both were originally derived from different theoretical analyses, nowadays it is increasingly the practice to choose the one which fits the trend of the data best. With suitable modifications, both of these two-parameter curves appear to cover most recruitment relationships adequately. A number of more complex curves have been described (e.g. Paulik, 1973), but they have involved fitting more than two parameters and have not been widely adopted.

The Ricker recruitment equation describes a family of humped curves with low recruitment at high stock levels, whereas the Beverton and Holt model covers a family of asymptotic curves exhibiting constant recruitment beyond a certain stock density. The Ricker curve implies strong density dependence, increasing geometrically over a certain range of stock densities. The Beverton and Holt curve implies an arithmetically progressive reduction in the recruitment rate as stock density increases.

Figures 6.4a and b illustrate a family of Ricker curves described

Figure 6.4: Diagrams to Show the Form of the Ricker Recruitment Curve, $R = \alpha S \exp(-\beta S)$. (a) β fixed, α varied. (b) α fixed, β varied. (c) Replacement-adjusted Ricker curves for the four values of α shown in (a) (see text for explanation).

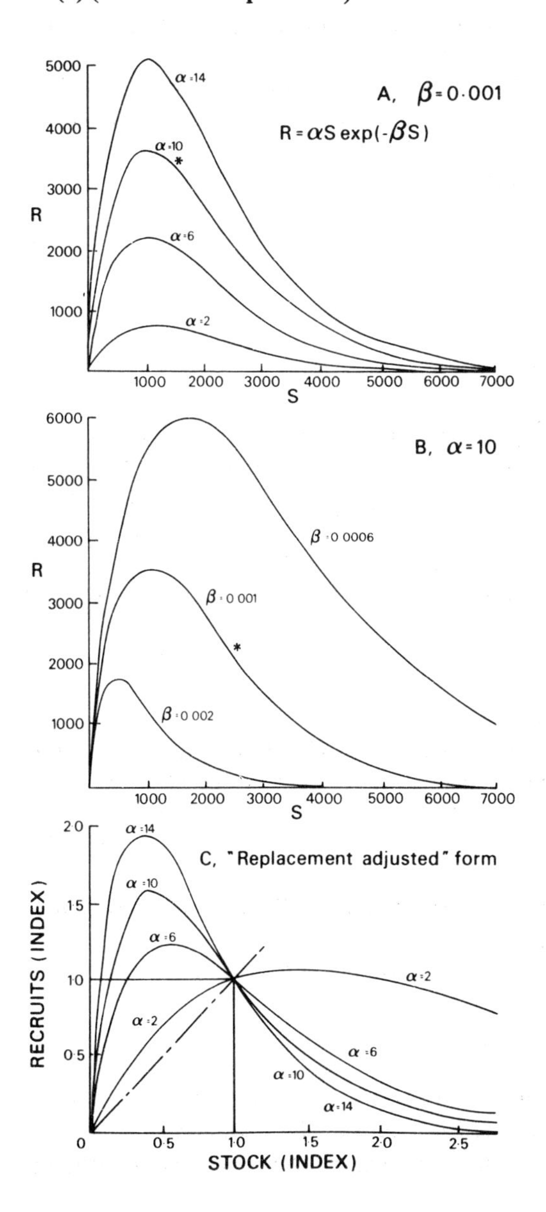

by the equation:

$$R = \alpha S \exp(-\beta S) \tag{6.1}$$

where R = recruits, S = adult spawning stock, α and β are the parameters of the curve (Figure 6.4a). Increasing α makes the peak higher and more humped, while increasing β decreases the height but makes the curve more peaked and shifts the maximum down at the same time (Figure 6.4b).

The maximum recruitment is given when

$$S_{\max} = 1/\beta \tag{6.2}$$

and the size of this maximum recruitment is

$$R_{\max} = 0.3679\,\alpha/\beta \tag{6.3}$$

β is therefore a density-dependent parameter in the Ricker model (Equation 6.2); α is density independent. The maximum of the hump can either be greater or less than the point at which recruits replace stock exactly (Figure 6.4c). The concept of replacement abundance is discussed below; for the moment we note that if S and R are in the same units, when $S = R = S_r$ we can substitute in Equation 6.1 and obtain

$$S_r = \ln(\alpha)/\beta$$

and so, writing $a = \ln(\alpha) = \beta S_r$, we can obtain the Ricker curve in a form in which replacement abundance, S_r, appears explicitly, i.e.

$$R = S \exp(a(1 - S/S_r)) \tag{6.4}$$

Figure 6.4c shows such replacement-adjusted Ricker curves differing in the single parameter a.

Beverton and Holt's asymptotic recruitment curve is

$$R = \frac{1}{\alpha' + \beta'/S} \tag{6.5}$$

(Note α' and β' are new parameters here.) Figures 6.5a and b illustrate a family of Beverton and Holt curves. Decreasing α increases the asymptote and reduces curvature, while decreasing β means that the

Figure 6.5: Diagrams to Show the Form of the Beverton and Holt Recruitment Curve, $R = (1/(\alpha' + \beta'/S))$. (a) α' fixed, β' varied. (b) β' fixed, α' varied. (c) Replacement adjusted Beverton and Holt recruitment curve for the four values of β' shown in (b). See text for further explanation.

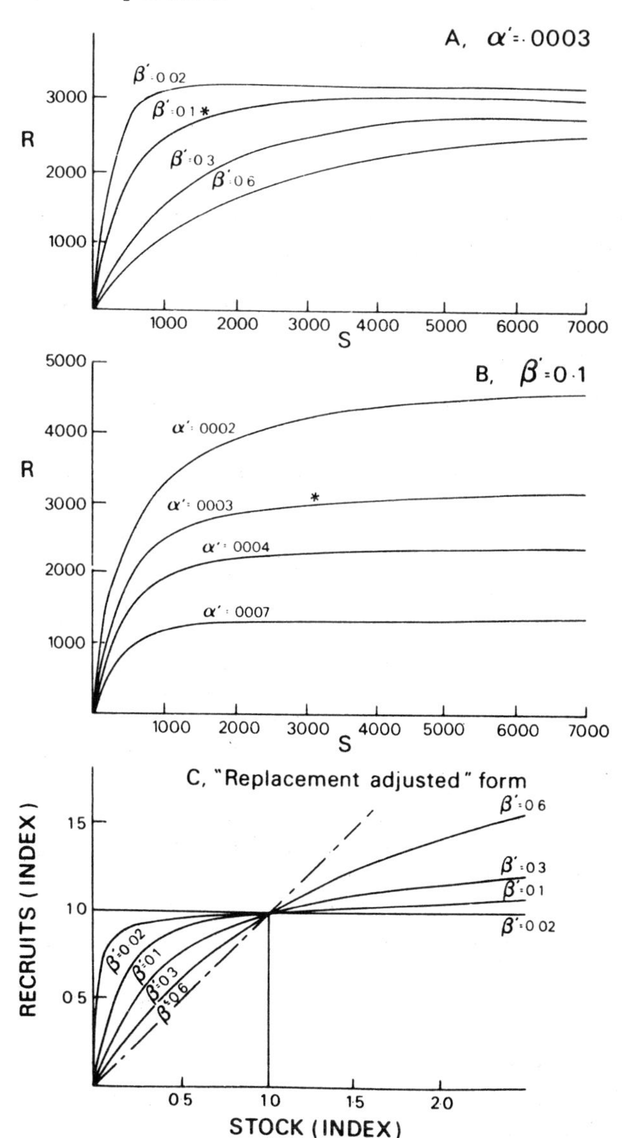

asymptote is approached more rapidly. As in the Ricker curve, the two parameters interact. The replacement-adjusted form equivalent to Equation 6.4 is

$$R = \frac{S}{(1 - a'(1 - S/S_r))} \tag{6.6}$$

where $a' = 1 - \beta'$, and $S_r = (1 - \beta')/\alpha'$. Maximum recruitment is given by

$$R_{max} = 1/\alpha' \tag{6.7}$$

produced at infinitely large stock size, $S_{max} = \infty$. So here α' is the density-dependent and β' is the density-independent parameter. Figure 6.5c shows the family of replacement-adjusted Beverton and Holt recruitment curves equivalent to Figure 6.5a but differing in values of β'.

The Ricker curve is said to be applicable when strong density-dependent mechanisms operate. These include: cannibalism by adults on the fry (as in some cod stocks, and in trout); when an increase in density of larvae means that they remain longer in a vulnerable stage; or when there is a time lag in the response of a predator to its prey's abundance so that a high initial density attracts (or generates) more predators without allowing for subsequent reduction of prey through exploitation. Highly-humped curves might be expected when there is 'scramble competition' for a limited resource (food, space or perhaps oxygen in a salmon spawning redd). In scramble competition everyone's ration is reduced, so in the extreme case mortality becomes high. Conversely, Ricker curves with low humps would be expected where 'contest competition' occurred (for example, juvenile trout feeding territories in riffles, safe refuges from predators on coral reefs). In contest competition successful individuals gain enough of the resource to survive and the losers die without affecting the winners' mortality rate. The Beverton and Holt curve is more appropriate when a ceiling of recruit abundance is imposed by available food or habitat, or when a predator continually adjusts its own attack rate to changes in prey abundance.

Fitting of either of these standard recruitment curves is best carried out on a computer by iterative least squares, i.e. in the case of the Ricker curve, by minimising $\Sigma\ [R - \alpha S\exp(-\beta S)]^2$. The major advantage of this method is that confidence limits can be estimated. An example is shown in Figure 6.6 from Cushing's work on Arcto-

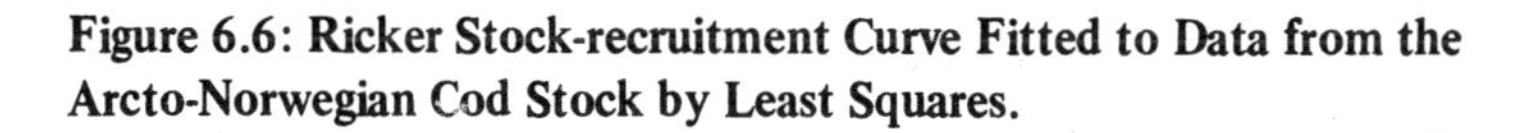
Figure 6.6: Ricker Stock-recruitment Curve Fitted to Data from the Arcto-Norwegian Cod Stock by Least Squares.

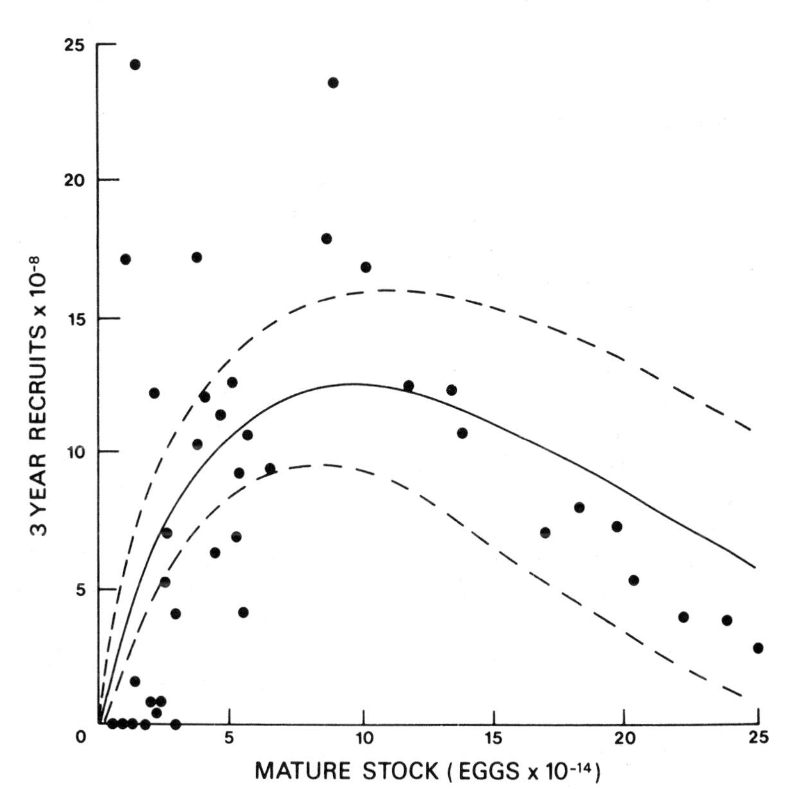

Source: Modified from Garrod and Jones (1974).

Norwegian cod. The shaded area is the zone outside of which there is only a 1 in 20 chance of finding the true curve. A second advantage is that least squares avoids the bias caused by scaling (Ricker, 1975).

The relative scales of the axes are crucial. Numbers of recruits may be adequate in some stocks, e.g. salmonids, but more satisfactory relationships are often found if recruits and stock are converted to equivalent egg numbers, as in Figure 6.6. The replacement versions of the curves are of considerable utility in calculating sustainable yields for some stocks, as we shall see below, but this really only applies for numbers where stock and recruits each consist of just one age class, the former being the most stringent requirement. This is the case for Pacific salmon, where Ricker curves have been most widely used in management. Since all adults die after spawning, the recruits replace them in a very direct sense. It is not quite the same in most fish populations, especially long-lived and slow-growing species like cod and halibut. In these cases it is not very clear what exactly constitutes replacement. Since there are many age classes of adults, the complete loss of one or even more cohorts may make little difference to the overall population and fishery yields (see Chapter 8). So the concept of replacement becomes a long-term statistical mean to a much greater extent than in short-lived species, and consequently any relationship becomes much harder to deduce against environmental 'noise'. Breeding stock and recruits expressed as egg equivalents are generally easier to use when attempting to fit a stock recruitment curve.

Beverton and Holt derived their asymptotic recruitment curve from the density-dependent mortality rate of the logistic population growth model which varies linearly (Chapter 3). By comparison, Ricker's humped curve is based on density dependence operating at the beginning of the interval between generations. Both models are continuous approximations of what is essentially a discrete process, but this seems to be of no disadvantage. The derivation of both models from the logistic growth curve has been made explicit by Erberhardt (1977). He demonstrates how Equations 6.1 to 6.6 above can all be expressed in terms of r and K of the logistic equation. For example, he shows the following (logistic parameters are on the right in each case):

Ricker curve: $\alpha = e^{r};\ \beta = r/K$
Beverton and Holt: $\beta' = e^{-r};\alpha' = (1 - e^{-r})/K$

So we can see that the density-independent parameters in the two models are almost equivalent, i.e. $\alpha = 1/\beta'$. However, whilst the density-dependent parameters are approximately the same when r is small, they diverge by up to about 20 per cent for larger r; $\beta \simeq (1 - e^{-r})/K$ when r is small.

Erberhardt considers that the Ricker humped curve describes recruitment in the classic r-selected species with rapid population growth, where marked changes in population size are an advantage in a fluctuating environment. This model represents situations where the population size can overshoot the asymptote, K, and then oscillate about that level. The magnitude of the oscillations will depend upon the degree of humpiness in the Ricker curve. Beverton and Holt's model on the other hand, describes K-selected species in stable habitats where the population tends to remain close to an asymptotic level. This model cannot overshoot and generate fluctuations in the same way. This contrast on the *r* and *K* spectrum between the two models is borne out by the finding that the extent of the hump in fitted Ricker curves correlates quite well with the cube root of the specific fecundity of the species (Cushing and Harris, 1973).

It therefore seems that humped curves could quite easily be fitted generally, species like the plaice having a low flat hump (e.g. Figures 6.4a and c, $\alpha = 2$). The assumption made by Beverton and Holt that recruitment would not be reduced even at extremely high adult densities does not seem realistic. Nevertheless, Harris (1975) clearly shows that a humped curve can only occur if mortality is dependent upon the initial stock density in some way; an asymptotic curve of the Beverton and Holt type results if mortality only depends upon the *current* density of the larvae. Stock density dependence or both types acting together can create a humped curve. Cushing and Horwood (1977) proved essentially the same thing empirically by running several different variants of a low-level simulation model in an attempt to mimic the stock/recruitment relationship for Arcto-Norwegian cod. Again, only stock dependence plus larval density dependence can generate growth and mortality curves similar in shape to the real data. Ware's (1980) model, described briefly at the end of this chapter, incorporates both factors.

When the spawning stock is actually replaced by recruits and fishing is concentrated on the mature fish, as in the Pacific salmon, then a simple Ricker curve can be used to manage the fishery for maximum sustainable yield. The slope of the Ricker curve is given by the differential of Equation 6.1, which is:

$$\alpha(1 - \beta S)\exp(-\beta S) \tag{6.8}$$

When this slope is 0 we have maximum recruitment, S_{max}, as in Equation 6.2, but when the slope is 1 we have the maximum recruits surplus to those needed for exact replacement of stock. The population size at which this maximum surplus of recruits occurs, S_s is found by solving iteratively,

$$\alpha(1 - S_s)\exp(-\beta S_s) = 1 \tag{6.9}$$

We can then calculate the maximum sustainable catch, C_s, a value related to *MSY*. (Note that strictly speaking this is not the same as a

Figure 6.7: One Use of Ricker Curves in Fishery Management. (a) Three replacement-adjusted Ricker curves of different α. Arrows indicate positions of maximum surplus recruits. (b) Catch per generation for curve 1 of (a). The exploitation rate is envisaged as increasing by 10 per cent generation; catches are shown as thick lines. Catches resulting from exploitation rates held at the values indicated are drawn as thin lines. Maximum catch is obtained at 42 per cent exploitation. (c) As (b) for curve 2 of (a). (d) As (b) for curve 3 of (a).

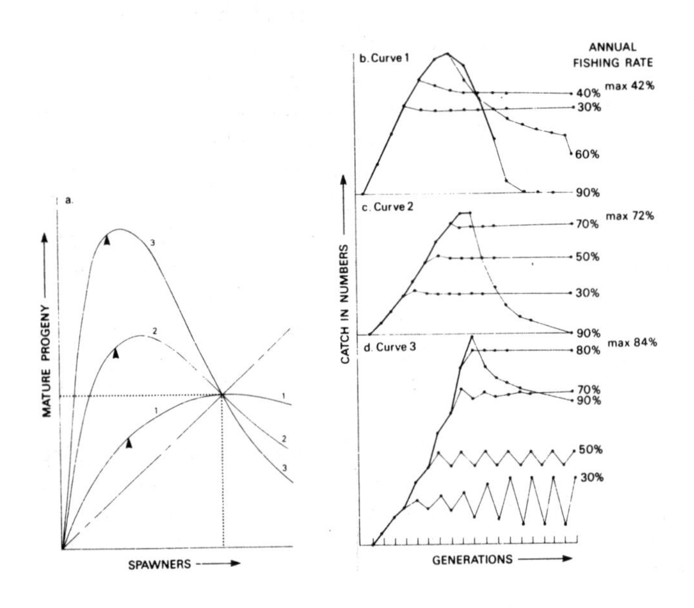

Source: Simplified from Ricker (1975).

yield in weight, although it has sometimes been used as such.)

$$C_s = S_s(\alpha \exp(-\beta S_s) - 1) \tag{6.10}$$

In addition to this estimation of C_s, the Ricker model can be used in a simple simulation of the effects of fishing on a stock, where its requirements hold true. The thick lines in Figure 6.7b to 6.7d, simplified from Ricker (1975), indicate what happens when the fishing mortality rate is progressively increased by 10 per cent in three stocks with Ricker curves differing in shape and magnitude as indicated in

Figure 6.7a. Fishing is considered to start on the virgin stock. The thin lines show catches which would result when fishing is held steady at the per cent rates indicated. C_s for each stock is shown on Figure 6.7a. We can see that as the hump of the Ricker curve increases, so the per cent exploitation rate at C_s increases also. Ricker curves with a high hump generate oscillations when the stock is near to replacement level, but these are damped at high fishing rates when the stock level is reduced. (In fact all Ricker curves show some degree of oscillation as equilibrium is approached, even in the unfished stock. Fishing just accentuates the tendency.) The lower the hump in the Ricker curve the more likely extermination of the stock becomes, curves I and II being eliminated by a 90 per cent fishing rate. Such a possibility of extinction becomes even greater when one considers random scatter about any real Ricker curve, a point we will return to at the end of Chapter 8.

One important point shown by Figure 6.7 is that there will be a peak of catch in the early years of the fishery, even at the optimum fishing rate which gives C_s. The difference between this early peak catch and C_s is greater the lower the hump of the Ricker curve; in the case of curve I in Figure 6.7b, the peak catch is 30 per cent greater than the ultimate stable level attained, at 50 per cent fishing. Such a high yield early in the exploitation of a new fishery can create management problems. The difference between early peak and sustainable catches is less in more highly humped curves, but here the penalties for getting the optimum rate wrong are much greater.

Ricker (1975) quotes an example of a salmon stock of 1,000 tons with a five-year generation time in which exploitation increases gradually by 10 per cent per generation. When described by a curve similar to I in Figure 6.7, after 25 years the catch will have risen to 440 tons, a third larger than the maximum sustainable catch of 330 tons, had anyone been able to calculate it. The slight drop in catch-per-unit-effort would hardly have been noticed against the background of environmental noise, and with more fishermen joining the fishery it would be expected anyway. There would be pressure to expand the fishery still further (see Chapter 9). Once the fishing rate reaches 60 per cent, the catches begin to fall and soon reach a level some two-thirds less than the peak catch in the early years (see Figure 6.7b). Catch per unit effort is now only 14 per cent of what it was at the early peak, and the number of adult spawning salmon in the rivers is only 8 per cent of the level in the virgin stock. Such a situation would certainly lead to pressure on management to 'do something about it', but it would be too late for the optimum action. Had the danger been recognised when the fishing rate was only up to 40 or 50 per cent, severe depletion need not have arisen. It is worth emphasising that, even at the fishing rate for C_S, the peak catch experienced in the early years can never be attained again; this general point is discussed further in Chapter 11. It can be seen that the Ricker curve could be a powerful management tool in stocks where its assumptions hold.

Using similar simulation experiments Ricker (1975) makes three further important points about the dynamics of this type of fishery. As far as we can see, the results will apply in general to any stock which could have a humped stock/recruitment curve. Shallower humps would only mean that the findings would apply less strongly, and a larger amount of random scatter would give them less rigour. First, a given stock size produces a greater catch when exploitation is increasing than when it is decreasing. Secondly, a given fishing rate will produce a greater catch when the rate is increasing than when it is decreasing. Both of these findings are important because they lead to spuriously high expectations which may cause overexpansion in the early years of a new fishery. The third finding concerns the exploitation of a stock consisting of a mixture of several species (or discrete stocks) each with a differently humped Ricker curve, as is the case in some Canadian Pacific salmon fisheries. The problem here is that the less productive stocks tend to reduce the total catch level below the optimum which could be achieved if all stocks had the same curve. Ricker's counter-intuitive finding is that it is better to spend one's effort enhancing the less productive stocks (through hatcheries, spawning channels, etc.) than improving the best ones. Ricker shows how the mixture of curves creates the situation where the combined optimum can actually be reduced if the best stock is improved. An elegant example of management recommendations based on this technique of multiple Ricker curves is described in Ricker and Smith's (1975) work on the Skeena river sockeye salmon fishery in British Columbia.

Referring back to the stock-recruitment data of Figure 6.1, we can conclude that a humped curve probably exists for many stocks with, however, great year-to-year variability superimposed. Ware's (1980) general recruitment model is at first sight satisfactory since it can assume humped or asymptotic shape. Unfortunately, its parameters depend on a detailed knowledge of the energy budget of the species. For the moment, Ricker's curve is as good as any other in describing the relationship since it can assume a variety of shapes and has the advantage that theory is available to allow it to be used in management. Even so, the scale of random scatter about the curve and the resultant uncertainty in fitting it argues for caution in interpreting the details of management recommendations. In some stocks, for example the Pacific salmon, Ricker curves can be used directly, but perhaps the most general use for the model in most fisheries is as part of a simulation of the fishery to determine the likely outcome of various alternative management strategies. The random scatter can be realistically incorporated in the simulation model, as we shall see at the end of Chapter 8.

Survival of Larvae

A summary of the changes in recruitment in North Sea plaice over 26 years is given in Figure 6.8 (Bannister, 1977). There is a cyclical trend in the data with random scatter superimposed, including one exceptional year class in 1963. The trend may reflect cyclical changes in primary and secondary production similar to those documented for the western English Channel (Southward, 1980). Garrod and Colebrook (1978) analysed similar but more complex links between recruitment fluctuations and meteorological and oceanographic phenomena such as the surface temperature anomaly in the Gulf Stream Drift in the North Atlantic. Hart (1974) correlated greater survival of North Sea sandeel larvae with incursions of north Atlantic water into the North Sea; increased survival was thought to be a consequence of the richer plankton community associated with this water. All these instances suggest that recruitment may be determined by major ecological factors; if we can discover what these are we may be able to predict recruitment. Referring again to Figure 6.1, it is apparent that no empirical model from the previous section could give a good prediction of recruits from the stock in any one year simply because of the amount of scatter, however well the model reflected the trend in the stock/recruitment data. Figure 6.8 and these further examples lead us into a different approach.

Variation in recruitment from year to year results directly from changes in the growth and mortality of the fry and larvae, so that an understanding of the ecology of these stages is necessary if we are to make any improvement on the empirical models. This is a most important point since recruitment often has a greater influence on catches and effort than any other single factor in the fishery (e.g.

Figure 6.8: Two-year-old Recruits to the North Sea Plaice Fishery Illustrating a Cyclical Trend.

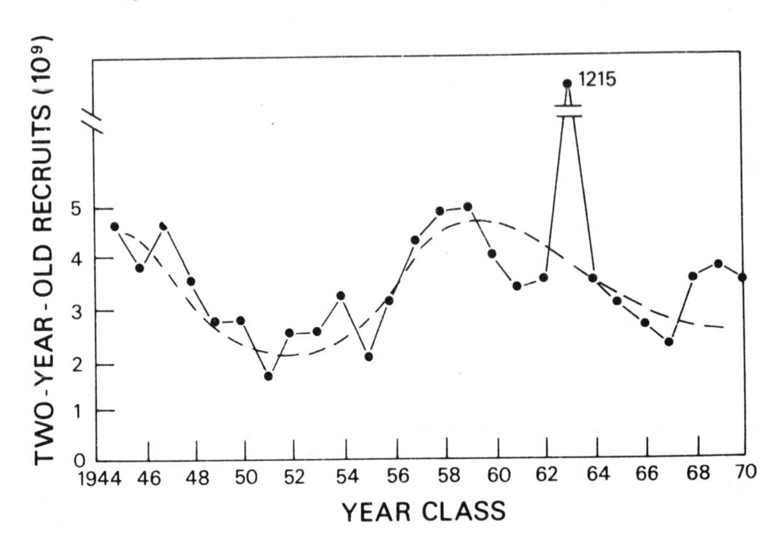

Source: Bannister (1977).

Bannister, 1977, for North Sea plaice). A detailed understanding of such factors important at the recruitment stage will enable one to test and build the predictive models needed to manage a particular stock (Chapter 8). Although correlation analyses or even sophisticated principal-component analyses linking recruitment to environmental variables may be quite successful (e.g. Smith and Lasker, 1978, on anchovy), only direct experiments can confirm the causal links suggested by such analyses. Therefore we are going to consider next some of the experiments on the survival of pelagic fish larvae which have recently clarified the first stages leading to recruitment.

Important experiments on the survival of pelagic fish larvae have been carried out by Houde (1975, 1977a, 1978) on tropical species of sea bream, lined sole and bay anchovy, and by Werner and Blaxter (1980) on Atlantic herring. In these experiments fertilised eggs collected from the wild or fertilised in the laboratory, are stocked at a range of densities in small tanks (1 to 20 litres). Algae remove toxic metabolites from the water in the tanks. Planktonic food for the developing fish larvae is maintained at predetermined densities by adding new food organisms two or three times a day as necessary. Suitable food, such as

Figure 6.9: A. Visual Field of a *Coregonus wartmanni* Larva, Illustrating Stereoscopic Fixation of Prey. B. Prey Capture by *Coregonus wartmanni* Larva.

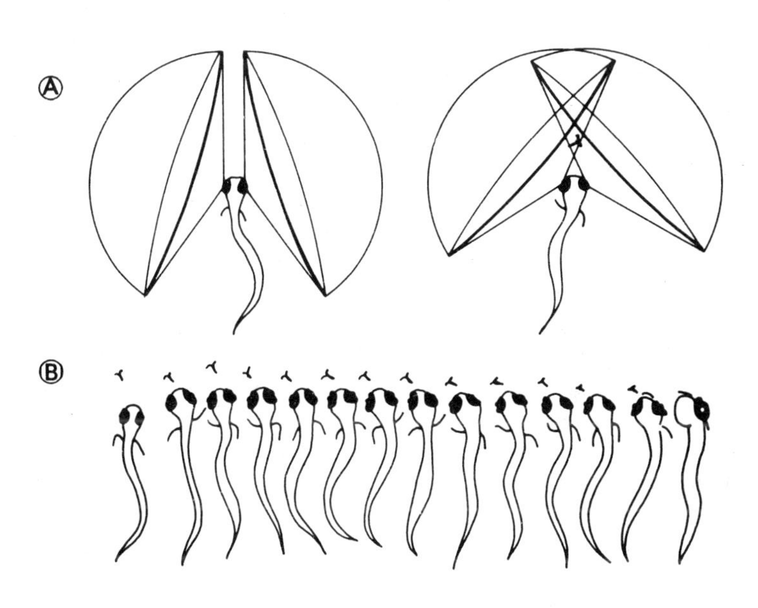

Source: Braum (1978).

copepod nauplii within the size range available to the young fish, is filtered from water samples collected from the wild, or alternatively nauplii can be raised in the laboratory from brine shrimp cultures. As the larvae grow, larger food has to be provided. Fish deaths are recorded and the larvae culled at intervals, giving estimates of survival and growth under each food regime. Such experiments are arduous and difficult to run, lasting in the case of herring for seven weeks. Tropical fish experiments can be run at 25°C and last about 16 days. Sea bream larvae begin to feed 36 hours after hatching at 2–3 mm long, once pigmentation around the eye cup has developed. Their yolk sac is completely resorbed at around 50 hours, so they have 14 hours in which to switch from endogenous nutrition to catching and digesting their own food. Balon (1975) terms this stage the eleutheroembryo. The larvae began to change into juveniles at about 12 days of age when they were 8 mm long. Herring also began to feed before the yolk sac was lost at about one week after hatching.

When a suitable small prey is within visual range (3 mm for herring and whitefish larvae), the fish tenses into an S-shape and then darts forward in a similar fashion to the lunge employed by adult pike (Chapter 1 and Figure 6.9). The larva's first attempts at prey capture are not very effective; success ranges from 1 in 50 attempts in herring to 1 in 3 for pike. After a week or so, approximately every other attempt is rewarded. The area searched for food by the developing larva can be calculated from its visual field (Figure 6.9) and swimming speed. For example, young herring can search 8 l/hour (Blaxter and Staines, 1971), and larval whitefish search 15 l/hour at 4°C (Braum, 1978).

The percentage survival of the larvae in these four experiments is summarised in Table 6.1. Each species exhibits a food threshold below which the fish larvae are unable to survive. Beyond this level survival increases but in general survival does not improve much above 500 plankters/litre. Inspection of the table shows that below the food threshold not only are the chances of survival very low for a larva, but they are also much more variable. This latter we consider to be an important point, for in most publications we have found that average survival values are quoted. The average does not reveal the variation between replicates of the same food and stocking density seen in the table. For example, at a food density of 300–500 organisms per litre, there is relatively little variation in survival between replicates, at 100 organisms/litre there is more, while at 30–50 plankters per litre between-replicate variability is very great indeed. (Analysis of variance (anova) techniques allow for such variability, but in fact are not often robust against a change of scale of variation across the experiment.) Table 6.1 shows that the design of such experiments needs to be very careful; Houde's 1975 work was not sufficiently replicated to allow analysis by anova, as were his 1977 and 1978 experiments. Even if the use of anova reduces the force of this criticism somewhat (as in Houde, 1978), it is clear from Table 6.1 that median survivals would be more informative than an average. For example, Houde's eight experiments on bay anchovy at 50 food organisms/litre gave an average survival of 5.9 per cent, but the median is 0.5 per cent! Since half of any results can be expected to fall above the median and half below it, the median is a more accurate reflection of the central tendency for these experiments than the average, which is biased by one high figure. Even use of the median obscures the fact that one replicate led to exceptionally high (40 per cent) survival, albeit with larval growth rather below normal. Similar variability at low food

Table 6.1: Percentage Survival of Pelagic Fish Larvae in Six Experiments. Replicates are included in each case where they were carried out. For discussion see text.

Stocking Density larvae/litre — Food Concentrations (Plankters/litre)

Sea bream (Houde, 1975)

	100	500	1500	3000
2	12	46	74	62
4	0	67	25,39	20,36
8	0	22,7	61	76
16	0	4.2	31	44
32	0.1	4.7	3.6	8.3

Sea bream (Houde, 1978)

	10	25	50	100	500
0.5	0,2.9,26	0,5.7,23	2.9,17,23	40,43,80	54,60,89
1	0,1.4,2.9	0,1.4,13	10,12,27	1.4,36,43	47,77,83
2	0,0,2.1	0,0.7,22	1.4,7.1,14	16,16,64	72,79,91

Atlantic herring (Werner *et al.*, 1980)

	30	100	300	1000	3000
8	4.1,0	3.7,0.8	12.2,14.8	17.2,18.0	11.2,5.4

Bay anchovy (Houde, 1977)

	50	100	1000	5000
0.5	6,40	11,0	49,20	63,74
2	0.7,0	2.9,4.3	66,59	57,62
8	0,0.7	0.2,0.9	18,34	75,48
32	0.3,0.1	9,10	55,43	63,68

Lined sole (Houde, 1977)

	50	100	1000
0.5	0,5.6	27,11	78,67
2	0,0	10,4.3	59,14
8	0,7.1	4.6,4.6	56,45
16	0,0	3.2,0	15.4,46.1

Haddock (Laurence, 1974)

	10	100	500	1000	5000
26	0	0	1.1	7.9	13.9

concentrations occurred with Werner and Blaxter's herring, some replicates again leading to reasonable levels of survival.

Could whatever happened in these occasional replicates also happen in the wild? It is possible that we have to look no further than to discover the sources of variation in these simple experiments to account for a large proportion of the observed variation between year classes in the wild.

Growth of the larva in the experiments showed a similar pattern to survival. In general, growth rates increased rapidly from low food densities to about 100 plankters per litre, then levelling off at higher food densities. Despite the fact that the experiments cover very different species in tropical and temperate waters, the general feature of the survival and growth of these planktonic fish larvae are remarkably similar.

Werner and Blaxter's herring were stocked at 8 per litre, a value within the natural range of up to 10 per litre which is encountered in the wild. However, the experiments which covered a range of stocking densities showed considerable contrasts between species. Larval density up to 32 per litre did not affect anchovy survival, whereas fish densities above 16 per litre reduced survival of lined sole and sea bream, probably as a result of competition for food amongst the juveniles towards the end of the experiments. Although the higher fish densities employed would be uncommon in the wild, the differences seem to reflect the ecology and behaviour of the species. Anchovy are pelagic schoolers as adults, while the bream and sole shoal less often. It would be interesting to see the results of this type of experiment carried out on a wider range of species. Other work has been done, but it has nearly always been at a limited and high range of food densities in work aimed at aquaculture.

At unnaturally high food densities, growth was generally neither enhanced nor reduced, but by vital staining a batch of *Artemia*, Werner and Blaxter recorded a more rapid movement of food through the tubular gut of the herring larvae (Figure 6.10). At food densities of 300 organisms/litre and over, food passed through the gut so quickly that the nauplii were barely digested, some even swimming away after defecation! The probable explanation is that larvae are adapted to feed at low densities so that a capture attempt will be made on anything which comes in range and presents the right sized stimulus. At high rates of encounter with suitable food organisms, the capture behaviour continues, although inappropriate, so that new food items are pushed into the gut too fast. Adult fish are usually able to regulate their food

Figure 6.10: Rate of Movement of Brine Shrimp through the Gut of Herring Larvae at Different Shrimp Densities.

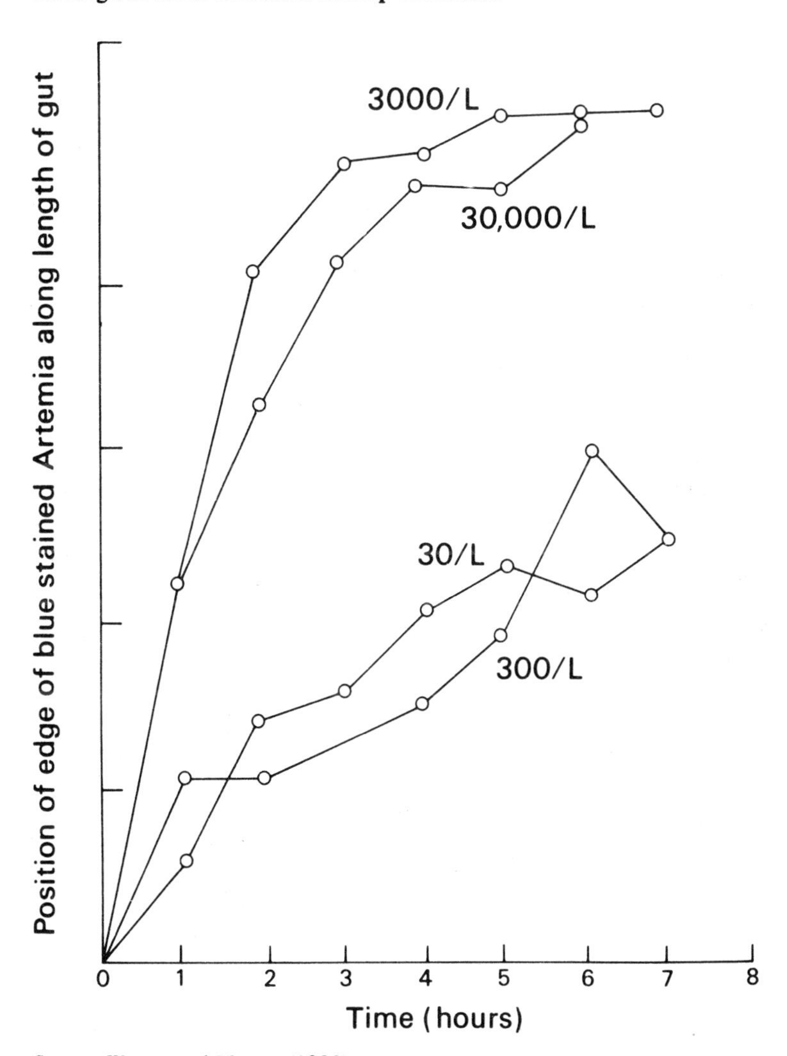

Source: Werner and Blaxter (1980).

intake but the larvae do not seem to be able to do this, a failure which emphasises the high evolutionary premium placed by selection on successful capture behaviour by these pelagic fish larvae.

The timing of the deaths of the larvae is of great significance. Werner and Blaxter carried out a control with no food at all and nearly all the fish survived until two weeks of age. Beyond this time massive mortality set in, so that all the fish with no food were dead by four weeks. Even at the optimum food densities, most deaths of larvae occurred during this period, from two to three weeks after hatching. Similar phenomena were noted in the tropical species, on a more rapid time scale of days rather than weeks. Why should this be? The deaths are almost all due to starvation resulting from failure to feed adequately in the crucial period of change-over from internal to external nutrition. If the larvae do not encounter enough planktonic food (of the right size) at this early stage they will starve and die some short time later. As mentioned (Chapter 5) young larvae in poor condition, apparently starving but not dead, have been found at sea. This raises the question – would larvae which have not been able to feed early on, recover if they encounter food later, perhaps in a plankton patch of higher density? Or is there a 'point of no return' beyond which they are doomed irrespective of their subsequent fortunes?

Laurence (1974) performed six-week experiments on larval haddock in relation to food density and the point of no return. Haddock begin to feed 2 days after hatching, but the yolk sac lasted until 6 or 7 days of age. Laurence determined the last point after hatching when food had to become available for the larvae to feed successfully. Batches of haddock were kept without food for 2, 4, 6, 8, 10, 12 and 14 days post-hatching, and then presented with abundant food at 2,000 organisms/litre. More than 50 per cent of haddock larvae deprived of food until 2, 4, and 6 days grew and survived well, but all of the fish deprived of food for 8 days or more died. Laurence concluded that 8 days was the 'point of no return' for haddock at 7°C. In a further experiment, groups of larvae deprived of food until day 6 were presented with food at different densities; the subsequent survival was similar to that found in the main experiment where food was provided immediately after hatching (see Table 6.1). The work suggests that the haddock must already be feeding when yolk sac nutrition ends. Furthermore, it is significant that more than 50 per cent of the 8- and 10-day deprived fish died later with food in their gut, and had therefore been able to initiate successful prey capture behaviour, but to no avail. This suggests that some irreversible change in the ability to digest food occurs if the larvae have not fed prior to the point of no return. Similar experiments by Lasker *et al.* (1970) indicated that at 22°C anchovy reached such a danger point at 3 days old; beyond this they starved and died even if food was available. Blaxter and Hempel (1963) demonstrated a similar phenomenon in Atlantic herring.

Starving anchovy larvae exhibited cell shrinkage and separation, loss of intercellular fluid in the musculature and liver and pancreas degeneration. O'Connell (1976) demonstrated that these histological criteria were better than purely morphological criteria in discriminating larvae according to their chances of survival. In larval cod, the RNA/DNA ratio provides a useful index of nutritional status which can be easily measured. Starving fish have a low ratio (Buckley, 1979).

O'Connell's work with anchovy larvae suggests that degeneration of the pancreas may be critical, even when other signs of emaciation were present. Pancreas in starved anchovy up to two days old were in good condition, implying that these fish had a reserve of zymogen which enabled them to begin digestion of any food consumed. Although adult fish can mobilise lipid depots (Chapter 4) for energy when starved, larvae have no such lipid reserves and protein mobilisation starts immediately. It seems that this affects pancreas function first, and if the pancreas degenerates once the yolk sac has gone, the animal cannot begin to digest food even if it is encountered and eaten. The lack of lipid food reserve in the young larva is interesting. The yolk supply represents the investment in larval survival that the parent of the species has been evolved to make; this investment will tend to be minimised (for any life history type) provided the species persists (somehow) through time (see Chapter 5). To provide larvae with more reserves imposes an extra cost on the parents. In this case enough pelagic larvae can, on average, manage to build up their own reserves to save the parents this extra investment.

Early work on larval survival suggested that actual prey densities in the wild were as much as two or three orders of magnitude lower than those found necessary for survival in laboratory experiments; Hunter (1972) invoked the 'plankton patch hypothesis' to explain the discrepancy. Natural processes amplifying small differences resulting from ocean currents lead naturally to patchiness in the plankton (Steele, 1976; Harris, 1980), but plankton sampling commonly expresses results as average densities. So it is suggested that the 'wild' food densities quoted are spurious in the sense that high food concentrations will be averaged with low ones. Fish larvae will encounter patches of plankton above their critical threshold for survival even though the measured average plankton densities are below this level. The implication is that a further hazard of larval life is for a larva to find itself fortuitously in such a plankton patch above the critical density.

Plankton sampling can sometimes be very inadequate. For example, in the thread herring survey off Florida analysed by Houde (1977b), there was no correlation whatsoever between herring larvae density and plankton density because the plankton nets used had mesh too large to catch food eaten by thread herring! In addition, the early laboratory experiments may have been misleading and experimental protocols have now been improved. (Houde notes that often too many larvae were stocked for extrapolation to the wild.) In the most reliable experiments, growth, condition factor (Chapter 4), and mortality rates closely mimic those found in the wild (Laurence, 1974). Houde (1978) has also recently improved the method of calculation of critical food densities in such experiments. This is done by calculating a log regression of an arcsin transformation of per cent survival on log food density. The level at which only 10 per cent of larvae survive can be read off the regression; this is taken as the critical threshold. Since some larvae are going to survive even below this, the 10 per cent level may be thought rather arbitrary, but Houde's method does have the advantage that data from all the replicates can be included in the regression. The variability in the data leads to the width of the confidence limits which may be attached to the critical food density estimate. Table 6.2 lists some critical prey densities calculated in this way.

How do these values compare with densities of suitably sized plankton when accurately sampled? In the case of the three Florida species assessed by Houde himself, densities of suitable copepod nauplii and copepodites were measured at 50 to 300 per litre, occasionally up to 1,500 per litre. If in addition the larvae ate the much smaller tintinnids these measured oceanic food densities could support adequate larval survival without the need to invoke the plankton patch hypothesis. In the case of the Atlantic herring, densities in the Firth of Clyde were only 0.3 to 2 suitable prey per litre and this is 100 times too low for survival. Werner and Blaxter concluded that only patches could sustain the herring. It is instructive to note that patches may be dynamic since they may alter in density and location with time, just as the fish density does over the period of high mortality. In Werner and Blaxter's experiments the ratio of densities of herring larvae to their brine shrimp prey fell from 25:1 to 200:1. In the wild it is possible that grazing by the larvae may reduce plankton patch densities locally.

Other work, including work on lake-dwelling coregonids (Dabrowski, 1975), seems to fit the same pattern; in some cases the fish larvae seem to need plankton patches, in others they may not be necessary.

Two general comments seem appropriate; first, even if survival is

Table 6.2: Critical Prey Densities Calculated According to the Method of Houde (1978). Confidence limits calculated from equations published in Houde (1978) and Werner and Blaxter (1980).

	Food Densities	
	Plankters/litre	Approx. 95% Confidence Limits
Atlantic herring	171	50 to 250
Bay anchovy	107	46 to 256
Lined sole	130	72 to 234
Sea bream	34	25 to 46

adequate in normal plankton densities, it is only marginally so, and most experiments indicate that growth is more rapid at higher densities, before levelling off at food levels above 500 organisms per litre. Growth may therefore be more rapid in a plankton patch even if the average plankton density is just about adequate. Larvae may therefore be less vulnerable to predation if they spend less time at a small size in the planktonic phase. For example, Houde (1975) calculated that larval sea bream took five days longer to reach juvenile size with food at 500 organisms per litre than at 1,500 per litre, thereby prolonging the danger from predation. It is also worth noting that although many authors consider that predation on the larvae is heavy, few quote any quantitative evidence for this opinion. Secondly, since natural processes in the plankton community naturally lead to patchiness, it is not necessarily a disadvantage to feed only in a patch, in contrast to the way in which the plankton patch hypothesis is usually quoted as a 'bad thing'. Indeed, patches may be so reliably formed that it may in fact be an adaptation to rely on patches. In terrestrial ecology many experiments on predator-prey-grazer universes have shown that long-term persistence is only possible in an environment composed of a mosaic of patches (Hardman and Turnbull, 1974; Huffaker, 1958; Takafuji, 1977). Field work has suggested that similar rules apply in aquatic ecosystems (e.g. Vince *et al.*, 1976; Macan, 1966). So a possibility is that only larvae which survive best in patches persist over evolutionary time.

As we have seen, there is sufficient variation in survival from experiments at one food density to account for observed variation in year-class strengths. The range of plankton density and patchiness can be added to this, so that it hardly seems necessary to look for further sources of variable mortality. However, many consider that predation by invertebrates such as arrow worms during the larval stage could be a density-dependent factor, even if it were not the major cause of mortality. The change in behaviour of larvae from dispersed food-searching to schooling towards the end of larval life may be instructive. Schooling is no protection against a short-range invertebrate predator, but does help protect against the more mobile fish predators to which the juveniles become attractive once they have grown. Discussion of the role of predators on fish larvae remains speculative because of the lack of data and experiments. Foraging strategies of these invertebrates have not been studied and we do not know, for example, whether they are attracted to aggregations of fish, or whether the functional response (Holling, 1954) reduces their impact at high densities. There is clearly a need for large-scale experiments on larval survival which somehow include predators as part of the system, to parallel similar experiments on terrestrial organisms.

If there is no single critical period, pelagic fish larvae certainly experience a number of hazardous processes. Developing young fish encounter behavioural problems in catching prey efficiently, physiological problems in switching from yolk sac to external nutrition before starvation destroys their pancreas, and ecological problems in the form of predators. There is also their own luck in finding themselves in a patch of prey of suitable size and density. Even with adequate food, in experiments many larvae die; O'Connell could not find out why – certainly all the deaths could not be accounted for by genetic defects.

To summarise the evidence on pelagic larval fishes, there seems to be a critical time just after the yolk sac is finished when the fish must encounter sufficient prey of suitable size. If this does not occur, irreversible degeneration of the digestive system takes place and they cannot live beyond this time even if food is successfully attacked later. Subsequent survival and growth will be optimum if the larvae find themselves in a plankton patch, and rapid growth will mean less predation from invertebrates. All these sources of hazard combine to give great variation from year to year and place to place. They can easily produce the observed variation in year-class strengths. In some cases juveniles enter the fishery directly, while in others recruitment comes at a later stage. One or several of the processes involved could be critical (i.e. the key factor) for a particular species; perhaps it is surprising that there is not more variation than that observed in fact! So, having recognised the general nature of the problem, we arrive again at having to deal with the particulars of larval survival in each species stock to get any further.

Models of Larval Survival and Recruitment

There have been a number of attempts to make detailed models of recruitment in particular species, these exercises often failing for lack of realism, overambitious generality, or too much special pleading. We will describe two such models which have been reasonably successful.

A carefully constructed bioenergetic model for winter flounder larvae has been published by Laurence (1977). Data for the model came from an experiment similar to those discussed in the previous section on feeding rate and survival at five prey densities. The energy values of the larval prey and the gut contents were determined, and the bioenergetic model constructed using equations similar to those discussed in Chapter 4. Laurence noticed that the flounders fed only in the light and incorporated the length of the feeding period in the model. Winter flounder spawn in spring when the photoperiod is about twelve hours in their latitude, and the temperature is 8°C. Some of the results are shown in Figure 6.11, which plots the critical minimal prey densities below which the larvae could not gain energy rapidly enough to offset metabolic losses during the twelve-hour feeding period. Below these densities the young flounders die. As expected, the critical densities are highest during the first feeding period after the yolk sac is lost. After about eight days of age they fall as the larvae grow and can search faster. When the flounders attain 75 μg in weight at about four weeks of age, the critical food density increases again until they begin to metamorphose at approximately 500 μg at seven weeks old. The requirement then falls again, probably because the efficiency of prey capture increases as the metamorphosing fish begin to adopt the classic flatfish 'lurk-and-lunge' technique, in place of constantly searching for prey. Laurence's model emphasises the limits of applicability of the critical period hypothesis. Some periods during larval life are more 'critical' than others, but all could be hazardous if food density is lowered. Once metamorphosis is reached, the young flatfish are much safer, and can begin to withstand temporary food scarcity by calling on reserves, although in some species there appears to be strong scramble competition for food during their first year on the bottom.

The aims of a model of gadoid recruitment by Jones and Hall (1973) were: first, to determine if realistic mortality and growth could be generated by considering the encounter rate of larvae with food; and

Figure 6.11: Size of Larvae of Winter Flounder Plotted Against Critical Prey Density (as Energy). The critical density was that below which there was not enough time in the day to feed.

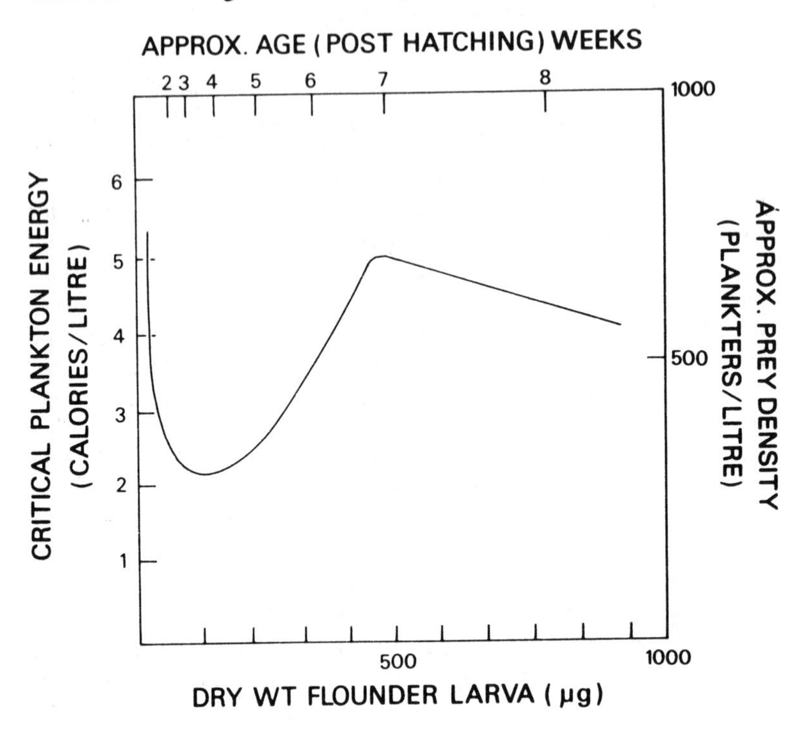

Source: Laurence (1977).

secondly, to see whether realistic variation in year-class strengths resulted from variation in larval food supply. This latter was achieved by combining the larval model with a full simulation of haddock and whiting populations. The larvae of these gadoid fishes increase their food intake as they grow by switching to successively larger prey. The prey themselves grow and can form the food of the same cohort of haddock as they become older. Each cohort of fish can then be considered to grow up with its own private food supply. Several cohorts of larvae may start in any one year partly because the spawning season is quite protracted and partly because of fluctuations in the copepod prey.

Larval survival was modelled by increasing the search rate for prey with age. Mortality of larvae was related to the ratio of the number of prey encountered over an estimate of the critical food density. Prey

Figure 6.12: Some Results from Jones and Hall's Recruitment Model. (a) Examples of stock recruitment curves for different initial food densities. (b) Coefficient of variation of simulated year-class strengths. (c) Frequency distribution of haddock year-class strengths compared with a set generated by the simulation.

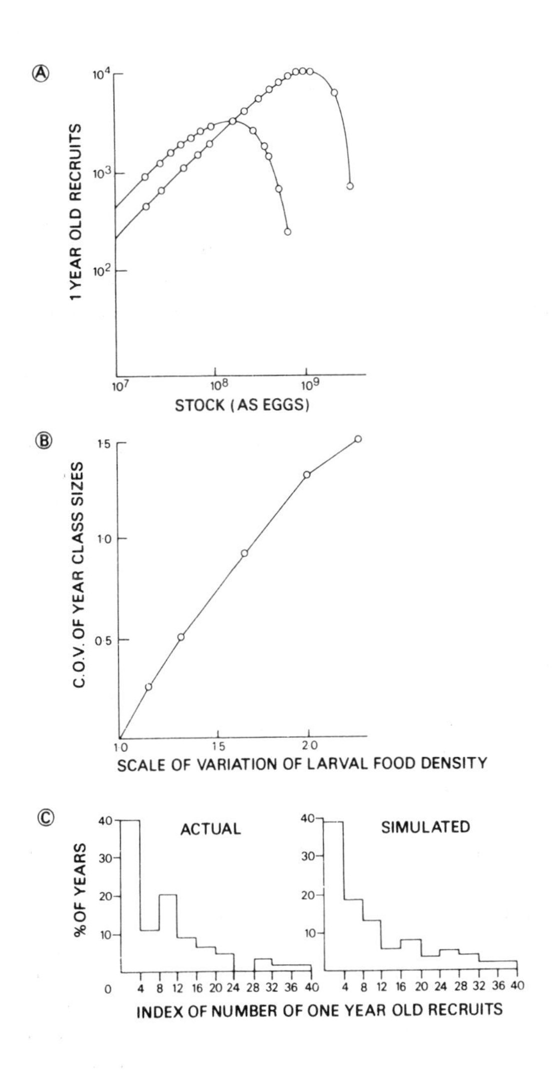

Source: Redrawn from Jones and Hall (1973).

mortality was just the number eaten by the fish. Both the initial fish density and the critical food density could be varied in different runs of the model and the outcome at 45 days of age recorded. This process gave a series of recruitment curves. The intercept of each curve was then altered to force the curve to pass through a fixed point based on a calculated long-term average of stock and recruitment for each of the two species obtained from field data. This was to make the model more realistic. Two such curves for haddock representing different conditions in the model, are illustrated in Figure 6.12a.

The larval portion of the model was incorporated into a full life history simulation in steps of 1/100th of a year, based on growth, reproduction and mortality of each age class (see Chapter 8 for examples of models of this type). The model was run with initial food density varying at random between certain limits for each simulated year, i.e. in each simulated year the total number of eggs generated in the previous year were allowed to develop with an initial food density picked at random from within the limits allowed. The point of the exercise was to compare runs resulting from different values for these food limits.

Year-class fluctuations virtually uncorrelated with stock size (as eggs) resulted from the model, closely mimicking the real situation (compare Figures 6.6 and 6.1). Over many simulation runs, the coefficient of variation (COV = s.d./mean) of the year-class strengths was plotted against the range of random variation allowed in the food supply (Figure 6.12b). The results showed that variation in year-class strength as great as that observed in the wild could be generated in the model by a food supply which varied at random by a factor of two to three times. Furthermore, the distribution of simulated year-class strengths closely resembled the actual ones, lending confidence to the assumptions of the model (Figure 6.12c).

Both of these models confirm that larval food is a key factor in the ecology of recruitment. However, as we have seen, predation can be important in salmonids (Larkin and Macdonald, 1968), cod (Cushing, 1977), and may interact with food supply through growth during the planktonic phase in many species with planktonic larvae. Food may be a key factor and predation a regulating factor in the special ecological sense of these terms (Ricklefs, 1979). We feel that the discussion in this chapter has shown that there are only a few general principles in recruitment other than the need for consideration of the exact details of the ecology of each species. When this can be done we can improve upon the empirical Ricker curve as a management tool.

Recruitment and Life History Strategies

On a more theoretical basis, however, recruitment characteristics may be of more fundamental significance than their use in management models may suggest. To conclude this chapter we will consider two recent pieces of work. The first attempts to relate recruitment and life-history characteristics to production; the second links recruitment to the way that life-history parameters are affected by density. Garrod and Knights (1979) have presented an interesting story which ambitiously tries to explain the characteristic features of commercial fish yields through the recruitment and population growth of the species involved. Dividing global fisheries into six zones based on habitat, they claim that the zones with more variable environments generate higher fish production because they have a higher proportion of pelagic shoaling species, such as the clupeids. Garrod and Knight's zones are shown in Figure 6.13a together with a scatter plot of production against the percentage of pelagic shoaling species based on the figures given in the paper (Figure 6.13b). Although the correlation is statistically significant (at the 5 per cent level), it is brought about by the three points to the right of the graph which represent exceptionally productive upwelling areas over deep water, where it is not surprising that demersal fisheries are low. So the data do not support Garrod and Knight's assertion very strongly. Nevertheless, pelagic shoaling species are adapted to persist in these variable environments through a suite of life-history characteristics whose result is a higher intrinsic rate of population growth (Chapters 3 and 5). These are the classic r-species more familiar to students of insect population dynamics. Species with a higher r have greater harvesting potential, since they potentially have a greater surplus over and above the numbers of fish needed for replacement of the current generation by the next (see discussion of surplus yield models in Chapter 7). Small size, rapid growth, early maturity, high fecundity and a short generation time are r characteristics (Chapter 5). It is therefore intuitively reasonable to attribute high yields to these species. However, many fisheries in high latitudes depend upon large, long-lived demersal species (see Chapter 2) like the cod, which nevertheless at first sight cannot be classified as K-strategy species because of their very high fecundities. Garrod and Knights demonstrate that despite their high fecundity, such species have in fact low r values and characteristics if one calculates r on the basis of replacing recruits. The high fecundity can be considered the special adaptation of a K-type species to the planktonic larval mode of life. Presumably this is itself

Figure 6.13: (a) Potential Fish Yield Per km² in Various Regions. Shaded bars represent the contribution of pelagic shoaling species. (b) Potential Fish Yield Per km² Plotted Against the Percentage of Pelagic Shoaling Species in the Fishery. Spearman rank correlation is 0.42, which for n = 24 is just significant at the 95 per cent level. For further explanation see text.

(a)

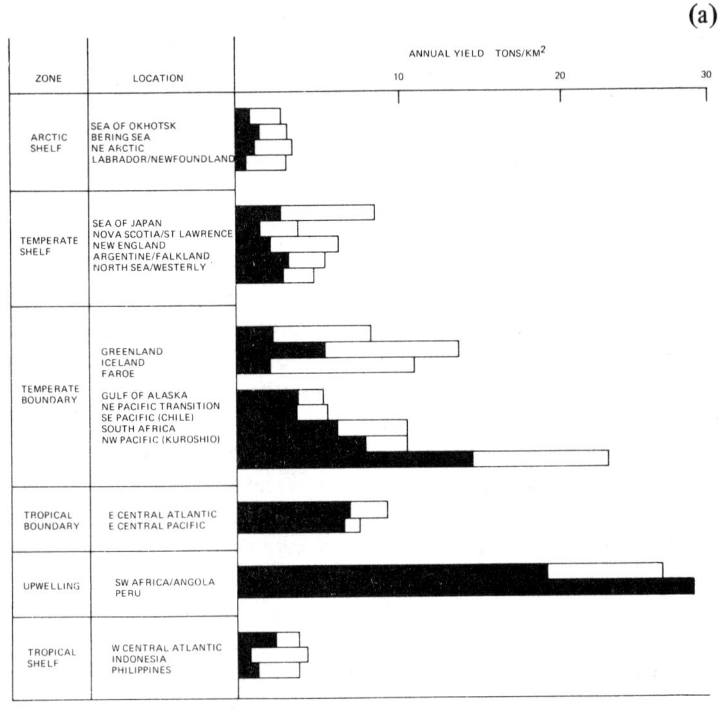

Source: Redrawn from Garrod and Knights (1979).

an adaptation partly for dispersal and partly to exploit the profitable energy source which the plankton represents to a small predator (see also Chapter 5). We could perhaps call these species 'pseudo-K' strategists. The elasmobranchs conform to the usual K-strategist pattern with their small number of advanced offspring, and their trend towards viviparity, evolved through the possession of internal fertilisation (see Chapter 1).

Garrod and Knights link intrinsic population growth values, calculated by their recruitment method which is discussed in detail below, to global patterns of fish production. They argue, for example, that

(b)

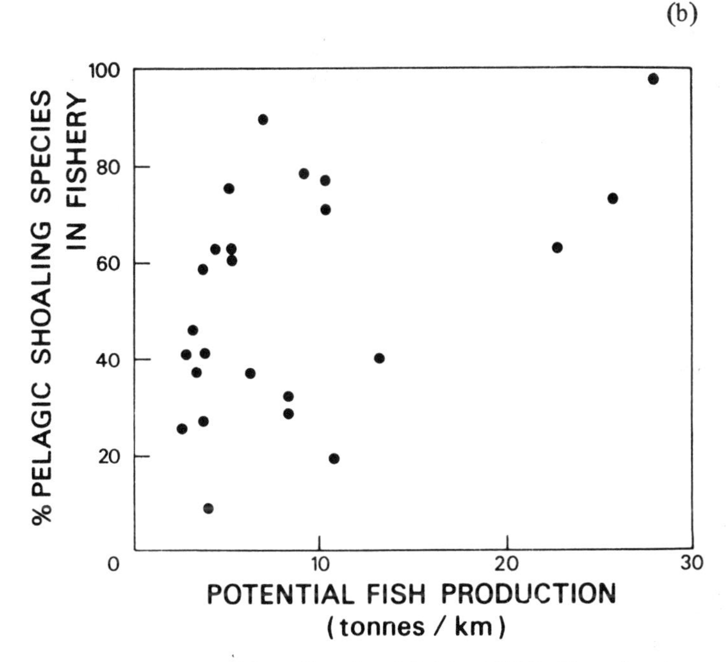

Source: Data extracted from Garrod and Knights (1979), Figure 1.

differences in the type of fisheries between the North Pacific (mainly pelagic fisheries) and North Atlantic (mainly demersal), which at first sight ought to be very similar steep-shelf high-latitude zones, can be accounted for by greater variability of the North Pacific primary production, making that zone more suitable for the shoaling pelagic species.

The conceptual framework for Garrod and Knights' speculations relates fish production and rate of population growth to three axes of diversity: temporal adversity, similar to the 'durational stability' of Southwood *et al.* (1974); biotic adversity, reflecting the lowered levels of primary production moving out from shelf to deep oceanic waters; and environmental adversity, reflecting the harsher climates of higher latitudes. According to this framework, the highest production (in the long term) should be from low *r*-types in stable areas and high *r*-types in unstable ones.

The main theme of these arguments, that fish production reflects the life-history strategy which a species has adopted to persist in the long term, is illuminating and unexceptionable. However, the specific

Figure 6.14: Variability in Recruitment in Selected Stocks of Four Species. Percentage scale calculated here is approximately 4σ to give the 95 per cent confidence interval. Median variability for each species is indicated by the solid bars.

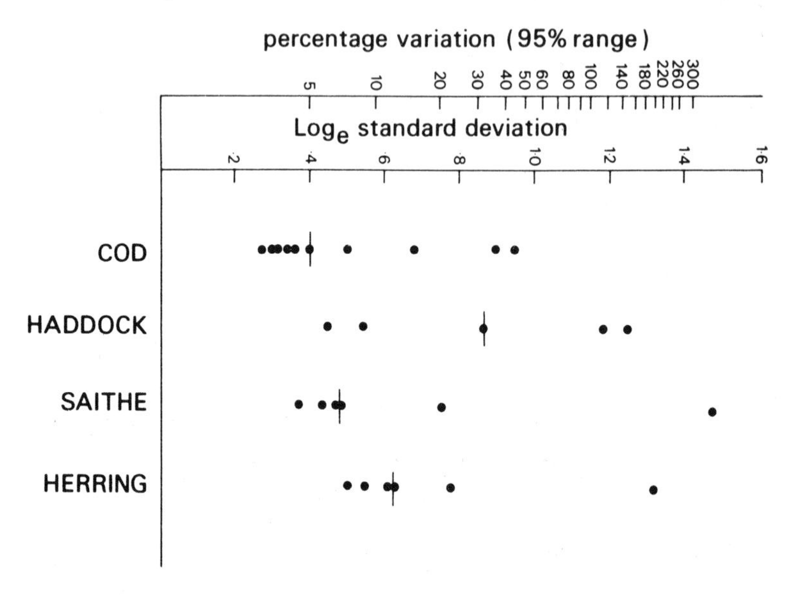

Source: Data extracted from Garrod and Knights (1979), Table 3.

prediction that higher yields and stock sizes are a consequence of variability in recruitment is less rigorously supported by the evidence assembled in the paper, and there are three main criticisms of the work as published.

First, the data Garrod and Knights assemble on variability in recruitment in different stocks and species do not really convince that pelagic shoaling species are more variable. Their values are plotted in Figure 6.14 together with medians for each species. Each species itself exhibits a wide range of variability in different stocks, not all of which can be explained by latitude. Differences between the values are unlikely to be artifacts, since all are based on a long series of data. Amongst the gadoids, saithe are the strongest schoolers (Pitcher *et al.*, 1976) yet the median for saithe is scarcely different from that for cod and considerably less than that for haddock. Based on the medians, herring are only twice as variable as saithe. This point is, however, not a very serious criticism since it remains true that year-class variation may be high in open-water species. Nevertheless, Chapter 5 has explained that these species can be buffered by age-class structure more often than is usually assumed, in a similar way to more long-lived demersal stocks, some of which can be extremely variable in recruitment, as Figure 6.14 shows.

The second group of criticisms requires a more detailed explanation of the recruitment model employed by Garrod and Knights. They consider that the population parameters of the stock will be adjusted to each other to ensure replacement in the long term. The number of eggs per recruit is taken as a measure of survival, because it predicts, over evolutionary time, the numbers needed to counteract the various sources of adversity. They devised the following method to get an estimate of the intrinsic rate of natural increase for each of their species. The number of eggs per recruit, E_r, was calculated from fecundity, growth, sex ratio, age at maturity and mortality over a cohort with methods similar to those discussed in Chapter 3. Data from five species gave a common average of 1.009×10^6 eggs per recruit. A scaling factor was then calculated for each species by dividing the common E_r by the species-specific E_r. The scaling factor was needed to adjust the eggs needed to produce one recruit to the arbitrary standard for any stock size. When this was done, data on stock and recruitment at one year old can be validly combined for many species; Garrod and Knights did this for the seven stocks shown in Figure 6.15. A log/log regression describes the scatter plot in Figure 6.15 well; Garrod and Knights obtain the equation $R = 7.24\,S^{0.574}$, relating stock (as eggs) to recruitment. The exponent 0.574 implies weak density dependence where an increase in stock will produce a somewhat smaller increase in recruitment. Stock recruitment curves for each of the individual species can be obtained by retransforming the equation back to the absolute form with the scaling factor. The next stage was to simulate the population growth of each species using the stock/recruitment relationship just calculated along with published values for the other essential life-history characteristics (age structure etc.). Each species was run for 200 years of population growth and the population sizes plotted against year. These population growth curves were asymetrically S-shaped (probably due to the choice of log axes for the recruitment formula), most species reaching equilibrium numbers within 20–100 simulated years. Garrod and Knights then fitted the Richard's growth equation to the curve; this is just an asymmetrical form of the logistic with an extra parameter to describe the degree of asymmetry as used by Pella and Tomlinson in their general

Figure 6.15: Combined Standardised Stock-recruitment Relationship Used by Garrod and Knights (1979). For further explanation see text.

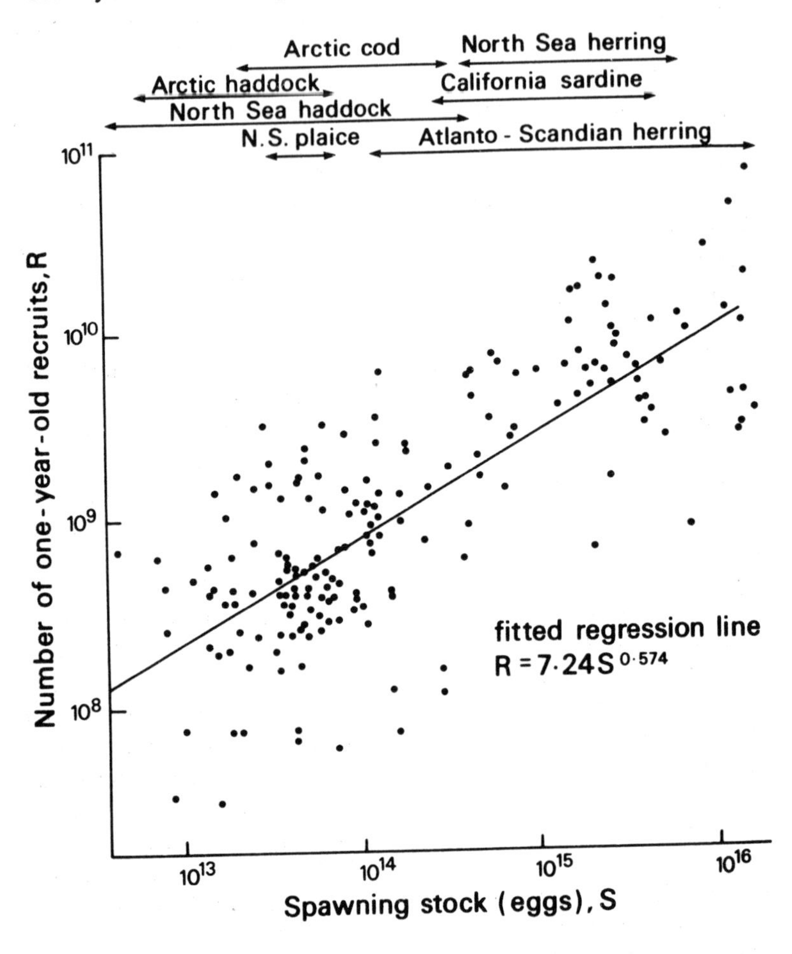

surplus yield fishery model (see Chapter 7). This growth equation gave as one of its parameters an estimate of the intrinsic rate of natural increase, r'. Values of r' obtained in this way are listed in Table 6.3. The consequences of these rates of increase for the yields of fisheries were then discussed as above.

The regression-based recruitment relationship used in the exercise can be criticised on several counts. For many there is evidence of low recruitment from high stock sizes from the abundant stocks of the early

Table 6.3: Values of r', the Rate of Population Growth Obtained by Fitting the Richard's Growth Equation to Simulated Population Growth Curves generated by Garrod and Knights' Recruitment Model.

	r'
Dogfish	0.026
Pacific halibut	0.035
North Sea plaice	0.036
Arcto-Norwegian cod	0.037
Atlanto-Scandian herring	0.044
North Sea herring	0.053
North Sea haddock	0.061
Californian sardine	0.094
North Sea sprat	0.131

years of exploitation prior to depletion. Species such as haddock and cod are generally considered to exhibit humped curves for this reason, as in the classic example of Arcto-Norwegian cod in Figure 6.6 above. Surprisingly, the data for this very stock are some of those used by Garrod and Knights, who have superimposed their own regression and a Beverton and Holt curve for comparison (Figure 6.16). This is carried out on a data set which appears to have been truncated (compare Figure 6.6 with Figure 6.14), leaving out data for the humped part of the curve. However, even if the missing data had been included, and this had been true of each of the species concerned, plotting the data overlapping all on a common graph, such as that in Figure 6.14, would obscure the curves and produce the scatter observed. For this reason the main criticism of this method is that it eliminates any possibility of a humped recruitment curve for the individual species. There is therefore little reason to have much confidence in the calculated values of r'. The authors plead that they are dealing with rates of increase now, under heavy exploitation, but then this is not how they actually use the values in their long-term simulation. The value for haddock, calculated from conventional life table methods in Chapter 3, is $\bar{r}_m$ = 0.180 for the years 1957 to 1967, about twice the rate here. Much of this is due to the 1962 year class with r_m = 0.873.

The third, and perhaps most important criticism centres on Garrod and Knights' claim to have demonstrated that higher variability in recruitment leads directly to higher average stock size and higher average yields. The authors show a graph (their Figure 6) which illustrates

Figure 6.16: The Arcto-Norwegian Cod Stock-recruitment Relationship Used by Garrod and Knights (1979). Compare with Figure 6.6.

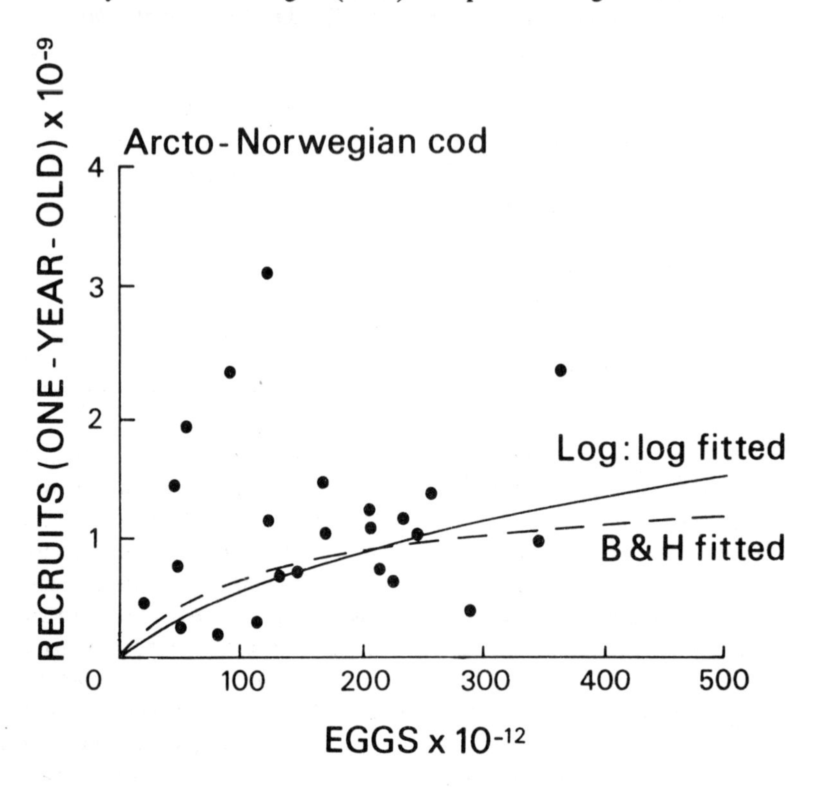

a greater enhancement of yield by this mechanism in stocks with a high r'. This finding, apart from being intuitively surprising, is different to the results of any simulations which we have run, including those described in Chapter 8. The reason seems to be that a logarithmic standard deviation of stock size has been employed. The arithmetic average of values from a log normal will, when transformed back to the arithmetic scale, be higher than the original arithmetic transformation of the log mean. Such an artifact would account almost exactly for the stated twofold increase in yield when the standard deviation of recruitment approximately equalled the mean.

These criticisms do not denigrate Garrod and Knights' fascinating attempt to relate global fish production to the ecological strategies evolved in response to the 'templet of habitat' (Southwood, 1977), providing many ideas for further development and research.

Ware (1980) has devised a general recruitment model based on the two separate processes of density-dependent pre-recruit mortality and stock-dependent reduction in population fecundity. These were the two density factors recognised by Harris (1975) and shown to be necessary for a humped recruitment curve by Cushing and Horwood (1977). In Ware's model the stock-dependent factor is brought about partly by density-dependent growth and partly by reduction in specific fecundity, and here he differs from the ideas of earlier workers like Cushing and Horwood (1977) or Ricker (1975) who thought that cannibalism producing density-dependent mortality reduced population fecundity. Ware cites good evidence for the reduction in specific fecundity in high-density populations, but his evidence for density-dependent growth in marine fish is less convincing. The model is based on the surplus energy which is available for either somatic or gonadal growth. Ware calculates this surplus energy from detailed life table and growth data. The maximum size of the fish is governed by a limit at which all surplus energy is used up in gonad growth.

The model generates a family of curves which may be either humped or asymptotic, the two extremes being found in fish which have either strong or weak stock-dependent fecundity limitation. This factor is high in K-type fish like cod, who can grow large before reaching their size limit. Such large long-lived fish eat large food organisms which tend to have long generation times. Dense populations of these fish can overcrop their food supplies and so we can expect selection to have favoured mechanisms which avoid this. Small short lived r-type fish like herring feed on planktonic prey with short generation times and are unlikely to reduce food availability by feeding, but face large year-to-year fluctuations. Population fecundity therefore tends not to be limited, and only larval mortality operates.

Ware's general model is a useful advance, but as yet the data needed to fit its parameters are available for only a few species of fish.

Recruitment is an artifact caused by man's fishing. In this chapter we have described the variable nature of the components of the fish's ecology which may be included in recruitment in different species. Recruitment must be estimated for effective fishery management. An expensive but effective way to do this is by monitoring the pre-recruits; models are cheaper, but difficult to build. We have described the more successful of the general simple models of recruitment, especially the Ricker curve, and have identified large environmental scatter as the main difficulty. We will describe in Chapter 8 how such models are used in fishery management.

The most important component of recruitment in sea fisheries, and the only one to have been subjected to experimental study, is survival and growth in the pelagic larval stage. We have reviewed evidence that the food supply is the single most important factor in predicting larval survival. Lasker (1978) has shown that in anchovy, a survey of the larval food organisms themselves may be a useful and practical predictor of year-class strength. In the wild, it seems likely that larvae survive better if they find themselves in a plankton patch.

Finally, we have discussed two attempts to link recruitment to the suite of life-history characteristics evolved in *r*- and *K*-selected fishes.

7 PREDICTION OF FISHERY YIELDS: SURPLUS YIELD MODELS

How many fish can be taken without destroying the stock? More precisely, how can the harvest of fish biomass be maximised without impairing the prospects of exploiting the fishery again in the future? This apparently simple question forms the basis of all fishery yield analysis; failure to answer it adequately has contributed to mistakes in fishery management in the past. The rest of this book is concerned with the methods which can be used to predict yields from managed fish stocks. Because of the importance of this topic to the next generation of fishery managers, we address ourselves to two related questions which, in the light of past events, we feel must be asked in addition to covering the essential methods of quantitative prediction of yields. The questions are: in what ways has the advice of fishery biologists been misleading? And why has biological advice so often been ignored?

The basic problem is illustrated from first principles in Figure 7.1, which shows the change in biomass with time in a hypothetical fish population under different harvesting regimes. After a crop is taken at time A, by gill nets for example, the biomass grows back to equilibrium level along an S-shaped curve of a type familiar to population biologists. The precise form of this S-shaped curve is not important at this stage, although the 'logistic' curve which we will consider below is often considered appropriate. It is sufficient to note that at low stock levels the rate of growth of biomass increases with stock size, but that after reaching a maximum, the rate begins to decrease again as the carrying capacity, represented by the upper limit of biomass, B_∞, is gradually attained. In practice both the growth rate and B_∞ will vary with time, but we will put aside this complication until later in the chapter. The figure shows regeneration of biomass taking place at a slower rate along the upper part of the S-curve after the smaller crop taken at time B. From time C, we show an example in which a series of regular crops keep biomass oscillating between the levels indicated by the dotted lines. It is worth noting that cropping regime C exploits the part of the S-curve which gives the maximum rate of biomass regeneration: in fact for the 'logistic' curve this is at $B_\infty/2$. Beyond time D, crops are being taken too frequently for the

Figure 7.1: Biomass Plotted against Time for a Cropped Population which Generates Biomass According to a Logistic-type Model. B_∞ represents the carrying capacity of the environment.

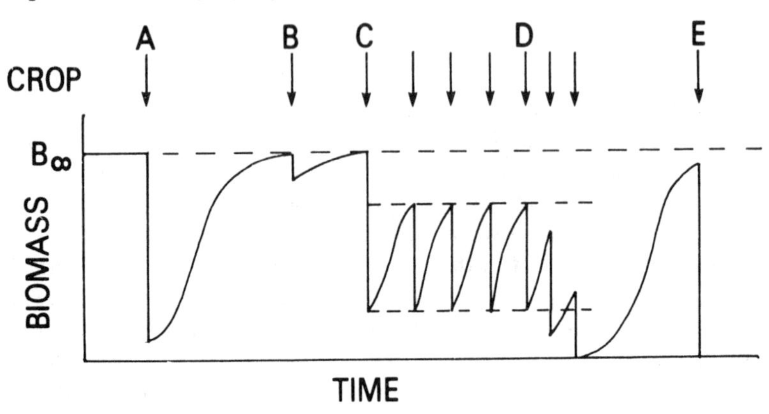

replacement of biomass to keep pace, and eventually the population is reduced to non-viable levels (too scarce to find mates, perhaps). In our Figure 7.1 this point is not quite reached and the stock recovers, only to be exterminated by a massive crop taken at time E.

From the biologist's viewpoint the correct fishing strategy is apparently one which provides the maximum sustainable yield (MSY) of fish, option C in Figure 7.1. As we will see later in this book, fishery managers nowadays have to consider a much wider range of biological, economic and even social factors in predicting an optimal cropping strategy for real fisheries. However, even if fishery scientists in the past have been naive in considering the steady-state MSY as the goal of fishery management, we believe that MSY remains a useful concept. We cover it in detail in this book for two reasons. First, the history of management which aimed purely at MSY provides useful object lessons about both the inadequacy of the classical fishery models and the misleading way in which mathematical advice can be presented to decision makers. Secondly, despite its deficiencies, MSY must be included as one factor in any management plan. At the very least, it sets a limit to the size of the industry.

Classically, two different types of mathematical models have been used to predict MSY. Biomass regeneration, considered as a single indivisible (and essentially unexplained) process, is the basis of the first and simplest type of model, called surplus yield, surplus production or Schaefer models after one of their early proponents (although the surplus yield approach can be traced back to Graham in 1935). It is a 'black box' approach to modelling concerned only with inputs and outputs to the biomass of the population. Most biologists will realise that 'biomass regeneration' subsumes a great many real population processes such as tissue growth, mortality and increase in numbers through reproduction. These processes describe phenomena going on in individual fish: population changes are considered as the sum of the effects on individuals, which of course can be grouped in convenient subsets of ages, sexes or even genotypes. The second type of fishery model, termed dynamic pool, analytical, or Beverton and Holt models, have traditionally attempted to deal with some of these processes explicitly. The earliest approach of this type was by Baranov as long ago as 1918. We will return to the 'dynamic pool' models in the next chapter. In the rest of this chapter we will be concerned with the surplus yield approach, and some of its recent extensions.

Surplus Yield Models

When integrated over time, the changes in biomass illustrated explicitly in Figure 7.1 become production, a population parameter studied directly by many ecologists. If fishing extends over time we can therefore ask what proportion of production can be taken as a total crop, or yield. The left-hand diagram in Figure 7.2 is an Allen curve (see Chapter 4) showing production between two times; in the right-hand diagram we have drawn an Allen curve when three discrete crops are taken. On purely empirical evidence it has been suggested that something like 8 per cent of the total production can be taken as MSY. Since rough Allen curves are quite easy to sketch for fish populations with relatively little field work, ostensibly it is attractive to use the Allen curve to predict likely yields from unexploited fish stocks, as usefully carried out by Balon (1972, 1974) in the Lake Kariba fisheries, for which wildly optimistic forecasts of yield had been made prior to his analysis. In fact, of course, this approach, beyond providing a rough 'order of magnitude' value, cannot help further since there is no way of predicting how quickly the population moves down the Allen curve. Mathematically, this is because the time information has been lost at the integration stage. In fact, both Figure 7.1 and Figure 7.2 are poor representations of the real situation for another reason. Relatively few fisheries take large discrete crops, and fishing is usually best considered as a continuous process, at any rate within a season. In Figure

Figure 7.2: Allen Curves, Giving Production from a Plot of Numbers against Mean Weight. In the right-hand graph, three discrete crops are shown.

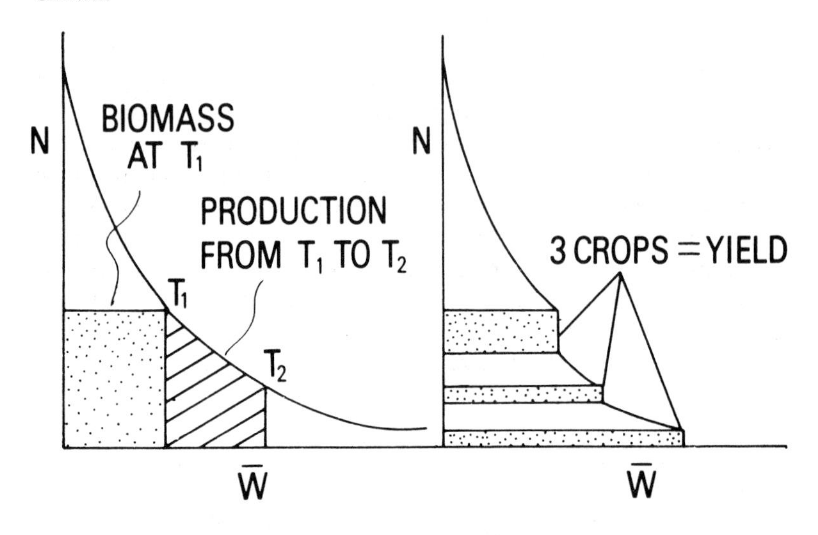

7.2 one could imagine continuous fishing being represented by stippling under the whole Allen curve, but while this might provide a neat visual representation, it could not be used for predictions. So, although the Allen curve clarifies the relationship between production and yield, it cannot help us model the fishery.

The continuous process of biomass regeneration taking place at the same time as fishing is the central feature upon which all surplus yield models are built. Biomass produced over and above that needed for exact replacement is regarded as a 'surplus' which can therefore be harvested. We will examine criticisms of this basic assumption later in the chapter, but for the moment will let it stand. When the quantity of biomass taken in the fishery is exactly equal to the surplus produced, the fishery is assumed to be in equilibrium, providing an equilibrium yield, Y_e. This too can be a dangerous assumption if the surplus yield model is applied in an unsophisticated way. Note that the 'surplus' does not actually have to be produced as it is only the *rates* which are important. A third kind of danger in this type of model can lie in a lack of biological realism in the equations which predict these rates.

Derivation

Just as in Figure 7.1, at some level between 0 and B_∞, the rate of

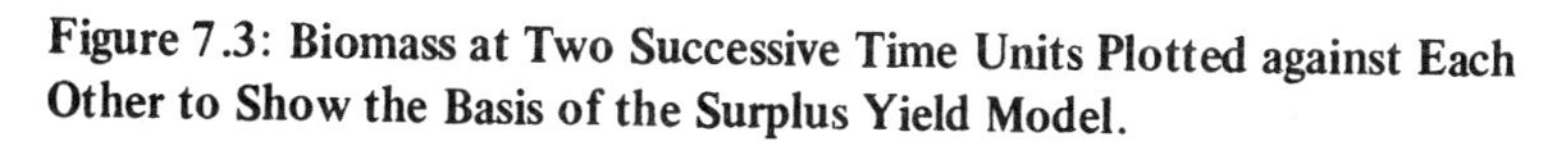

Figure 7.3: Biomass at Two Successive Time Units Plotted against Each Other to Show the Basis of the Surplus Yield Model.

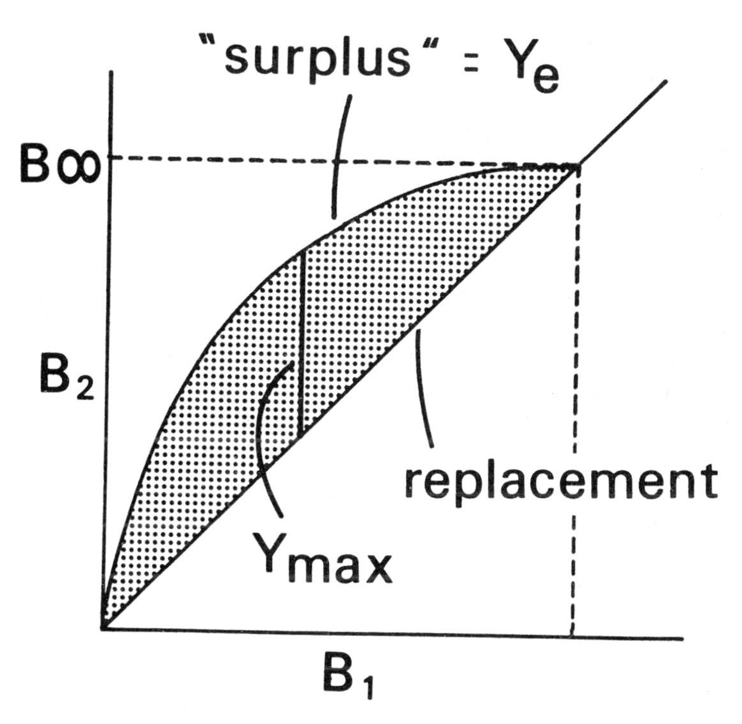

production of surplus biomass will be at a maximum, giving the MSY, Y_{max}. At this stage we will begin to write the model down in simple algebra, since mathematical representation is going to get more helpful as the models get more complicated. As far as possible, equations are stated in words as well.

Figure 7.3 is a kind of phase diagram of biomass plotted before and after regeneration has taken place, B_1 and B_2. You can think of B_1 and B_2 as values a year apart, but they could be on any time base. Logically enough the 45° line represents the exact replacement of biomass: the surplus biomass above this line is stippled. The regeneration process we will represent by $G(B)$, which could be any mathematical function but in our example would be the logistic growth curve. Other forms which may be taken by $G(B)$ are described later; for the moment we will just take the most common type.

It is instructive to derive the simplest form of the surplus yield model from first principles. The rate of change of biomass in the population is given by $G(B)$, any function of biomass which is appropriate, i.e.

$$\frac{dB}{dt} = G(B) \qquad (7.1a)$$

Fishing the population will reduce $G(B)$ by some function of fishing effort, which we will write for the moment as $H(f)$. So now

$$\frac{dB}{dt} = G(B) - H(f) \qquad (7.1b)$$

To complete the basic surplus yield model, it only remains to find reasonable forms for these two functions that fit available data. For example, one could simply suppose that increasing fishing increases $H(f)$ proportionally. If f is the instantaneous fishing rate and q is the efficiency or 'catchability' using this particular gear, then we can write this assumption as:

$$H(f) = qf \qquad (7.2a)$$

From this we can see that yield (Y) can be given by

$$Y = H(f)B = qfB \qquad (7.2b)$$

Similarly one might assume that the logistic curve could apply to the biomass generation function (Chapter 3), so that writing k for the biomass growth rate (equivalent to r_m for numbers), and B_∞ for the maximum biomass (equivalent to N_{max} for numbers) which could be supported by the environment,

$$\frac{dB}{dt} = G(B) = B\,k\,(B_\infty - B)/B_\infty \qquad (7.3a)$$

So the complete model including fishing becomes

$$\frac{dB}{dt} = B\,k(B_\infty - B)/B_\infty - Bqf \qquad (7.3b)$$

At the equilibrium we said above that the yield must be equal to the regenerated biomass, i.e.

$$Y_e = B\,G(B) \qquad (7.4)$$

(Note that the actual regenerated biomass is given by the stock biomass times its *rate* of growth.) We will get Y_e when the two rates, of fishing and biomass growth, are equal, i.e.

$$G(B) - H(f) = 0 \tag{7.5}$$

This is true at all points along the curve drawn in Figure 7.3. The general surplus yield model of Equation 7.3b can now be solved for all kinds of conditions, including values of Y_e when appropriate.

One value which is of obvious use in fishery management is f_{opt}, the rate of fishing which will produce the MSY, Y_{max}. We can do this simply by adapting Equation 7.4 specifying the conditions as in Equation 7.5.

$$\begin{aligned} Y_{max} &= B\,G(B) \\ \text{when}\quad G(B) - H(f_{opt}) &= 0 \end{aligned} \tag{7.6}$$

Classic Surplus Yield Models

Versions of the surplus yield model differ in their choice of exact form of $G(B)$: the three most common forms are listed in Table 7.1, where $H(f)$ is normally represented by Equation 7.2. Schaefer's (1954) classical model used logistic growth, a symmetrical S-shaped curve; Graham's (1935) model was similar. Fox (1970) used the Gompertz curve, the asymmetry of which is more realistic for growth in weight (see Chapter 4). An ambitiously general version was invented by Pella and Tomlinson (1969), whose $G(B)$ was continuously variable in shape, at the expense of having to fit an extra parameter, m. When $m = 2$, Pella and Tomlinson's model is the same as Schaefer's; when $m = 1$ it is the same as Fox's. It is crucial to realise that predictions of Y_{max} and f_{opt} from these types of models depend entirely on the exact form of $G(B)$, so it is very important to choose the one which best resembles the growth of the stock in question. The snag is that it is sometimes quite difficult to do this, especially with Pella and Tomlinson's model. As we will see later, even sophisticated modern versions of the surplus yield model suffer from this disadvantage. Paradoxically, use of the model can be made almost too easy by the widespread availability of a computer program which fits Pella and Tomlinson's model, even to inappropriate data.

So far the model may not look immediately helpful. The brilliant insight of Schaefer was that, by a mathematical trick, the whole model could be expressed in terms directly useful to fishery managers,

Table 7.1: Biomass Growth Functions for the Classical Surplus Yield Models.

Schaefer:	$G(B) = \mathrm{d}B/\mathrm{d}t = kB(1 - (B/B_\infty)) - H(f)$
Fox:	$G(B) = \mathrm{d}B/\mathrm{d}t = kB(-\ln(B/B_\infty)) - H(f)$
Pella & Tomlinson:	$G(B) = \mathrm{d}B/\mathrm{d}t = kB(1 - B^{m-1}/B_\infty) - H(f)$

and at the same time it could be fitted using real data easily and routinely gathered by the same managers. We will now see how this was done, although it is not important to remember the exact details since they can always be looked up in the standard fishery texts.

As a first step in fitting the model to real data, the surplus yield under equilibrium conditions, Y_e, is written as a function of stock biomass (i.e. the 45° line in Figure 7.3 is laid down flat on the horizontal axis). For the logistic growth curve in Schaefer's model we have, from Equation 7.3b;

$$\frac{\mathrm{d}B}{\mathrm{d}T} = Bk\,[(B_\infty - B)/B_\infty] - Bqf \tag{7.7a}$$

where $G(B)$ now equals the unrestricted intrinsic rate of increase of biomass, k reduced by the factor in brackets, which gets smaller as B approaches B_∞. Krebs (1979) has aptly termed this factor the 'unutilised opportunity for population growth'. This rate equation can be integrated with respect to time to give, in the absence of fishing ($f = 0$), the familiar S-shaped curve drawn in Figure 3.3. There is in fact a variety of different-looking equations for this logistic curve depending on which constants are thought to be important. However, we do not actually need the explicit version of the equation, since in a fishery we have Y_e when $\mathrm{d}B/\mathrm{d}T = 0$ (i.e. Equation 7.7a is zero) and we can move Bqf to the other side as Y_e to give

$$Y_e = Bk\,[(B_\infty - B)/B_\infty] \tag{7.7b}$$

which is the curve drawn in Figure 7.4b. Calculus can be employed to show that Y_{max} occurs when $B = B_\infty/2$ for the logistic curve. For the asymmetrical Fox model we get a similar equation for equilibrium yield,

$$Y_e = Bk\,[\ln(B_\infty) - \ln(B)] \tag{7.7c}$$

Figure 7.4: The Schaefer and Fox Surplus Yield Models Compared. Equations are shown above the curves. (a) Biomass plotted against time for the two models. (b) Equilibrium yield against biomass. (c) Catch per unit effort against effort. (d) Equilibrium yield against fishing effort.

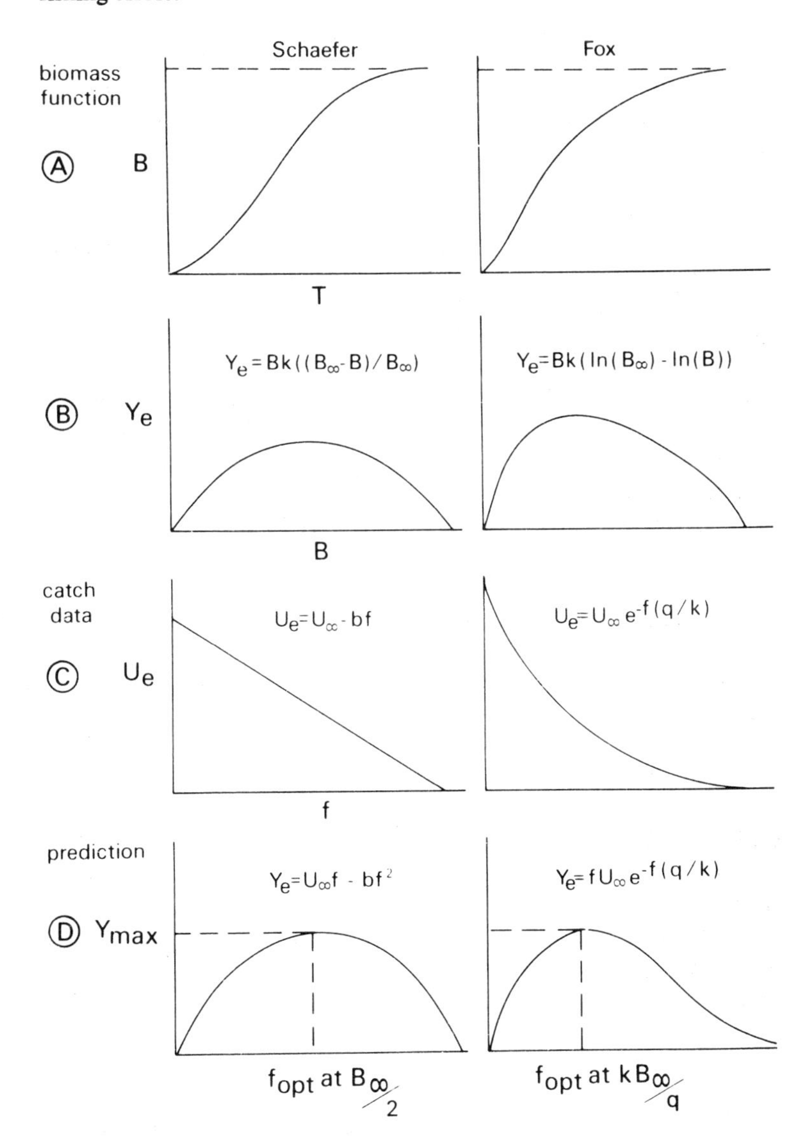

which is also drawn in Figure 7.4b.

The next stage is to define U_e as the fish catch per unit of effort at equilibrium yield, i.e.

$$U_e = Y_e/f \tag{7.8}$$

The units chosen for expressing effort do not matter as long as they remain consistent. For example, units of kg landed per boat per day are often used, and these are the sort of simple data which have been accumulated in all but the most anarchic fisheries for very many years, even when any biological data on the stock have been completely lacking. Although the arbitrariness of the units is convenient for mathematics, in practice there are difficulties. Fitting the model properly depends upon many years of back data from the fishery and during this time the fishing gear, size and power of the boats etc. are quite likely to have changed. Modifications to the units have been employed to take account of some of the alterations, for example 'landed kg per horsepower per boat per day'. The details of the problem of estimating effort are fully discussed in Gulland (1974), but it is worthwhile remembering that it is not always straightforward to quantify fishing effort consistently.

Since we can define yield as biomass times rate of fishing times catchability as in Equation 7.2b, we get

$$Y_e = fqB$$

we can substitute for Y_e in Equation 7.4 to get $U_e = fqB/f$ and can therefore write:

$$B = U_e/q$$

Curiously, this now expresses stock size as catch per unit effort (CPUE) over catchability. Of course this cannot really be used to predict stock size as the relationship is true only at equilibrium; it just expresses the fact that CPUE will change at different biomass levels. The trick recognised by Schaefer is to substitute this new value for B and divide by U_e in Equation 7.7b to get, for the Schaefer model,

$$f = (k/q)\,[1 - U_e/U_\infty]$$

which can easily be expanded and rearranged to solve for U_e giving

Figure 7.5: Summary of the Analysis for Surplus Yield Models.

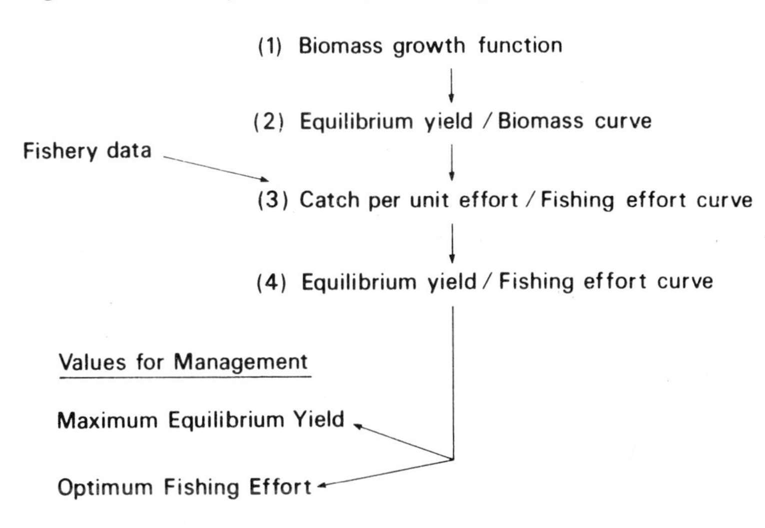

$$U_e = U_\infty - [U_\infty(q/k)]\, f \tag{7.9}$$

where $U_\infty(q/k)$ is the slope 'b' of the straight line in Figure 7.4c. The point now is that we have a useful expression describing the relationship between CPUE and our fishing effort. As can be seen in Figure 7.4c, this produces a straight line when U_e is plotted against f. For the Fox model we get a similar function in the same way, i.e.

$$f = (k/q)\,[\ln(U_\infty/q) - \ln(U_e/q)]$$

which can be rearranged and expanded to solve for U_e giving

$$U_e = U_\infty \exp(-\frac{q}{k}f) \tag{7.10}$$

which is a declining exponential curve when U_e is plotted against f, as shown in Figure 7.4c.

Classically, the lines in Figure 7.4c are fitted by regression to past data from the fishery. The results of this process were used in the final stage, which purported to give the fishery manager exactly what he wanted by relating fishing effort directly to yield. Since $Y_e = fU_e$ by definition, we can get, for the Schaefer model from Equation 7.9,

$$Y_e = U_\infty f - bf^2 \tag{7.11a}$$

and for Fox from Equation 7.10,

$$Y_e = fU_\infty \exp(\frac{q}{k}f) \tag{7.11b}$$

If we plot these curves on a graph, as in Figure 7.4d, we can read off directly the fishing effort, f_{opt}, which is needed to achieve the MSY, Y_{max}. Alternatively, these equations can be differentiated to find Y_{max}. The process is summarised in Figure 7.5. Problems raised by this simple fitting procedure are discussed later in this chapter.

Problems in Applying the Classical Models

By way of example, Figure 7.6 shows the surplus yield model fitted by Schaefer himself to data from the Peruvian anchovy fishery in 1970. The yield taken in 1971 prior to the collapse of the fishery in 1972 (Idyll, 1973) is indicated on the plot. The fact that it is only 25 per cent greater than the predicted MSY, and well within the apparent 'safety zone' should give cause for unease about these simple surplus yield models. Even now, several years after the crash, the anchovy fishery has scarcely recovered. In 1971 this single fishery produced 15 per cent of the total world catch and well over half of the total world clupeid catch. Fish meal from Peru was hugely important in Western agriculture as a source of cheap high-lysine protein; the loss of the fishery has had significant effects on the whole Western economy, exacerbating the increased energy costs of the Middle Eastern oil crisis. The world as a whole could ill-afford the loss of ten million tons of protein a year. Furthermore, the collapse of the anchovy fishery has had indirect ecological consequences which may turn out to be irreversible. First, there is the loss of 4–5 million guano-producing seabirds who depended on the anchovy stocks alongside man (Schaefer, 1970). This has both economic and ecological effects. Secondly, as with the Californian sardine fishery, the whole ecosystem may have swung over into another mode (judging from the catches in 1973 and 1974, perhaps favouring the sardine over the anchovy), which would make full recovery of the anchovy unlikely. Although the destruction of the anchovy fishery was not a direct consequence of reliance upon the surplus yield model, much of the blame must be attached to the misplaced confidence that allowed (or failed to halt?) such large catches, even after the first warning signals of recruitment problems in late 1971 (Paulik, 1971). We will now begin to probe the models in more detail in an attempt to discover what went wrong.

Several variants of the classical fitting process for surplus yield

Figure 7.6: Schaefer Model Fitted to the Total Catch of Peruvian Anchovy – Data Taken from the Fishery in the 1960s. Note that estimated catch by seabirds is included in the total. Estimated MSY is about 10^7 tons in this model.

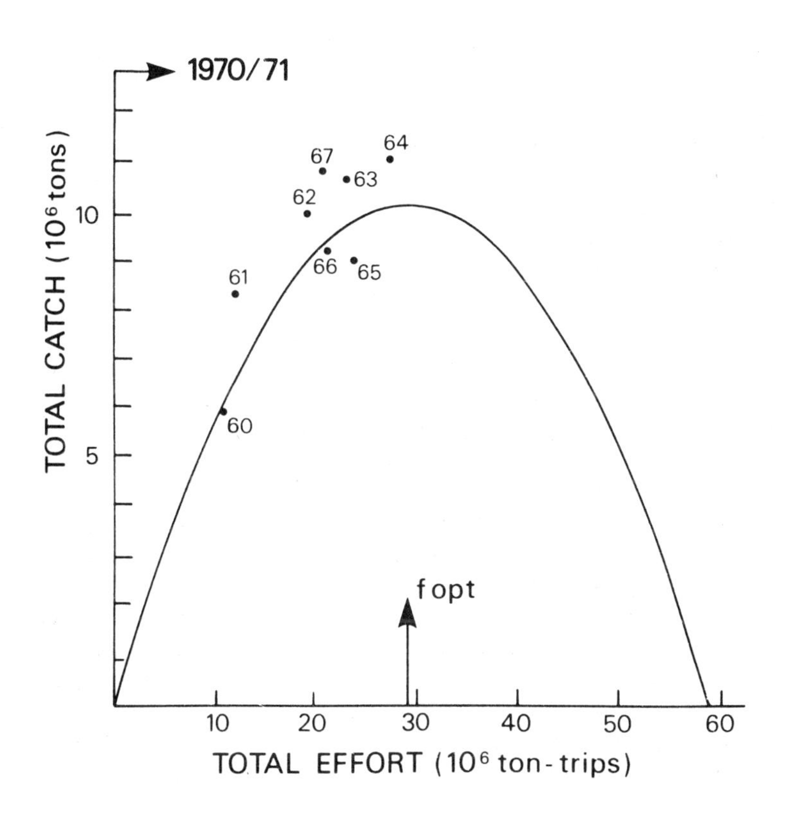

Source: Redrawn from Boerema and Gulland (1973).

models exist, depending on what sort of fishery data are available. Ricker (1975) gives an excellent technical account. It is important however to realise that the position of Y_{max} in relation to B_{∞} is fixed for a particular version of the model. For Schaefer, Y_{max} is *always* at $B_{\infty}/2$, the middle of the parabola in Figure 7.4d. For Fox it is at $B_{\infty}k/q$. A particularly dangerous consequence of this is that it is possible by rearranging and solving a couple of simultaneous equations, to estimate MSY and f_{opt} from two or more equilibrium yields taken in past years, without knowing q or the stock size (Ricker, 1975,

p. 316). Whilst this use of mathematics is elegant, the crucial point is that there is no way of knowing if the yields which were assumed 'in equilibrium' were really so. If they were not, as we will see in a moment, the whole analysis could be disastrously wrong. Again, although fitting the additional parameter 'm' for the $P + T$ version may lend some additional confidence in terms of realism, remember that the only 'general' thing about this model is that the biomass regeneration function (i.e. the equivalent of Figures 7.4b, c, d) can assume a variety of shapes, but not *all* possible shapes, taking into account, for instance, a minimum viable stock size. For Pella and Tomlinson, once m is fitted, Y_{max} and f_{opt} are again fixed in relation to B_{∞}. We will return to one way of overcoming this problem at the end of the chapter.

The major practical advantage of surplus yield models is that they require only catch and effort data, the sort of information accumulated over many years in most fisheries. MSY is seductively easy to calculate, in fact no biologists need be employed in the fishery and managers do not even have to get their hands and feet wet in examining actual fish. However, just as the assumption of a single biomass regeneration function provides the surplus yield model with its attractive simplicity and practical advantages, it is also the root of its major and dangerous disadvantage. The model ignores the real biological processes which actually generate the biomass. Changes in biomass are made up of contributions from the separate but interacting processes of growth, recruitment and mortality. Even if the model worked well in a steady-state population, time lags in the response of one of these processes to altered fishing are quite likely to be different to time lags in another, rendering a single $G(B)$ inapplicable. Furthermore, the population processes themselves may be drastically altered by different age structures in the fish population, and age structure is also ignored by the surplus yield model (Jensen, 1973). Therefore not only is the surplus yield model unlikely to be serviceable for a fishery newly exploited from the 'virgin' state because age structures are altering rapidly, but it is also unlikely that the assumption of a single $G(B)$ will hold under a wide range of conditions. The consequence of this is that the model will tend to give falsely optimistic answers. For example, following a large increase in fishing effort (which is not treated differently from a small increase on the standard theory) the contributions of the more fecund older fish to recruitment could be greatly reduced.

Something of this kind happened to the Peruvian anchovy in the spawning season of October 1971, following a series of large increases in fishing effort. Two further mistakes depleted the anchovy: first, heavy fishing continued despite the known recruitment problem; and secondly, heavy fishing continued into the 1972 season despite the incursion of unproductive warmer surface waters into the normally highly productive cold upwelling on the coast, an event which happens with an irregular periodicity of 15–20 years off Peru, and well known to the Peruvians, who term these meteorological and oceanographic events *El Niño* (Valdivia, 1978; Craig, 1974). Murphy (1977) cites the 'rather reasonable variance' attached to the surplus yield predictions made for the anchovy fishery prior to 1972, demonstrating that although managerial blunders compounded the problem, the prime culprit was the misplaced confidence engendered by the models. For this reason we would suggest that unmodified classical surplus yield models should be employed only with great caution in fishery management; the stakes in terms of vital protein resources are just too high to gamble with inadequate mathematical tools.

Extensions of the Model

There have in fact been a number of recent developments and extensions to the standard surplus yield method which do now mean that these can be employed legitimately in some cases where biological details of the stock are scanty and/or very difficult to obtain, provided the results are treated with caution.

Pope has extended the Schaefer model to deal with the case of several species being caught in the same fishery at the same time. This problem of 'multiple stocks' is not dealt with adequately by any of the conventional methods, but the practical situation is common for example in tropical fisheries and in tuna purse seine fisheries. In its simplest variant, the joint MSY of the two species stocks is estimated by searching for a maximum yield on a three-dimensional version of Figure 7.4d where the surplus yield curve has become a parabolic dome extending over a base plane covering the combinations of stock size of the two species. With more than two species in the fishery, graphical representation becomes impossible, but the mathematics of searching for Y_{max} can still be carried out. In its present state, Pope's pioneering attempt still suffers from most of the criticisms levelled against the classical surplus yield models, but his approach can in principle apply to multiple stock versions of the recent extensions of the classical theory which we will outline below. The mathematics become very difficult however, as May *et al.* (1979) found with a rather different approach to the same problem, which we have outlined elsewhere in this book.

The classical variants of the surplus yield model give no indication of the length of time taken for the stock to reach a new equilibrium level after changes in fishing effort. Fletcher (1978) has given a rigorous analysis of these transient changes in yield and biomass for the Schaefer and the Pella and Tomlinson models. He also considers some of the conditions which can lead to extinction of the stock. However, Fletcher's analysis generally applies only to small changes in fishing effort and, since it is based on the standard SY model, it has no way of including the possibility of different time lags in response to changes in fishing effort in each of the various real processes which actually contribute to biomass change.

One of the most practical simple modifications to the surplus yield model has been carried out by Walter (1973), who very sensibly pointed out that rate of change of biomass now was likely to be influenced by past biomass just as much as, or even more than, by the current biomass level. Only the current biomass is included in the standard model. In the simplest case, Walter considered biomass at only one time delay, so $G(B)$ is given by

$$\left(\frac{1}{B}\right)\frac{dB}{dT} = b + a_1 \ln(B_t) + a_2 \ln(B_{t-1})$$

where B_t is biomass now and B_{t-1} is biomass at the previous time interval, usually one year ago. The constants, b, a_1 and a_2 are fitted by regression. In the more general case the equation is extended to include m delays:

$$\left(\frac{1}{B}\right)\frac{dB}{dt} = b + a_1 \ln(B_t) + a_2 \ln(B_{t-1}) + a_3 \ln(B_{t-2}) \ldots + a_m \ln(B_{t-m}) \quad (7.12)$$

The coefficients a_1 to a_m are fitted as before using a multiple regression, with the difference that this time the regression is carried out by a 'stepwise' method, which has the effect that only biomasses which contribute significantly to the current rate of biomass change are included in the final equation. In stepwise multiple regression one works through all the various combinations of the coefficients a_1 to a_m, discarding the regressions which fail to improve on the fit of the model (a standard programme is available in most computing centres). Even the current year's biomass may be discarded from the final

Table 7.2: Comparison of Predictions of f_{opt} with Actual Values of f Using the Fox and Walter Surplus Yield Models.

	f_{opt}		f
	Fox	Walter	
Californian sardine			
1934	1,175	1,800	800
1944	1,175	940	1,500
1954	1,175	310	550
Atlantic yellowfin tuna			
1960	9.94	12.5	5.3
1963	9.94	6.0	13.0
1966	9.94	6.4	9.1

equation in this way. The result is the best predictor of the rate of change of biomass which can be derived from this particular set of data. After fitting this delay equation, the MSY, Y_{max}, and the optimum effort, f_{opt}, are calculated in a way analogous to the standard procedure. A whole family of yield curves is generated by the Walter time delay model, so that f_{opt} can easily be different for different years, unlike the standard models which always give the same answer. The Walter method therefore, almost fortuitously, provides a practical way of overcoming the objection that the fishery must be in equilibrium before equilibrium yields can be calculated. The disadvantage is that fishery data for a long run of consecutive years must be available, the span of the time delay being determined by the length of the run.

Walter gave three examples of the utility of his model. First, he was able to demonstrate that the demise of the Great Lakes trout fishery was not due to fishing beyond f_{opt}, an encouraging result since we know that the invasion of sea lampreys was responsible. His next two examples are shown in Table 7.2. By Walter's analysis the Californian sardine *was* overfished prior to its collapse in 1955, contrary to most conventional analyses. Atlantic yellowfin tuna, along with other tuna fisheries currently showing signs of overfishing (Rothschild and Suda, 1977) was exploited beyond f_{opt} from the early 1960s according to Walter's model, whereas tuna fishery managers would have considered themselves safe by conventional analyses. The major criticism which can be levelled at the Walter model is that it works best as a *post hoc* analysis, and could not be safely employed for predicting the outcome of current large changes in fishing rates on next year's stock. It is a much more attractive and general method than the classical surplus yield models, however, and it is surprising that Walter's time delay method seems to have been virtually ignored since its publication. A very similar model has been derived quite independently by Orach-Meza and Saila (1978), successfully describing the Maine lobster fishery.

Incorporating Randomness

The single most important factor lacking in the analysis of the Peruvian anchovy, which meant that the model could not stand much chance of predicting the likelihood of collapse, was environmental randomness. To the general ecologist, failure to consider random factors is very surprising since the literature has always been full of such concern. Indeed in the 1950s 'density-independent' factors were held by many to be the sole check on natural populations of insects, and a debate raged between them and the supporters of 'density-dependent' controls. Perhaps the failure typifies the isolation of fishery ecology from the mainstream of ecological thinking at that time: fortunately this is no longer true and fishery work now draws on a wide range of concepts from ecology, economics, and optimality theory. In fact, it is in something of a state of flux, as we shall see in the final chapter.

In Peru, the pattern of offshore currents was the key environmental factor which determined the nature of the ecosystem in which the anchovy lived. The upwelling of nutrients brought by the cold northward current allowed a rich plankton community to flourish; the anchovy cropped this as food. Larval anchovy at the critical time of the first meal require food organisms of a particular size range and density (see Chapter 6). The sparsity of food organisms in the *El Niño* of 1972 caused larval death and compounded the loss of offspring which would have come from the older fish wiped out by overfishing in 1970 and 1971. During *El Niño* the change in surface currents and consequent alteration of the normal feeding opportunities also caused adult anchovy to concentrate in a narrow band near the coast, further exacerbating the effects of fishing. All these general effects of *El Niño* were known in Peru, if not by the fishery managers. When *El Niño* occurred in the 1950s and 1960s prior to the overexpansion of the industry, the stock had both the age structure and reproductive capacity to recover rapidly in the following season: overfishing destroyed this robustness and brought about the recruitment failure. A post-mortem such as this is all very well, of course, but we have to show that a fishery prediction model can take account of events like

El Niño.

In trying to do this, two problems cause difficulty: modelling and predicting the occurrence of the phenomenon itself, and then modelling its effects on the fish stock. Taking the first problem, although the oceanographic and meteorological events are doubtless inherently describable, a predictive model is still well beyond the capabilities of atmospheric science – in fact we cannot even forecast the weather five days ahead, let alone concomitant wave and current changes in the sea. We might hope to correlate *El Niño* with some apparently unrelated external events, such as the incidence of sunspots or earthquakes, and it is likely that we could draw up combinations of circumstances in which *El Niño* was more probable, as insurance companies do for various risks, but this would not be a predictive model as such. The second problem is even worse; although ecologists could set out a plausible general list of effects of the current change on the ecosystem, it would take an army of experimental ecologists to determine the quantitative details of the relationships with anchovy ecology and behaviour, and the anchovy's food and predators. (This lesson about the amount of effort needed in big ecological models was learnt the hard way by the ecological research community during the International Biological Programme.) Fortunately, both problems can be bypassed.

All that is necessary is to build a random perturbing factor into the model of the fishery; in the case of the surplus yield model this is the biomass growth equation $G(B)$. The frequency and magnitude with which the perturbing factor operates can be determined from the past history of the fishery, or even a sensible range of guesses can be made and the model run for each of them. The random influence can be made to operate 'catastrophically' by being switched on and off with the appropriate probability in any one year (i.e. a P of about 1/15 occurring in any one year for *El Niño*). The probability of the event occurring in each of a series of successive years can be reduced in the model if this is what we see happens naturally. In addition, the magnitude of the effect on the stock can be taken at just one level, or at several levels of seriousness (which could be selected from a Poisson distribution; see Jeffers, 1978, for further details of this type of modelling). Alternatively, the perturbing factor can be modelled by assuming that the magnitude of its effect is normally distributed about a mean value, large fluctuations away from the mean therefore being less likely than small ones. The magnitude and frequency of perturbations are linked by just one variable in this type of model, the standard deviation of the curve. This sort of model is termed a white noise

Figure 7.7: Example of White Noise – Random Numbers Selected from a Normal Distribution with a Given Mean and SD.

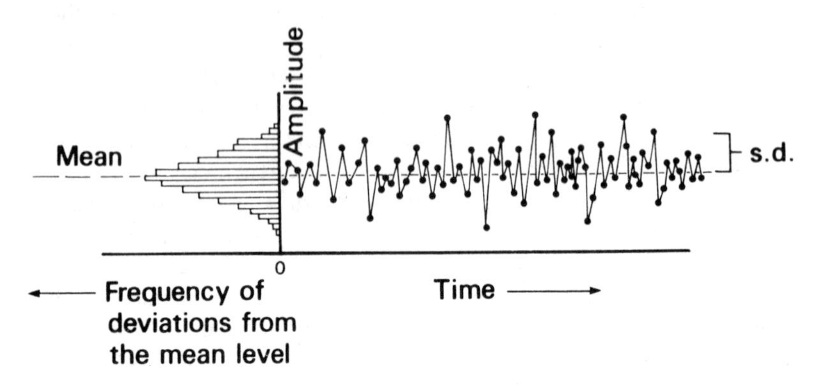

generator. Figure 7.7 illustrates the probability distribution and a typical set of results from a white noise generator.

Using the improved model, any management strategy for the fishery must then be proven to be robust against the 'worst 50-year case', just as bridges must be built to withstand the '50-year wind' and marine structures the '50-year wave'. In fact, as we will see in the final chapter, we can have an even more sophisticated approach if detailed information on the stock is available.

For a particular fishery the model would probably be run as a numerical simulation, but Beddington and May (1977) have dealt analytically with the inclusion of environmental randomness in a Schaefer surplus yield model, elegantly providing a number of valuable insights. To understand Beddington and May's model, we must first note that we can rearrange the general surplus yield equation into a linear density-independent term followed by a second density-dependent term linked to the square of stock biomass. For the Schaefer version (Equation 7.7a) we have

$$\frac{dB}{dT} = (k - qf)B - k(B^2/B_\infty) \tag{7.13}$$

Beddington and May incorporated randomness into Equation 7.13 by making the intrinsic rate of increase of biomass, k, a normally random variable. The mean value of k, $\bar{k}$, is put into the density-dependent term in the model. Randomly-selected values of k from 'white noise', k_r, are used in the density-independent term. This was thought to be realistic because random factors usually operate in a density-independent fashion in ecology. Beddington and May's model then is

$$\frac{dB}{dT} = (k_r - qf)B - \bar{k}(B^2/B_\infty) \tag{7.14}$$

The white noise generator for k_r has a mean of $\bar{k}$, and a standard deviation σ which Beddington and May have set to a value which they consider realistic, but which could easily be adjusted to an appropriate level for a particular fishery. The mathematical analysis of Equation 7.14 shows that as MSY is approached with increasing fishing effort, the variability of the yield increases, as shown in Figure 7.8a. Note that increasing variability is just another way of saying lower predictability. When effort is increased so far that the fishery moves out beyond the Schaefer MSY, variability of yield increases more rapidly, graphically illustrating the dangers of destroying the stock at high fishing efforts.

Beddington and May's analysis also provides a dramatic comparison between the alternative strategies of managing the fishery with a constant fishing effort (e.g. fixed number of boat days per year), or attempting to take a constant yield each year, as might be advocated under an unsophisticated quota system. In fact the 'constant yield' policy considered by the authors is somewhat more realistic, since a manager may accept a reduced yield at low stock levels because of high costs, but the general conclusions of their work apply to any constant yield policy. The response time of the fishery, T_r, is calculated from matrix algebra. T_r represents the time taken to recover from small disturbances, and can therefore be thought of as a measure of robustness of the stock – just the sort of general measure we need. T_r for the two harvesting policies is illustrated in Figure 7.8b, which shows that a constant yield approach is much more likely to lead to dangerously high response times.

A number of recent papers have followed Beddington and May's work and have emphasised that the surplus yield model has entered a new era of utility; recent work implicitly includes environmental randomness. There has also been much discussion of the bias caused by problems in fitting the model, and in estimating Y_{max} and f_{opt}. Some work has explored the consequences of using different methods of fitting the model. The general theme of all of the recent work is to approach the fishery as a problem in optimal control. A consequence of this is that much of the mathematics has become almost so abstruse as to lose sight of the biology!

Figure 7.8: (a) Yield and Variability of Yield Plotted against Effort for Beddington and May's Model. (b) System Response Times for Constant Effort and Constant Yield Fishing Strategies.

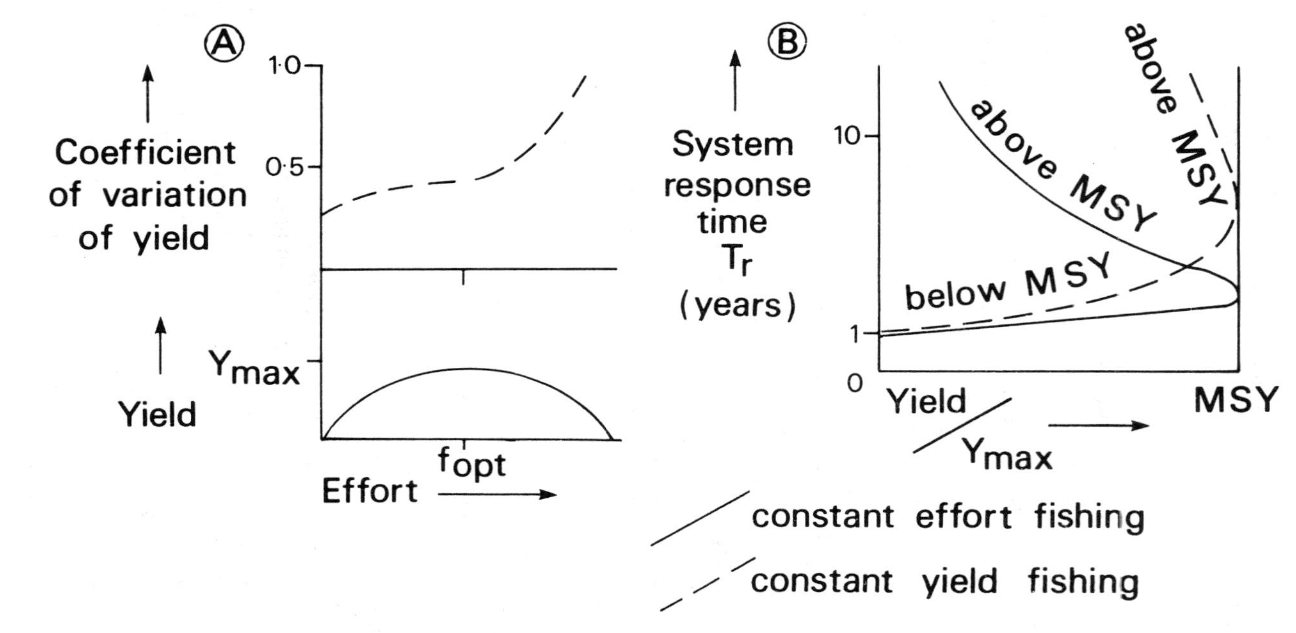

Source: (b) Redrawn from Beddington and May (1977).

Problems in Fitting the Models

The standard methods of fitting the surplus yield model to data are always subject to the danger that the data employed in fitting did not come from a fishery in equilibrium, as we have emphasised above. For example, in the all-too-common case of a fishery which has had a history of increasing yields, the plot of yield against effort will always lie above the true equilibrium curve, so that if these points are used to fit a model, the MSY will be overestimated. The danger is exemplified by considering that fishing at an f_{opt} for an MSY overestimated by 25 per cent could reduce a virgin fishery to less than 10 per cent of its potential in ten years, even without the exacerbating effects of randomness described by Beddington and May. If one includes the likely effects of randomness, then the risk of extinction is large for this degree of inaccuracy in estimating MSY, so it is important to use a model which includes randomness and does not include the assumption that past data came from a fishery in equilibrium.

Until quite recently two standard methods of fitting the surplus yield model were in use and had replaced the classical methods we described earlier in this chapter. We can rewrite the differential equation for the Schaefer model (Equation 7.7a) as a difference equation relating biomass in consecutive years t and $t + 1$, i.e.

$$B_{t+1} = B_t + B_t k(1 - \frac{B_t}{B_\infty}) - Y_t \qquad (7.15)$$

The yield in year t, Y_t, is given by biomass times fishing rate times catchability as before

$$Y_t = B_t q f_t \qquad (7.16)$$

It is possible to rearrange Equation 7.15 and include the yield for the subsequent year to obtain an equation,

$$\frac{[Y_{t+1}/f_{t+1}]}{[Y_t/f_t]} - 1 = k - (k/qB_\infty)(Y_t/f_t) - qf_t \qquad (7.17a)$$

which can be fitted by ordinary multiple regression to fishery data of Y and f. This will provide estimates of k, B_∞ and q, so that f_{opt} and Y_{max} can then be calculated as before. Note that Equation 7.17 is the same as

$$[U_{t+1}/U_t] - 1 = k - (k/qB_\infty) U_t - qf_t \qquad (7.17b)$$

where we have written U for catch per unit effort as before. A more realistic version of Equation 7.16 is

$$Y_t = B_t(1 - \exp(-q'f_t)) \qquad (7.18)$$

which implies that CPUE goes down with increasing effort. If this latter is true, then it is not possible to derive a linear equivalent of Equation 7.17a, and a non-linear procedure for fitting has to be adopted (see Hilborne, 1979, or Ricker, 1975).

The major problem with these methods is that the results are biased because the regressed variables in Equation 7.17 are not independent. The bias can cause serious overestimation of f_{opt}, which, as we have seen, could be dangerous in many fisheries.

Gulland's (1961) method of overcoming these problems was to regress U on a moving average of effort, instead of taking each year's data separately. Recently, Roff and Fairbairn (1980) have demonstrated that Gulland's method cannot be used because a spurious relation between U and f arises too easily in most data. The artifacts can arise in two ways. First, the presence of effort in both axes biases the plot towards an inverse correlation. Although Gulland thought that increasing the period of years in the moving average would correct for this, Roff and Fairbairn found that the second artifact (below) destroyed any improvement. Secondly, spurious relationships were brought about by high positive autocorrelations of effort with time (e.g. as the fishery developed), and by negative correlations of U with time (e.g. as stocks were depleted). Roff and Fairbairn used information from 19 stocks of four species each with at least 15 years of data. They found, for example, that there was temporal autocorrelation, significant at the 1 per cent level, in twelve of the stocks. Two possible artifacts were investigated in detail: the correlation between U and time, and between average f and $1/f$. By running all possible lengths of moving average (between two and ten years long), the authors tested to see if artifacts generated in the data affected the U/f relationship vital to the correct fitting of the Schaefer model. If the artifacts did not affect this important relationship, one would expect their incidence to be independent of the significance of U/f correlation itself. Table 7.3 clearly shows that this is not the case. The significance of the U/f correlation depends heavily on the significance of the correlations found in one or both of the artifacts. Roff and Fairbairn have concluded that the Gulland method of fitting the Schaefer model should no longer be used.

Table 7.3: Roff and Fairbairn's Demonstration that the Relationship of the Important Variables in Fitting the Schaefer Model is Affected by the Spurious Correlations. Simulations over all possible lengths of moving average. For further details see text.

	Significant Correlation in the Artifacts		
	Neither	One	Both
Significant Correlation in U and f Relationship			
Significant	1	38	47
Non-significant	17	19	4

From a practical fishery manager's point of view, Schnute (1977) has met a long felt need. Schnute solves the problem of fitting the surplus yield model in the presence of environmental randomness, and has developed an admirably clear modern method and nomenclature. An approximate version of the model can be fitted using multiple linear regression on a programmable hand calculator, but he also gives a method of obtaining an exact solution for Y_{max} and f_{opt} for each year, using a computer. Most important, Schnute derives confidence limits for the yield and fishing effort values, together with a way of predicting next year's catch with known precision from the fishery data. In addition to this, he introduces a useful 'failure index', I, which measure objectively how well the surplus yield model explains the existing fishery data. If I is close to 0, the model works well, but as I tends to 1, past and present fishing efforts become poor predictors of catch and the whole surplus yield rationale becomes suspect for that fishery. Although Schnute deals only with the Schaefer version of surplus yield, a more general analysis is in preparation.

Mohn (1980), Uhler (1980), and Hilborne (1979) have all compared Schnute's method with the standard estimation techniques using simulation. Although each of these authors has employed somewhat different simulation models and has asked slightly different questions, their general conclusions are similar. They find that the major sources of bias are the same as those discovered by Roff and Fairbairn. They also recommend that Schnute's method is the least likely to be biased, although all currently available methods can give dangerously wrong answers in heavily-exploited stocks, just where they are needed most. Mohn (1980) simulated a 30-year fishery starting at low exploitation, peaking after 15 years and then falling again. This mimicked the opening up of a stock, overfishing it, with subsequent collapse. Uhler (1980) based his work on catch and effort data for 33 years of the Pacific yellowfin tuna fishery, to which he added progressively larger amounts of white noise. The Schaefer parameters were then estimated from the data for each simulated year. Hilborne (1979) compared simulations of slow-growing fish with a fast-growing stock, where most of the production was due to recruitment. The different estimation techniques were brought to bear on simulated catches from the test populations. Both Hilborne and Mohn consider that the Schnute fitting method was less biased, but unstable. Schnute's 'failure index' was, however, a good guide. When this was low, the parameters of the model were likely to be accurately estimated, but when it was high, dangerously wrong answers were possible. Schnute's 'failure index' therefore can be taken as a useful management tool by urging caution in the interpretation of estimated f_{opt} and Y_{max} values. Hilborne has also considered the effects of several management control systems and we will return to his conclusions in Chapter 11.

Incorporating Recruitment

Walter (1978) has made a further important advance in considering the actual fishing effort which should be used when variable recruitment is incorporated into the Schaefer model. Walter is unhappy with treating variation due to recruitment as 'noise', although the lack of any clear relation between adult stock and recruiting young in many species suggests that it is not an unreasonable assumption (Cushing, 1977). Walter restructures the Schaefer model so that the biomass is regenerated *both* from recruitment *and* from tissue growth net of mortality. His basic equation is

$$\frac{dB_j}{dT} = B_j k(1 - B_j/B_\infty) - B_j qf \qquad (7.19a)$$

which is a slight rearrangement of Equation 7.7a and the symbols are as before, with the exception that B_j is now the biomass of the stock in year j, derived partly from recruits the previous year. Walter considers the case of Ricker-type recruitment (Chapter 6), so that with density-dependent recruitment of this type, the long-term equilibrium yield comes to lie on a parabola somewhat below the standard Schaefer curve. The approximate formula for the long-term average yield is derived as:

$$\bar{Y} = \frac{qf}{(1/B_{\infty}) + \alpha\beta} (\alpha + k - qf) \tag{7.19b}$$

where α and β are the coefficients of the Ricker recruitment curve that has been fitted to stock-recruitment data from this fishery. If recruits make up a large proportion of the stock, as they may do in short-lived fish like sprats, the approximation used to derive Equation 7.19b does not hold and the basic formula has to be solved for $\bar{Y}$ numerically. Recruitment functions other than the Ricker curve are easily incorporated into the model instead.

Optimal Control of the Fishery

However, the importance of Walter's model is that the basic equation can be solved in another fashion, this time pertaining to a particular year, j. The approximate solution gives

$$Y_j = B_{\infty} qf \ln(B_j/B_{\infty} + 1) \tag{7.19c}$$

which is a shallow curve relating the *attainable yield* (AY) to fishing effort for year j. The two curves, from Equations 7.19b and 7.19c, can be plotted on the usual surplus yield graph of yield against effort. Where the two lines intersect, we have an effort which represents the *attainable equilibrium yield* (AEY) for that year. For efforts above this AEY, the biomass will tend to decline in the next year, for efforts below AEY, stock biomass will tend to increase – all other things being equal and provided there are no wild fluctuations in recruitment. The situation is illustrated in Figure 7.9, which is taken from Walter's analysis of Atlantic mackerel stocks. Inspecting the attainable yield (AY) line for 1974, we can see that the AEY is given by an effort of about 0.4 million fishing days. From the graph, it is clear that the maximum value AEY can ever attain in one year is the same as MSY, the peak of the long-term yield curve. Walter suggests that the safest management strategy will be to try to approach this maximum as quickly as possible, even if recruitment fluctuations mean that it is never possible actually to get there.

We will describe here the general process of managing a stock using Walter's model; the procedure is summarised in the flow chart in Figure 7.10. We are assuming here that the aim of management is to set a total allowable catch (or quota) for each year. The TAC can of course vary considerably from year to year, although it may be prudent to note that the final level of TAC may be altered slightly if economic as well

Figure 7.9: The Walter Attainable Yield Model Applied to Atlantic Mackerel.

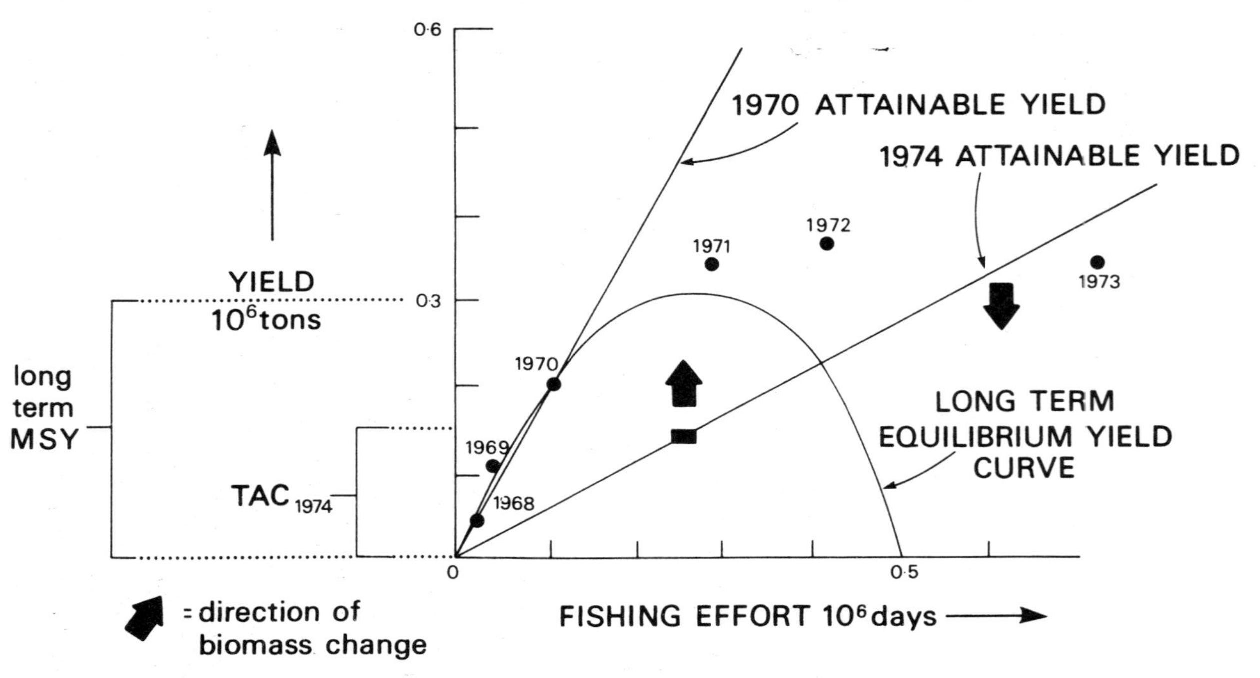

Source: Walter (1978).

Figure 7.10: Diagram Illustrating Passive Optimal Control or 'Real-time Management' Using the Walter (1978) Model. For further explanation see text.

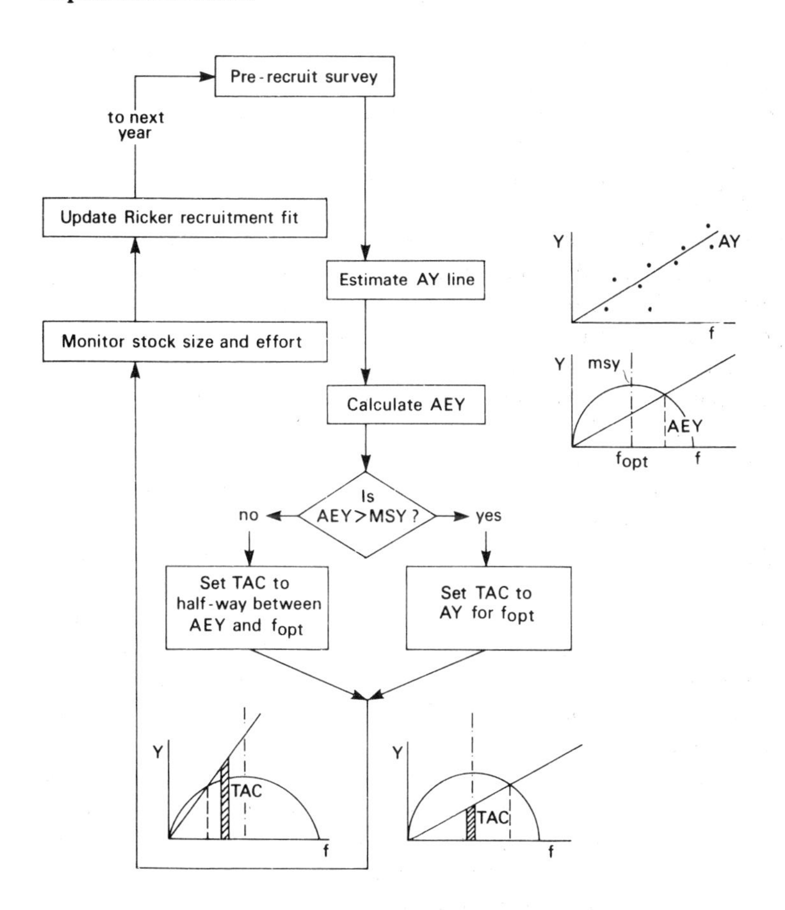

as biological considerations are brought into the final policy decision for the fishery. As in Figure 7.10, there are two versions of the procedure for setting TAC depending on whether the stock is currently fished above or below the maximum sustainable yield. First, for a stock which is being fished above MSY (and remember that this MSY is the one fitted by Walter's model, and is already *less* than in the conventional Schaefer analysis), the best policy is to take a TAC corresponding to the MSY effort, i.e. f_{opt}. If this is done, the stock biomass will be given a nudge towards better yields from the higher biomass in the following year. Figure 7.9 shows that for the mackerel in 1974 the TAC on this basis would have been about half of the full MSY, or about two-thirds the attainable equilibrium yield. Even if recruitment proves to be very poor in the following year, the new AY line calculated for this year will ensure that the stock is again nudged in the right direction, provided the total allowable catch is set to the level predicted by f_{opt} on the new year's AY curve.

Secondly, in an underexploited fishery, for example the 1970 line on Figure 7.9 and the below-MSY option in the flow chart in Figure 7.10, TAC could be set to the level predicted by f_{opt} in the same way as before. This is what is recommended by Walter himself. However, we can see two disadvantages in allowing such a large catch. First, such a quick reduction in stock size and changes in the associated age group structure might have unforeseen consequences for the stability of other species in the ecosystem, some of which might themselves be exploited or linked to exploited species. In turn, the recruitment of the managed species itself might be affected. Secondly, large catches in a new fishery might have unfortunate economic consequences which would be difficult to reverse, not only in lowering market prices, but also in encouraging an industrial capacity too large for the future lower yields at MSY. We will return to these points later in the book, but in the meantime we suggest that a TAC somewhere between f_{opt} and the AEY line would be a sensible tactic. In Figure 7.10 a catch halfway between the f_{opt} and AEY levels is indicated, although in a particular fishery another value might be adopted. It is worth noting the likely position of a conventional Schaefer curve on Figure 7.9, well above the fitted AEY line of Walter.

As well as the usual catch and effort figures, the Walter model calls for information on recruitment. For past years this can sometimes be deduced from stock data (see the original paper for details), but if the method is to be used in a fishery we would have to make a survey of pre-recruits such as are being carried out in many stocks today, such as the south-west UK mackerel stocks (see Chapter 3 and 6). Management using the Walter model is an example of *passive optimal control* (or 'passive real-time management') to which we will return later in the book.

Walter has separated two factors in biomass change instead of just one and has therefore moved one step towards the models we will consider in the next chapter which take this approach much further. For the moment we can see that the classical surplus yield models leave a great deal to be desired and may provide advice which can be misleading and dangerous, for the reasons outlined by Beddington and May and by more recent authors. However, the modern practical developments of Walter, Hilborne and Schnute may indicate a way forward when the massive survey costs required by the 'dynamic pool' approach are just not practicable. Uhler (1980) has emphasised that the Schaefer model will continue to be used for some time to come, simply because catch and effort data are all that is available. Even so, Mohn (1980) has warned that 'MSY estimation will be slow in changing from an art into a science'.

8 DYNAMIC POOL MODELS AND FISHERY MANAGEMENT

Two enhancements leading to greater realism are embodied in the dynamic pool approach to fishery management. First, the separate processes which alter fish population biomass are described explicitly as components in the model. Secondly, the age structure of the population is included. Before proceeding with our account of dynamic pool models in this chapter, we will briefly explain the importance of each of these points.

In his analysis of the problem of overfishing, Russell (1931) clearly identified the four main processes taking place in the fishery. Two of these, recruitment of new individuals (R) and tissue growth (G), added to stock biomass, whereas the other two processes, natural mortality (M) and mortality from fishing (F), reduced stock biomass. Russell's fishery equation was

$$S_2 = S_1 + (R + G) - (M + F) \tag{8.1}$$

where S_1 and S_2 stand for the biomass of the stock at the start of two successive time periods, usually one year apart. One can see intuitively from Russell's equation that when the two brackets are equal, losses will balance gains, and the fishery will be in balance, but when losses are greater than gains the stock will be depleted and overfishing will occur. In fact, Russell's simple equation summarises even the most sophisticated of contemporary dynamic pool models; nearly all of the complications arise because we have to obtain realistic sub-models for the four processes themselves. One problem is that Equation 8.1 does not identify correctly the rates at which the four processes are occurring, and in fact the symbols, G, M and F are conventionally taken as referring to the instantaneous rates of growth, natural mortality and fishing. Confusingly, R usually stands not for instantaneous rate, but for the number of recruits entering the fishery in one year.

Figure 8.1 is a diagrammatic model of a fishery with the loss and gain rates correctly identified. The solid lines represent flow of biomass and broken lines represent influences which alter the rates. Each of the four processes in the fishery could be broken down into a hierarchy of sub-models, and a major problem is deciding how far this decision should go. For example, the recruitment sub-model could be

Figure 8.1: Flow Diagram of a Fishery Illustrating the Four Main Processes Modelled by the Dynamic Pool Approach to Fishery Management. Each of the four sub-models can itself be broken down into more complex and realistic systems. Complexity and realism may also be increased by incorporating age structure. The diagram uses the MIT notation for system models.

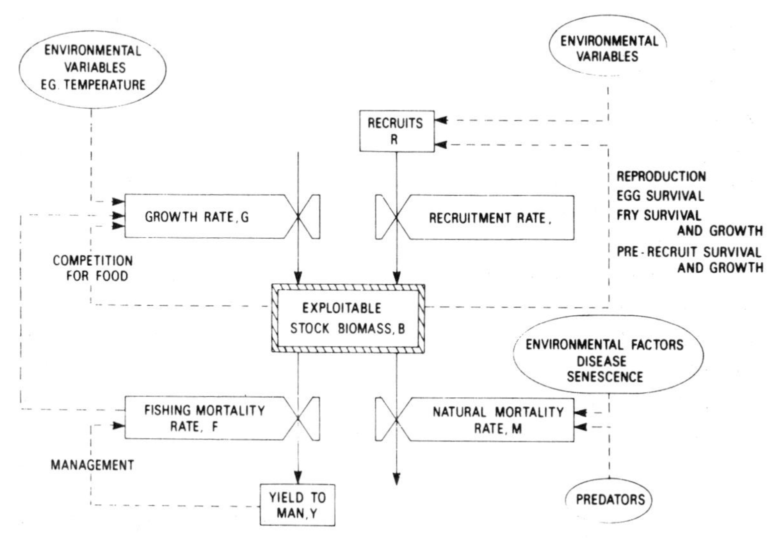

elaborated to include egg and fry survival and feeding and mortality of the pre-recruit stages. Natural mortality, representing losses to predators and to disease, could depend upon the foraging ecology of the predators. Feeding, assimilation and growth could also be modelled in great physiological detail. Similarly, the relationships between the catch recorded by the fishery manager and the gear employed are potentially of considerable complexity.

Fortunately, however, the four processes are modelled in a straightforward way in most dynamic pool models which are employed in fishery management. This is achieved by a compromise between realism and precision; the model must be a reasonable simulation of the real processes going on in the system, but it must also be tractable. It is quite easy for a biologist to build a detailed conceptual model of a system such as our fishery, but unfortunately it becomes increasingly difficult to obtain the data necessary to fit such a model as it gets more complex. It is quite easy too to conceive a model which takes an inordinate amount of computer time to run, even if the data were available to fit it. As a rough rule of thumb, sufficiently accurate answers can generally be obtained, compatible with computing time and tractability, by considering processes just one to two levels down in the hierarchy from the level at which the answers are required (in our case answers of yield and fishing effort). In practice, this requires little more than common sense and experience. For example, it would clearly be absurd to include the details of egg formation in the ovary or the molecular biochemistry of assimilation, in a model which aimed to provide a management plan for the North Sea herring fishery. So although the dynamic pool approach is hierarchical and splits up into sub-models the separate processes which were subsumed into just one factor in the surplus yield model, this operation need not go much further than obtaining functions, or empirical data, for the four sub-models themselves.

The majority of fish populations are composed of a number of discrete cohorts of age groups, each of which is the product of one breeding season. In most commercial stocks, one cohort is produced each year, but even in the tropics where the environment appears nearly constant, many species breed seasonally and in a co-ordinated way, so that the population structure remains essentially similar. When the age structure of the fish population is included in our fishery model, the four processes identified by Russell can operate realistically and independently on each age group. The consequences of considering age structure were illustrated vividly by Sir Alastair Hardy in his synthesis of marine biology *The Open Sea*, published in 1958, and we can do no better than to repeat his hypothetical example, which was based on cod. Assuming for simplicity that the fishery was in a steady state and that only fishing mortality was operating, Hardy neatly demonstrated that a fishing rate of 50 per cent could give the same numerical yield as 80 per cent fishing. Furthermore, 50 per cent fishing left more older fish in the population as illustrated by Table 8.1, and since old fish weigh more than young ones and in this example there are sufficient of them, a greater yield in weight was indicated with the lower of the two fishing rates. This effect is shown in Table 8.1b. If recruitment was postponed to age two in Hardy's example, there was an even more massive 81 per cent improvement in yield over the original 80 per cent fishing rate. Yield would decrease again if recruitment was postponed to much later ages or if the fishing rate was too low. The importance of Hardy's example is that it shows that there is going to be an an optimum combination of fishing and age of entry to the fishery which will extract the maximum yield from any particular age group structure and growth rate. This optimum may vary

Table 8.1a: Annual Catch in Numbers at 50 Per Cent and 80 Per Cent Fishing Rates from a Hypothetical Stock, Based on Cod.

Fishing Rate:	80% Annual		50% Annual	
	stock	catch	stock	catch
Age Group				
1	1,000	–	1,000	–
2	200	800	500	500
3	40	160	250	250
4	8	32	125	125
5	2	6	62	62
6	–	2	31	31
7	–	–	16	16
8	–	–	8	8
9	–	–	4	4
10	–	–	2	2
11	–	–	1	1
Totals	1,250	1,000	2,000	1,000

Source: From Hardy (1958).

Table 8.1b: Annual Catch in Weight at 50 Per Cent and 80 Per Cent Fishing Rates From a Hypothetical Stock, Based on Cod.

Age	Average Weight	Catch (Yield in Weight) 80% Fishing	50% Fishing
2	82	65,600	41,000
3	175	28,000	43,000
4	283	9,056	35,000
5	400	2,400	24,800
6	523	1,046	16,213
Total		106,102	161,138

Source: From Hardy (1958).

if the latter two factors are not themselves stable.

The point is explained by Figure 8.2, which illustrates the change in biomass of a hypothetical fish cohort as numbers decrease and the mean weight increases with age. In a steady state, the things which

Figure 8.2: Biomass Profiles for Fished Populations at Three Levels of Fishing. Broken line indicates start of fishing at progressively increasing ages (right-hand scale). Age groups in the population are shown on the left-hand scale. Vertical height above the base represents biomass of fish in an age group. Biomass profile of the unfished population is drawn to the right (shaded).

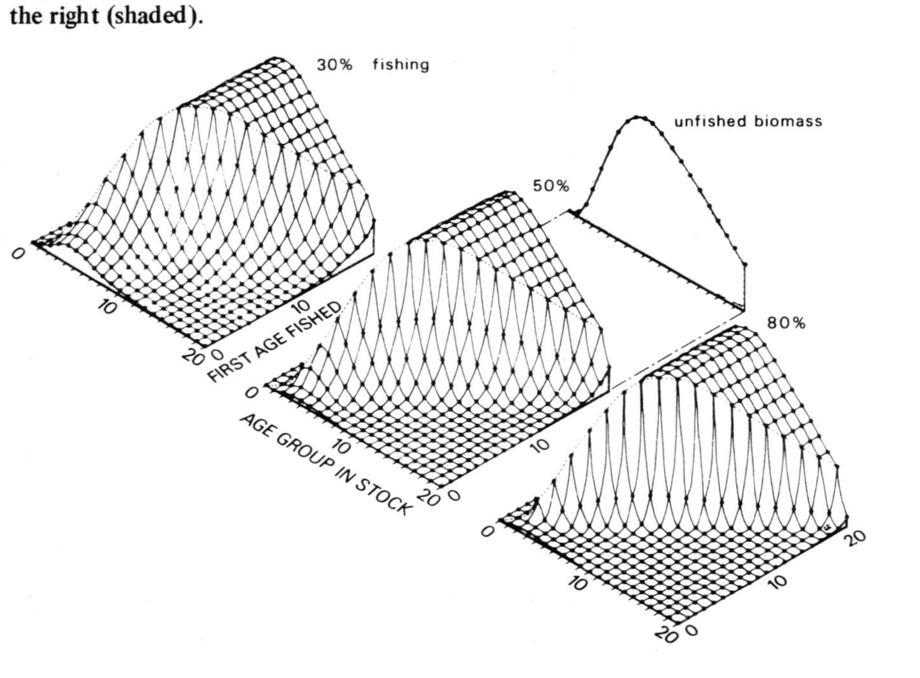

Figure 8.3: Total Yields Plotted against Age of Entry to the Fishery for the Three Fishing Rates of Figure 8.2. (Figures 8.2 and 8.3 were generated from a model in which $M = 0.2$, $W_\infty = 1{,}000$, $K = 0.1$, and $t_0 = 0$.)

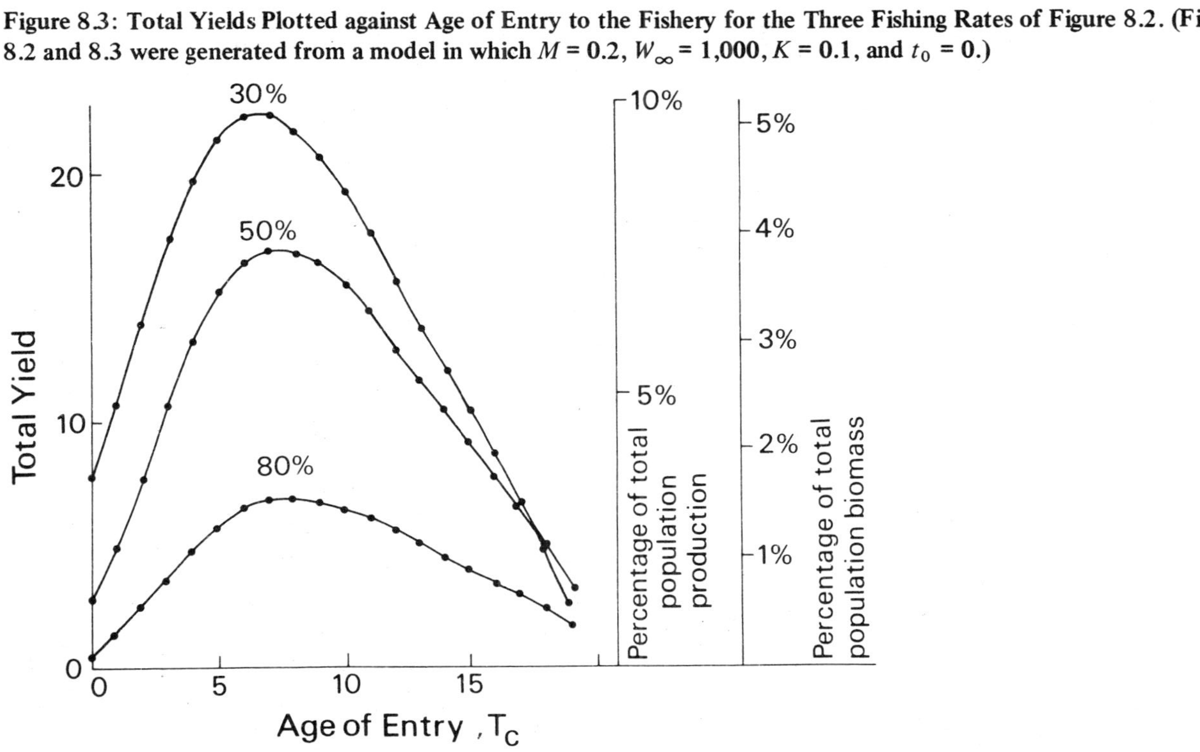

happen to one cohort as it gets older are the same as in a single cross-section of the population at one time. In any real fishery it is unwise to assume this of course, but we will stick for the moment with the steady-state model for the purposes of demonstration. Figure 8.2 illustrates the effect of fishing at 30, 50 and 80 per cent rates on the biomass of this steady-state stock for different ages of entry to the fishery. Figure 8.3 summarises the total yields under the three fishing rates. Clearly, the higher the fishing rate the lower the total yield and the later in life the optimum age of entry to the fishery. Differences between the three fishing rates, and the ages of entry, would be less for a fish which grew towards its ultimate size more rapidly than that chosen for our example because the peak biomass is reached earlier in life. The figure demonstrates that the aim of management using a dynamic pool model is to find the optimum combination of age of entry (t_c) and the fishing rate (F). A major research task is in obtaining sufficiently accurate information from the fish stock to enable such a detailed model to be fitted.

Dynamic pool models have also been termed explicit or analytical fishery models in recent texts and reviews; we prefer the older term dynamic pool because this seems to express better the contrast with the surplus yield methods. Most models of this type are now solved numerically rather than analytically, and any model can be termed explicit in one sense or another.

In the rest of this chapter we give an account of the general dynamic pool model and its classical solution by Beverton and Holt, followed by an explanation of how more recent developments in simulation modelling enable us to relax many of the assumptions and restrictions of the classical approach. As with surplus yield in the last chapter, we finish this chapter by showing how dynamic pool models can be used in the optimal control of fisheries, and discuss the range of options from which the fishery manager may select for the particular kind of information he requires.

Derivation of the General Dynamic Pool Model

The dynamic pool model is most conveniently introduced as a set of five integral equations, which lead up to a function giving the yield from the fishery. These equations are quite general and summarise formally processes which we can describe in ordinary language. Although they are simple equations, the presence of many integral signs can make them look formidable to the biologist, and this prompted Schaefer to tell Beverton in the 1950s, 'I like your book of flute music'!

The first equation merely states that the total numbers of fish in the stock, N, are given by the integral of numbers over all ages:

$$N = \int_{t_r}^{t_l} N_t \, \mathrm{d}t \tag{8.2a}$$

where the subscript t refers to each age group, t_r is the age of recruitment to the fishable stock and t_l the maximum age of fish in the stock. In practice, in stocks with discrete age groups Equation 8.2a can usually be replaced with the summation

$$N = \sum_{t=t_r}^{t_l} N_t \tag{8.2b}$$

A similar integral gives the numbers caught as

$$C = \int_{t_c}^{t_l} F_t N_t \, \mathrm{d}t \tag{8.3}$$

where t_c is the actual age at first capture by the gear used in the fishery and F_t is the instantaneous rate of fishing mortality on age t. The biomass of the fish stock can be calculated as

$$B = \int_{t_r}^{t_l} N_t \bar{W}_t \, \mathrm{d}t \tag{8.4}$$

where $\bar{W}t$ is the mean weight of fish aged t.

In all of these equations, weights and fishing mortality are relatively easy to obtain values for, but it is very difficult indeed to obtain estimates of numbers directly from a stock (see Chapter 3). However, this problem can be overcome if we have estimates of rates of natural mortality from age t_r onwards. We can then calculate the numbers at the start of each age from

$$N_t = R \exp\left[-\sum_{j=t_r}^{j=t} (M_j + F_j)\right] \tag{8.5}$$

The complicated-looking expression in Equation 8.5 merely expresses the total mortality rate, Z, up to the start of any age t in a mortality formula familiar to ecologists: $N_t = N_{\mathrm{start}} \exp(-Zt)$ (see Chapter 3). We are now in a position to write an integral to express the total yield from the fishery as

$$Y = \int_{t_c}^{t_l} F_t N_t \bar{W}_t \, dt \tag{8.6}$$

in which it is possible to find values for each of the main components.

Equation 8.6 is the general yield equation underlying all dynamic pool models. Partly for historical reasons, there have been two different approaches to its solution. The first of these derives from the classical and pioneering work of Beverton and Holt, who worked at the Lowestoft Fisheries Laboratory, UK. As good mathematicians, they solved Equation 8.6 analytically, after choosing suitable functions for F, N and $\bar{W}$, producing a single elegant equation for yield which could be calculated by tables or by using mechanical calculating machines. As we shall see, they created a major problem for themselves in their choice of a growth function. Even without this difficulty, exact integration places stringent constraints on the types of function which could be included. The second approach evolved with the general advent of computers in the 1960s; Equation 8.6 is solved numerically by literally adding up the area underneath the curve described by the equation, just as we could do by hand on a sheet of graph paper by laboriously counting the squares. The huge number of additions necessary for accuracy would be unthinkable if we had to perform them by hand, but computers and even programmable hand calculators are capable of performing the calculations in seconds. The advantage of this approach is that any combination of realistic functions or actual values from the wild, can be included in the yield equation, and we can easily build very flexible fishery models which simulate the effects of different management policies. First, however, it is instructive to examine in detail the classical yield model of Beverton and Holt.

The Classic Dynamic Pool Model

In order to solve Equation 8.6 analytically, Beverton and Holt had to make a number of simplifying assumptions; these are shown diagrammatically in Figure 8.4. Both fishing mortality and natural mortality were taken as constant and both were assumed to come into play immediately at their respective ages. This is termed 'knife-edge' recruitment. A von Bertalanffy growth curve was chosen to represent growth in length. Weight was taken as a simple cube of the standard equation (Chapter 4) so that the growth equation used is:

Figure 8.4: Diagram Illustrating the Assumptions of the Classic Beverton and Holt Dynamic Pool Model.

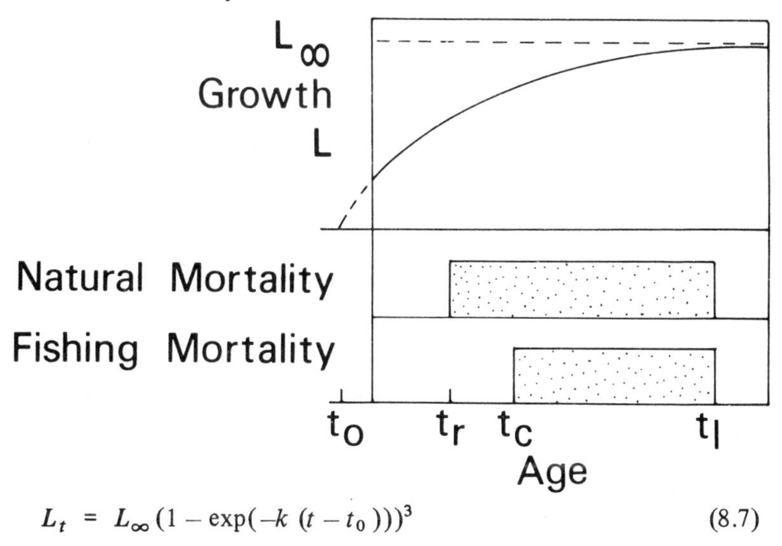

$$L_t = L_\infty (1 - \exp(-k\ (t - t_0)))^3 \qquad (8.7)$$

Subsequently, other workers published analytical solutions which allowed the weight exponent to vary from the value of 3. The choice of the von Bertalanffy curve for growth immediately constrained the yield equation to providing only annual totals, since the correct size of the fish is only indicated at one time per year with this model (Chapter 4). The main problem, however, was that Equation 8.7 cannot be integrated very easily. Beverton and Holt chose a method in which four separate components are added or subtracted to give the total weight of fish in the catch. This caused a further departure from reality as the contributions of the different age classes are lost in this process.

However with these simplifications, Beverton and Holt were able to integrate Equation 8.6 to give:

$$Y = \overbrace{FR \exp(-M(t_c - t_r))\, W_\infty}^{1} \sum_{n=0}^{n=3} \left\{ \overbrace{\left[\frac{U_n}{F + M + nk}\right]}^{2} \overbrace{[\exp(-(nk(t_c - t_0)))]}^{3} \overbrace{[1 - \exp(-(F + M + nk)(t_1 - t_c))]}^{4} \right\} \qquad (8.8)$$

where F = instantaneous rate of fishing mortality
M = instantaneous rate of natural mortality
R = number of recruits
W_∞ = ultimate asymptotic weight from von Bertalanffy curve
k = von Bertalanffy growth coefficient
t_0 = theoretical age at which fish have zero length
t_r = age at recruitment to fishable stock
t_c = actual age at first capture with given gear
t_l = maximum age of fish in stock
U_n = integration constants necessitated by use of the von Bertalanffy growth model, $U_0 = 1$; $U = -3$; $U_2 = 3$; $U_3 = 1$.

This equation (despite its formidable appearance!) can be calculated using a slide rule or tables, especially if the calculation is carefully set out in tabular form. Practice in the use of this equation is still part of the training courses in Fisheries run by the FAO since it remains useful in parts of the world where computers are not yet available. The meaning of each part of the equation, labelled 1–4 above, can be approximately described, bearing in mind the complication caused by the growth integral. The part labelled 1 merely expresses the number available for capture at age t_c as a proportion of the R recruits at age t_r. Parts 2 to 4 are mixed up in the summation which approximates the growth integral and are not easy to disentangle. Roughly speaking, however, part 2 expresses the proportion of deaths which are due to fishing, part 3 gives the proportion of recruited weight which survives to the start of fishing, and part 4 gives the proportion of weight dying from all causes during the fishery. Parts 2 and 4 together give the proportion of the ultimate weight, W_∞, which is caught in the fishery.

When Beverton and Holt derived Equation 8.8 it was difficult to predict recruitment, so that the results were expressed in terms of yield per recruited fish, which means taking the R across to the left-hand side of the equation. Most users of the classical dynamic pool approach have followed this practice, and so for a given set of values for growth, mortality and age of entry to the fishery, the usual procedure has been to construct a yield curve of yield-per-recruit plotted against increasing values of fishing mortality. Two such yield curves are illustrated in Figure 8.5. The maximum value of yield on such a curve represents the Beverton and Holt prediction of MSY. In this respect two types of Beverton and Holt curve have been

Figure 8.5: Yield Contours for the Beverton and Holt Model Applied to North Sea Plaice (Pre-war). This classic diagram shows yield per recruit plotted against age of entry, t_c, and fishing mortality rate, F. Yield curves drawn at the left refer to t_c values opposite the arrows.

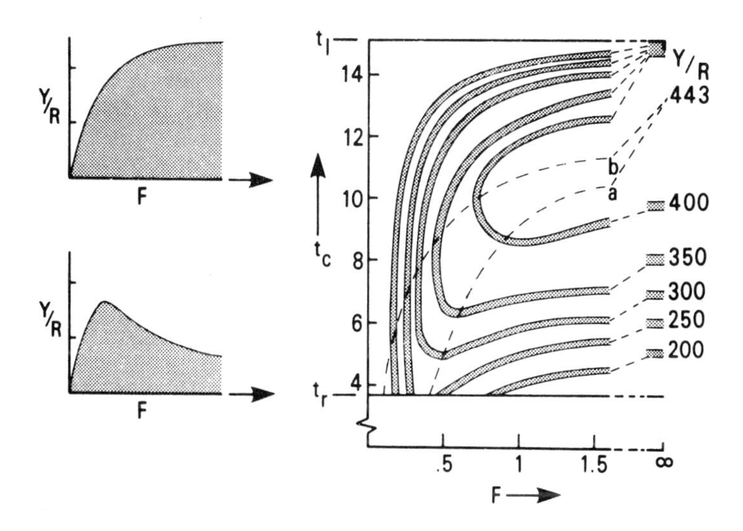

distinguished. In the first, as in the upper diagram on Figure 8.5, yield increases to a plateau with increasing fishing mortality. In this type, beyond the first part of the curve, a large increase in fishing mortality produces only a very small increase in yield. In the second type of curve, drawn in the lower-left diagram in Figure 8.5, a definite peak of yield is reached at relatively low fishing mortality. In fact, the 'plateau' type of curve does have a very shallow peak at high values of F, with one exception noted below, but is virtually asymptotic for most purposes.

At first sight the management policies suggested by these two types of curve are very different; the first seems to indicate maximum yields from heavy fishing, the second advocates low to moderate fishing mortality. However, this conclusion is misleading, as we shall see in a moment. The actual shape of the Beverton and Holt yield curve varies according to the relationship between growth rate and age at first capture, and in fact the two 'types' of curve grade into each other. Slow-growing fish (low k), which are caught early in life, have a lot of potential weight to put on, and this brings about a yield maximum at the low fishing rates which allow many fish to escape the gear and therefore realise some of their growth potential. The diagrams in Figure 8.2 were produced for a fish with $k = 0.1$, simulating relatively slow growth. Had a value even lower than 0.1 been used, the age of maximum biomass in Figure 8.2 would have shifted further right towards the older ages. A yield curve for such a fishery would exhibit an even stronger peaked effect over a wider range of t_c values. Conversely, in fast-growing fish the maximum biomass is reached relatively earlier, and would produce a yield curve of the plateau type at lower values of t_c. Notwithstanding this effect, all Beverton and Holt stocks will exhibit both types of curve depending on t_c. In fast-growing fish the peaked curves may be restricted to very young ages, perhaps even younger than t_r. This point will be returned to a little later when we have discussed a few characteristics of the yield-per-recruit curve.

An extremely useful way of employing the yield curves generated by the Beverton and Holt model is to examine the effect of different mesh sizes in the gear used in the fishery by drawing up a series of curves for increasing values of t_c. Since the larger mesh allows more younger fish to escape, the mean value of t_c will increase with mesh sizes. For our purposes this exercise also allows us to clarify the nature of the Beverton and Holt model. The series of yield curves are really slices through a three-dimensional surface and this is traditionally represented as contours joining points of equal yield per recruit. The yield contours are sometimes termed yield isopleths and can be considered analogous to a contour map. Such a contour plot is shown in Figure 8.5, based on Beverton and Holt's analysis of the pre-war North Sea plaice fishery. Contours are drawn at 50 g per recruit intervals (contours at less than 200 g per recruit are omitted in our figure for clarity – they follow smoothly the same outline as drawn). The yield contours delineate a yield surface shaped like a hill crowned by a plateau with a promontory extending down towards $F = 0.5$. The promontory covers the region of t_c values in which peaked yield curves are found. The lower diagram at the left in Figure 8.5 shows the peaked yield curve for $t_c = t_r = 3.8$ years, representing a slice horizontally across the yield surface. Similar slices taken at progressively greater values of t_c gradually become less peaked up to an age of entry of about eight years. Beyond this age, plateau type curves take over, as in the yield curve for $t_c = 10.5$ years illustrated at the top left in Figure 8.5.

Contours are drawn in Figure 8.5 for t_c up to 15 years and fishing mortality rates up to 1.5 (representing a 78 per cent annual mortality).

Beyond this value of F, the position of the contours does not alter much even if F is made infinitely large, as indicated on the figure. The maximum value of Y/R is attained at infinite F; the age of entry at which this occurs may be termed the optimal age, and for the pre-war plaice fishery it was 13.4 years. The Beverton and Holt model therefore predicts that we will obtain the MSY by employing gear which allows all fish younger than the optimal age to escape, but applies infinitely large fishing effort to all fish older than that (i.e. each and every fish older than this age must be caught). Quite clearly, this is an unattainable management goal, although this point is not made very clear in some textbook accounts of the Beverton and Holt model. It does mean however, that the fishery manager who attempts to use Beverton and Holt's method is ostensibly faced with a rather arbitrary decision about where he should aim for on the yield surface.

However, if we look more closely at the Beverton and Holt model, this is in fact a rather naive criticism. At the very least we can use the yield surface to indicate the region of the t_c/F space in which yields are within (say) 80 per cent of the maximum and then pick the minimum fishing mortality and the maximum t_c which would achieve it. Beverton and Holt themselves advocated using the model in this way by carrying out one of two eumetric fishing strategies. For example, if there was no hope of altering the mesh in the fishing gear (and resistance to such change is quite common in fisheries, for example fishermen in Kilkeel, Northern Ireland, resisted change from 15 mm to 17 mm mesh in February 1980), t_c is effectively fixed. By constructing a yield curve at this t_c (i.e. a horizontal slice across Figure 8.5), the manager could aim to alter F to the value required for peak yield. Such peak values of F lie on the 'eumetric' curve labelled a in Figure 8.5. When the value of t_c gave a curve which lay in the peaked region of the Beverton and Holt yield surface, the manager would be sensible to aim for an F which gave a yield about 80–90 per cent of the maximum on the lower F side of the slope. Above this he would gain diminishing returns for increasing effort. This policy also helps by preserving stock stability as explained below. For t_c values which produced curves in the plateau region, high F values are always indicated by Beverton and Holt, and the main decision would be to decide what percentage of the maximum yield attainable at the optimal age would be worthwhile trying to gain. An alternative eumetric fishing strategy applies to the manager who finds it difficult to alter fishing effort, but does have the option of altering the fishing gear. He can take vertical slices through Figure 8.5's yield surface and aim to set t_c to fit peaks of this new set of curves.

Figure 8.6: Yield per Recruit (Y/R) and Mean Weight of Fish in the Catch Calculated for the Blue Whiting (grams).

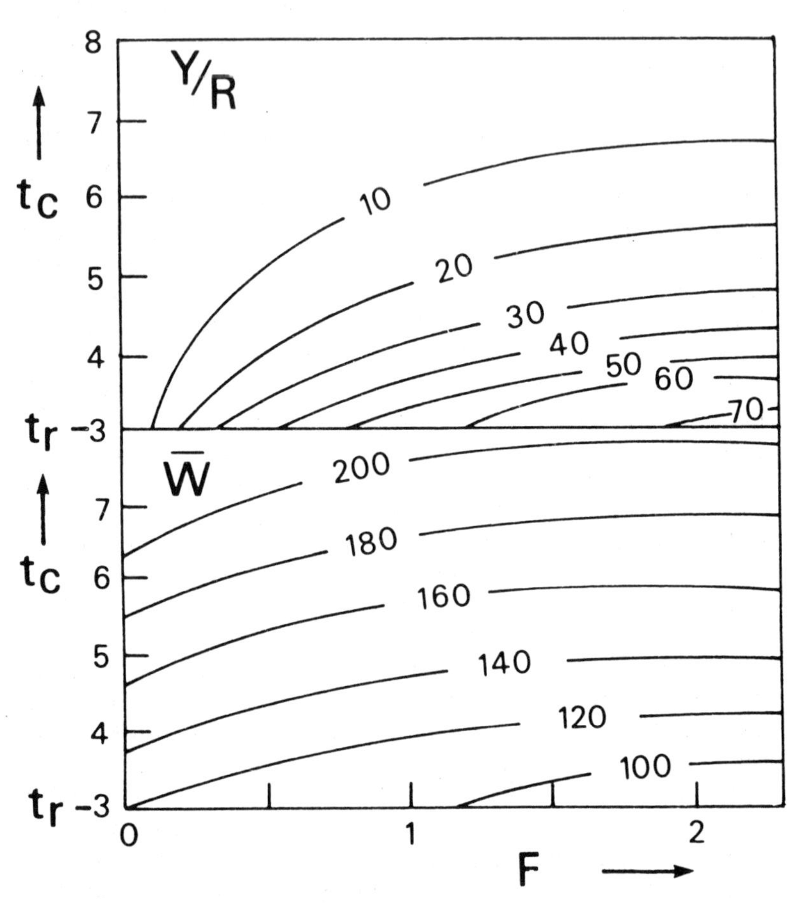

These peaks lie on the line labelled *b* on Figure 8.5.

The pre-war plaice fishery in the North Sea sustained an *F* of 0.73 (52 per cent fishing mortality per annum) and an age of entry of 3.8 years; Figure 8.5 indicates that this combination produced far from the maximum yield. Since the 1950s however, the fishery for plaice has altered with changes in the North Sea ecosystem and with changes in fishing pressure. Bannister (1977) has analysed more recent figures and come to the conclusion that fishing mortality is now lower, although North Sea plaice are still candidates for overfishing.

From Equation 8.3 it is possible to obtain predictions of the number of fish caught, from which the mean weight of fish in the catch can be calculated. This can also be conveniently plotted on contour diagrams against t_c and F. This weight may be of importance if the marketing of certain sizes of fish is difficult (e.g. big herring) or if the fish are to be processed by machine (e.g. blue whiting), or command a better price per unit weight (large cod). Figure 8.6 shows the Beverton and Holt model fitted to data for blue whiting, currently an underfished stock around the north western coasts of the UK (see Chapter 3). The lower part of the figure gives the contour surface for the mean weight of the fish caught. Under the heavy fishing recommended by the Beverton and Holt model, it seems that processing by machine of such small fish would be impracticable.

The yield per recruit contours in the upper part of Figure 8.6 are an example of the Beverton and Holt model in which the whole of the promontory of the peaked yield curves is masked below the age of recruitment. This is partly because the fish are relatively fast growing ($k = 0.23$), but is mainly the result of the blue whiting's high natural mortality rate ($M = 0.64$, equivalent to 47 per cent per annum).

This example therefore illustrates another vital feature of the Beverton and Holt model which has received little emphasis to date; this is that the yield surface is *always* of the same general shape. Stated in mathematical terms, the function in Equation 8.8 will always produce a similar-shaped surface whatever the values of the parameters. As the parameter values vary, the surface merely deforms linearly by stretching or contracting along the t_c or F scales. Variation in growth rate, k, stretches the contours in a horizontal direction on our plots. Variation in natural mortality stretches the contours, and the promontory, in a vertical direction. The value of t_r in relation to this stretching determines how much of the promontory is masked. High k and high M will both generate contours like those of the blue whiting in Figure 8.6, whereas low values of k and M tend to produce plots like that for the plaice in Figure 8.5. The important point though is that solely by virtue of choosing the Beverton and Holt dynamic pool model we have predetermined the shape of the yield contours, and therefore we have fixed both the maxima of yield towards which management should aim and the general type of management options likely to be employed. Whatever real data from the fishery may indicate, we can always fit Beverton and Holt yield contours, just as we can always fit linear regression lines to data however much a wide scatter or a clear trend away from the fitted line may render the model inappropriate. In this sense, the Beverton and Holt model can be just as misleading and dangerous as the simpler kinds of surplus yield models discussed in Chapter 7.

As with the Schaefer model, one of the attractions and dangers of the Beverton and Holt model has been that it requires relatively little information on the fish stock to draw up the yield contour map. In the case of the surplus yield model the information was of the kind which fishery owners and managers tend to gather anyway; in the case of the Beverton and Holt dynamic pool model it is of the kind which government-run organisations have excelled in organising and providing, i.e. routine sampling of fish markets at the major landing ports for each stock. Large quantities of impressive-looking figures are quickly gathered in this way and teams of technicians are kept busy reading otoliths in the government laboratories. We hope that by the end of this book we will have suggested to the next generation of fishery biologists that neither short cut can be a viable substitute for perceptive research on the fish stocks and the ecosystems of which they form a part, especially in an era of increasing exploitation of the world's natural resources.

To summarise the disadvantages of the classical Beverton and Holt model:

(1) The yield contour surface is always the same shape irrespective of whether this is appropriate or not.

(2) Despite Beverton and Holt's eumetric fishing strategy, a prediction of MSY at infinite fishing rates is not, to say the least, easy to deal with in practice.

(3) Although clearly expressed by Beverton and Holt as yield per recruit, there has been an understandable tendency to treat the results as though they were total yield. The advantage of Y/R is that where there is no clear relationship between adult stock and recruits it should provide the best long-term strategy. Unfortunately the 'best long-term strategy' may not help much when your stock has just suffered a recruitment failure. On the other hand, should there be a relationship between stock and recruitment, albeit with a wide degree of scatter, it will be much safer to incorporate recruitment into the model rather than assume it is constant. A further consideration is the psychological tendency to slip from considering something constant to considering it irrelevant, which is easily done if the Beverton and Holt model is unthinkingly applied.

(4) The model applies only to the steady state, whereas most real

fisheries, especially those most in need of management are in a considerable state of flux from year to year in many stocks subjected to contemporary economic, environmental and political pressures.

Extensions to the Beverton and Holt equation to deal with a dynamic situation have been published (e.g. Kutty, 1968; Paulik and Gales, 1964), but are little used because a much better general approach is now available. A similar qualification applies to the many published modifications to Beverton and Holt which have allowed some of the classical assumptions outlined in Figure 8.4 to be relaxed. To give them credit, Beverton and Holt (1957) considered a number of such relaxations, such as density-dependent growth, but the more recent flexible approach to dynamic pool modelling has superceded their work. In the 1980s, the contribution of Beverton and Holt can be seen to have been more in preparing the ground for modern fishery models rather than in the universal utility of the particular form of the model they produced. In fact, despite its widespread reproduction in fishery texts, Beverton and Holt's model was actually used in the management of very few real fish stocks on a global scale, probably in fewer cases than Schaefer's.

Modern Dynamic Pool Models

Modern dynamic pool models are built around summation of yields from each age class in the stock, instead of integrating over all ages. They are therefore much easier to understand, apply, and with the general availability of computers, easy to calculate. Unlike the Beverton and Holt model, the yield equation is completely general and does not specify the shape of the yield contour surface in advance. At this stage we will examine the components of a general yield equation in detail before giving some recent examples of its use.

$$Y = R \sum_{i=t_r}^{t_l} \left\{ \overbrace{\left[\frac{F_i}{F_i + M}\right]}^{1} \overbrace{\left[1 - \exp(-(F_i + M))\right]}^{2} \underbrace{\left[\exp\left(-\sum_{j=t_r}^{i} F_j + M\right)\right]}_{3} \underbrace{\bar{W}_i}_{4} \right\} \tag{8.9}$$

The major part of Equation 8.9 consists of a summation over all age classes in the stock, i.e. t_r to t_l. The first thing to note is that fishing mortality and mean weight are subscripted, i, so that different values can be used for each age class. The result of each stage of the summation will give the proportion of the number of recruits R which are caught, in each age group corrected to their mean weight. Unlike Beverton and Holt, the equation is quite easy to understand when dissected. The part labelled 1 in Equation 8.9 gives the proportion of total deaths during age i which result from fishing. Part 2 gives the proportion of the fish dying during age i as in Chapter 3, Equations 3.6 to 3.12. Part 3 gives the proportion of recruits that are alive at the start of age i as in Equation 8.5 above. Part 4 is the mean weight of fish in age i. The total yield from Equation 8.9 applies to the steady state and is made up of the contributions from each age group.

The important point about Equation 8.9 is its great flexibility: it can easily be modified to suit almost any situation or fishery (Beverton and Holt's model is in fact a special case of Equation 8.9!).

Perhaps the most significant modification to Equation 8.9 enables it to be used to predict yields in a fishery with changing exploitation. Equation 8.10 below gives the general dynamic pool model in which separate and successive years may be included.

$$Y_k = \sum_{i=t_r}^{t_l} \left\{ \frac{F_{ik}}{F_{ik} + M_{ik}} [1 - \exp(-(F_{ik} + M_{ik}))] N_{ik} \bar{W}_{ik} \right\} \quad (8.10)$$

$$\text{where } N_{ik} = R_{k-i+t_r} \exp\left[-\sum_{j=t_r,\, m=k-i+t_r}^{j=i-1,\, m=k-1} (F_{jm} + M_{jm}) \right]$$

Although this equation now looks rather complicated, all that we have done is add the subscript k to the variables to refer to the year. For example, W_{ik} is now the mean weight of fish aged i in year k and F_{ik} is the fishing mortality this group of fish is actually subjected to, rather than a long-term average. The complication in the second line of the equation enters because in order to calculate the current numbers in an age group we have to trace that group's history back in time until the year in which it recruited. This means that the actual recruits in any year can be incorporated into the model which now follows the fate of each cohort through time, as illustrated in the diagram in Figure 8.7 (see also the life tables in Chapter 3). In practice, not all of the information required may be available and some other information or guesswork may have to be used, especially for the first year the model applies to. However, the fishery model is no longer restricted to

Figure 8.7: Age Structure in the General Dynamic Pool Model. Diagram shows one cohort of recruits (R) followed through a sequence of years in the fishery.

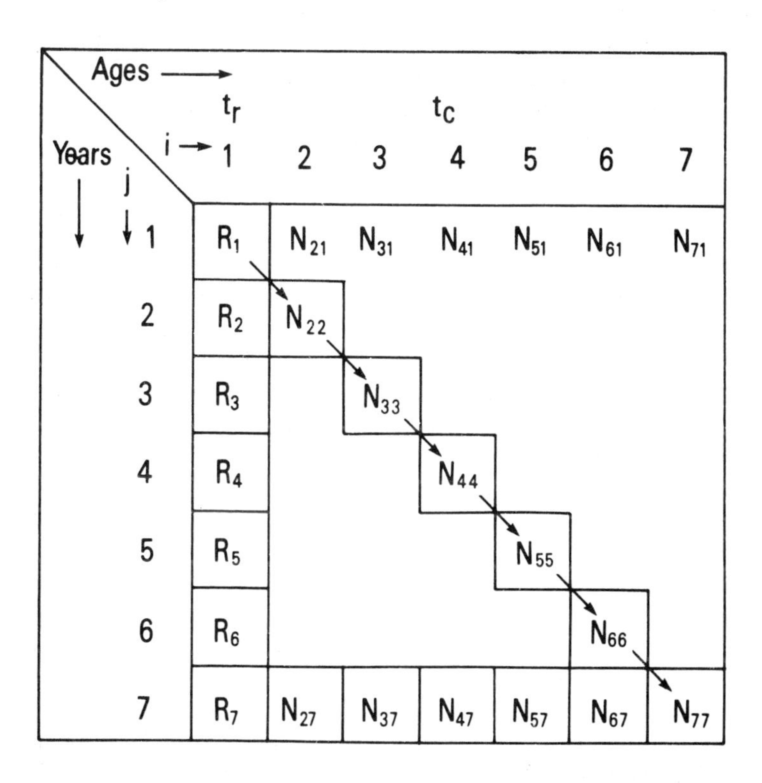

describing the steady state; it may be employed to simulate the actual exploitation history of a fishery. Even more importantly, it can be used to extrapolate the age classes into the future, given a knowledge of recruitment, and examine the consequences of changes in fishing rates, as we shall see in detail later in this chapter.

The flexibility of this general dynamic pool approach extends to each of the four sub-models. Any desired degree of generality or realism can be accommodated. Fishing mortality, assumed constant for ages greater than t_c in the Beverton and Holt model, can be set to any required values in the general dynamic pool equation. Unfished ages younger than t_c can be represented by zero F_i, older ages which are only just vulnerable to the gear to a low F_i, and ages which are fully recruited to the stock set to the standard value for F_i. In some types of gear, small purse seines for example, very large old fish are less vulnerable because of their faster swimming speed. In other cases large fish shift their feeding habits and no longer stay in the same areas as the rest of the stock. All of these changes can be easily accommodated in the model by altering the set of F_i values. A simple way of achieving this effect in practice is to use a standard value for F_i modified by a set of 'gear coefficients' ν_i. This was the method adopted by Clayden (1972) in a model of ten North Atlantic cod stocks, one of the earliest examples of the modern dynamic pool approach.

The vital feature of this type of model is its ability to simulate changes over a number of years. For example, fishing mortality might be linked in a density-dependent way to population density, as might occur in a fishery in which constant effort was applied. The model will mimic the behaviour of the fished population as it moves towards the steady state. This period of change is called the transient response and it is vital in the present world of rapidly changing economics, politics, and fishing technology where few fisheries are likely to experience the same fishing regime for more than a few years at a stretch. Again, a major problem is obtaining accurate values for fishing mortality and we will return to this problem later in the chapter.

Natural mortality, usually assumed constant, is not really so (see Chapter 5). Although the model contains the possibility of using different M_i values, the difficulty of obtaining accurate estimates of natural mortality means that this option is rarely exercised in practice. A simulation in which M is density dependent is likely to be a more realistic management tool than one which assumes M to be constant over all years. Provided that the exploitable part of the stock is beyond the age at which natural mortality rates decrease sharply, it is usually safe to assume that the same natural mortality will apply to all fished ages.

So far we have considered annual values of the main components of the dynamic pool model. These annual values are only approximations in most fish stocks, since both growth and mortality vary tremendously through the year from season to season. In temperate zones the growth of many fish is zero in the winter months (see Chapter 4). In addition, many fisheries are exploited seasonally as the fish undergo part of their spawning or feeding migrations. For example, Figure 8.8 illustrates the pattern of seasonal growth and mortality in a small temperate freshwater fish. The unrealistic nature of the annual rates is very clear. Very few examples of this type have been published for marine fish; an

Figure 8.8: The Pattern of Seasonal Growth and Mortality in a Small Freshwater Cyprinid. Solid lines show rates calculated on a four-weekly base; broken lines join the equivalent annual rates, which obscure the seasonal pattern.

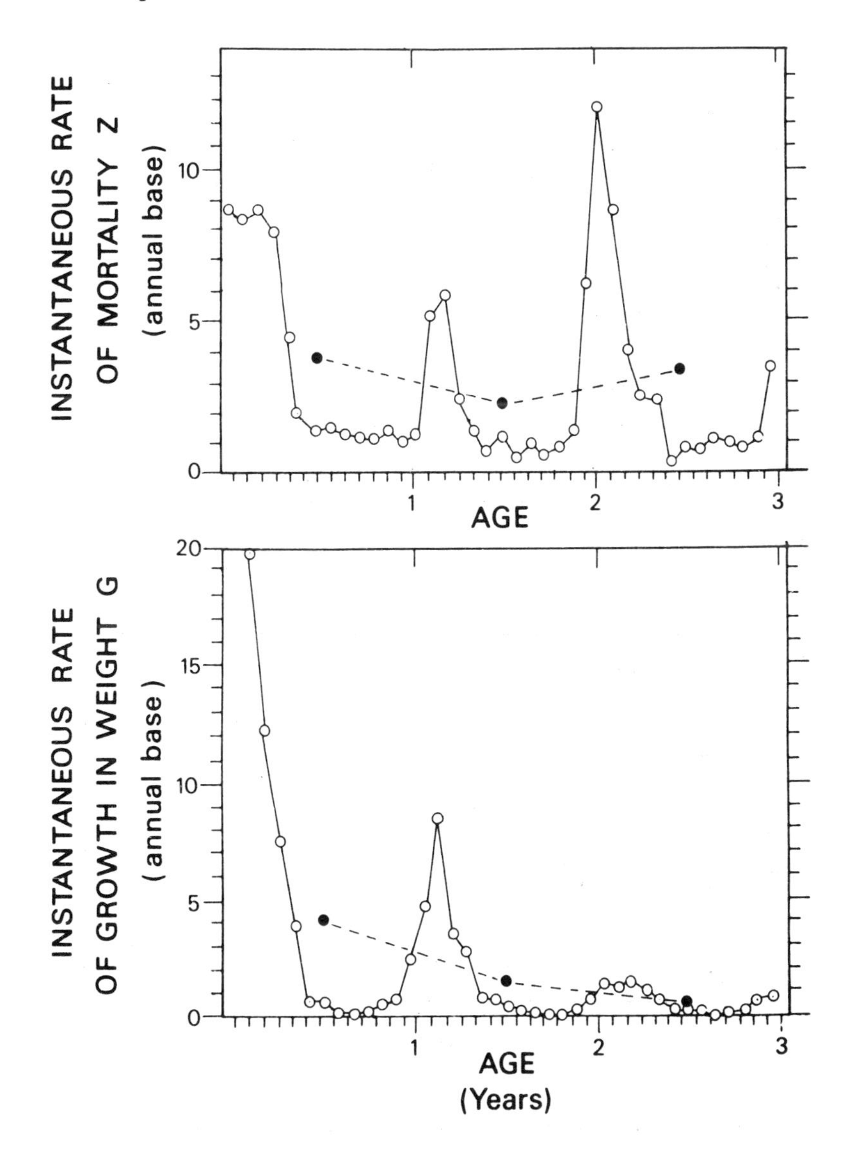

Figure 8.9: Two Examples of Seasonal Growth in Length. A. Minnows in the Seacourt Stream, River Thames, UK. B. Juvenile plaice from the Solway Firth.

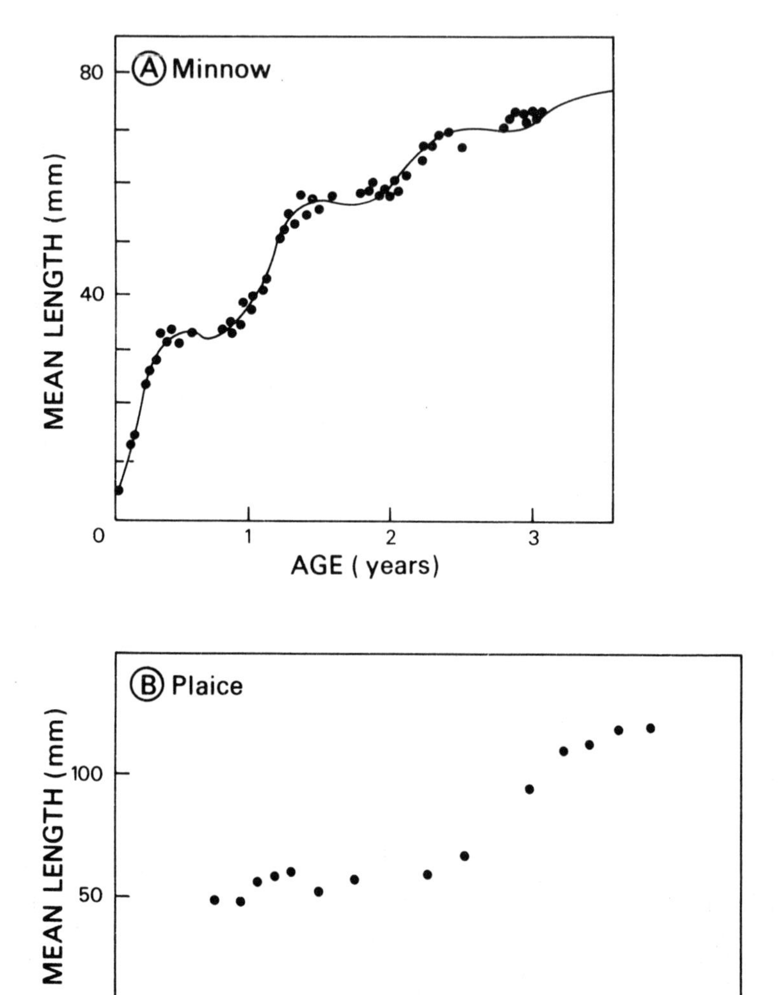

Source: A. Pitcher and Macdonald (1973a). B. Lockwood (1974).

example of seasonal plaice growth is given in Figure 8.9. Seasonal variation in natural mortality has not yet been described for marine fish, but more research effort may soon be directed towards this end with the advent of work on sea cage culture of sprats and salmonids.

A fully realistic dynamic pool model might therefore need to include seasonal variation in some or all of the main components. A modification to do this is easy to carry out on the general equation:

$$Y_k = \sum_{i=t_r}^{i=t_l} \sum_{s=1}^{s=y} \left\{ \left[\frac{F_{iks}}{F_{iks} + M_{iks}} \right] \left[1 - \exp(-(F_{iks} + M_{iks})) \right] N_{iks} \bar{W}_{iks} \right\} \tag{8.11}$$

Symbols are as before, but each year has been divided into y seasons. For example $\bar{W}_{iks}$ is the mean weight of age group i in year k for season s. Empirical values for weight can be used, or a seasonal growth model such as the one outlined in Chapter 4. The numbers in each age class are slightly more complicated to work out, but the principle is the same as in Figure 8.7. Care has to be taken to adjust mortality rates to the appropriate within-year values. The computer program only requires an extra loop to deal with this and can be used to explore the effect of altering the time of year of exploitation relative to the season in which maximal growth occurs.

In summary, the general dynamic pool model can incorporate any desired pattern of growth, natural mortality, and fishing mortality. The values which are employed can be empirical ones from the fishery itself, or can be generated by a suitable sub-model, such as the seasonal version of the von Bertalanffy, when this is more appropriate. Results may be expressed as yield-per-recruit as in the Beverton and Holt model, but it is much more valuable to use the potential of the dynamic pool model to the full and include recruitment directly. In this way predictions of yield for specific years can be made, but we have to overcome the problem of estimating recruitment mentioned in Chapter 6. Recruitment for the coming year can be measured by pre-recruit surveys (Chapters 3 and 6), and recruits for past years can be calculated using cohort analysis as described in Appendix 1. Alternatively, recruitment may be generated by a suitable model relating recruits to the parent stock, such as the Ricker model described in Chapter 6. Both of these approaches have their own deficiencies, and we will illustrate these by giving an example of each.

Management Applications of the Dynamic Pool Model

A pioneering approach to simulation by dynamic pool modelling was carried out by Clayden at the UK Government Fisheries Laboratory, Lowestoft, in 1968. Unfortunately his work was not published until 1972, by which time political changes in Europe had altered the situation he was describing.

Clayden's ambitious work simulated the fishing effort and catches of 15 nations' fleets on the ten main north Atlantic cod stocks (see Chapter 3) over 23 years from 1946 to 1968. In effect, ten dynamic pool models were run, each of which had 15 sub-models representing the fishing effort of each of the fleets. The growth of the cod was taken from empirical data on each cod stock, and natural mortality was assumed constant. Cod can be considered to recruit at about one year old when they leave the surface waters and become 'codling', living nearer the bottom. Recruitment was obtained by taking the estimated long-term average number of one-year-olds for each stock and modifying it by a factor based on the relative year-class strengths pertaining to each cohort of fish. This was done separately for each stock. Fishing mortality was assumed proportional to effort and constant over all ages in the fishery; gear selection was simulated by using gear coefficients for different ages according to the type of gear characteristic of each fishing fleet.

In the model, the fishing fleets were allowed to move in a dynamic fashion between the stocks from year to year. The attention of a particular fleet was determined by a set of economic factors which weighted the value of each stock to that fleet by the distance to the fishing grounds, the size of boat, the value of the cod, and the relative value of species other than cod which might be caught there. Further constraints according to the type of vessel were added; for instance, half of the Norwegian fleet, which consisted of quite small boats, was fixed as fishing on the coastal Norwegian cod stock.

The results of the simulation were encouragingly similar to the actual history of fishing on the different stocks. An example, giving the CPUE for the Bear Island cod stock, is shown in Figure 8.10. As well as simulating past years, Clayden made predictions for the then future years up to 1975 by using average recruitment values and an estimated 4 per cent per year increase in fishing effort. Despite the fact that his specific predictions were overtaken by political changes, his general conclusions that total cod yields could not be increased much further, and that catch-per-unit effort would fall, have been dramatically borne

Figure 8.10: Example of Simulated and Actual CPUE for One of the Ten North Atlantic Cod Stocks Modelled by Clayden (1972).

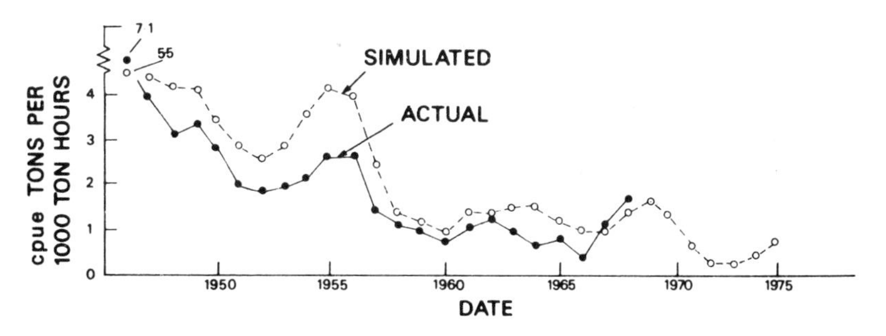

out by events.

As might be expected, Clayden's results were most sensitive to recruitment and he concluded that increasing accuracy in the estimation of year-class strengths would pay worthwhile dividends in improving the predictions from his model. As explained in Chapter 6, a great deal of research effort is now being put in just such a direction, especially in pre-recruit surveys.

It is important to recognise the historical significance of Clayden's work; this was one of the first published examples in which a true simulation model was employed, and management recommendation for the fishery predicted actual total yields rather than yield per recruit.

A similar model to Clayden's was published by Garrod and Jones in 1974, this time for the Arcto-Norwegian cod fishery. This is the largest single cod stock in the north Atlantic, capable of yielding about three-quarters of a million tons of fish per year if managed correctly. Garrod and Jones described growth by a conventional von Bertalanffy curve, constant natural mortality was assumed, and a series of gear selection coefficients were employed to modify the fishing mortality for partially recruited ages. The main difference from previous work was the use of a Ricker recruitment curve. The Arcto-Norwegian is the most northerly of the Atlantic stocks, growing relatively slowly (k = 0.12), maturing quite late at seven years old and recruiting to the stock between three and five years old. Recruitment is quite variable from year to year (Chapter 6) unlike the more southerly cod populations. Using the age-group structure of the late 1960s Garrod and Jones ran their simulation model to predict trends in the fishery according to different fishing mortalities and different mesh sizes in the trawl. Some of their results are shown in Figure 8.11, and they clearly show that an increase

Figure 8.11: Garrod and Jones' (1974) Predictions for the Arctic Cod Stock under Three Levels of F and Two Mesh Sizes.

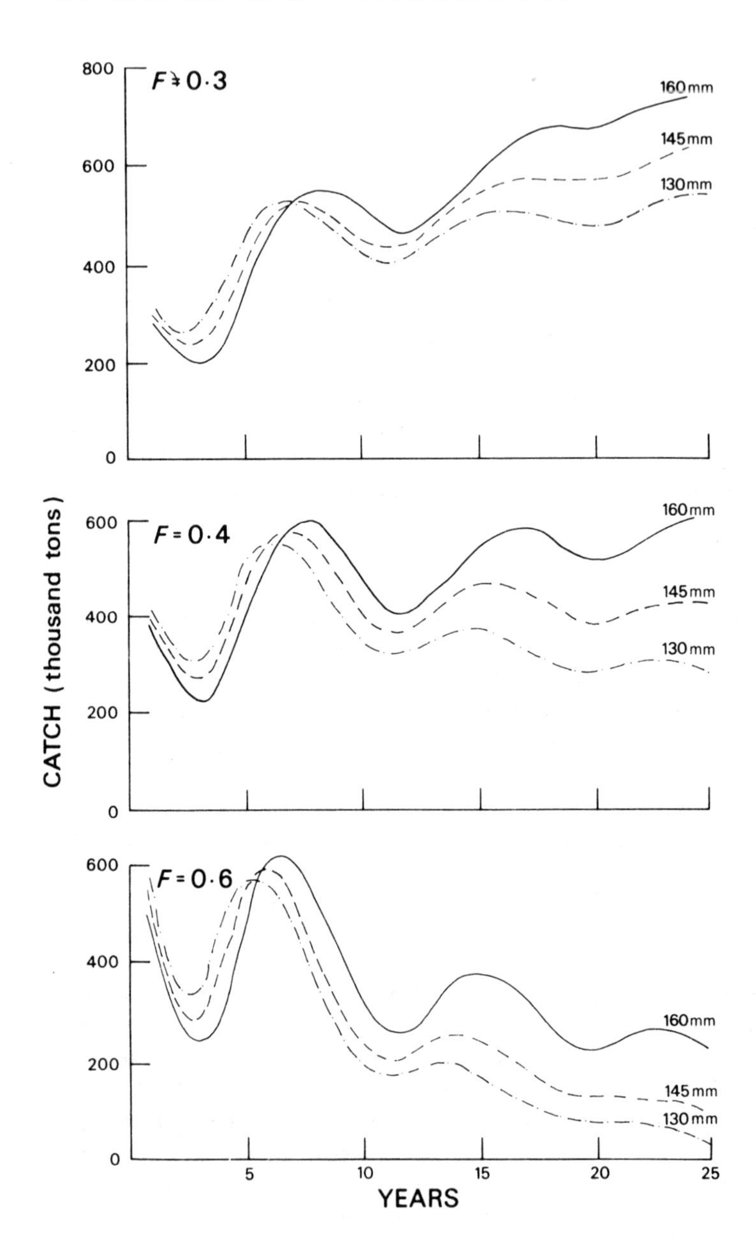

in mesh size to 160 mm, together with a reduction in fishing mortality to 0.3, provide the best chance of sustained yields in the future, although initial yields might be a little lower. The bulge in the predicted yields in Figure 8.11 is caused by the very large 1969 year class working through the fishery. Higher fishing rates with mesh around 130 mm would overexploit the stock.

Sadly, Garrod and Jones' more pessimistic forecasts have been borne out in practice. Mesh sizes have only recently been increased (1979), but from 120 mm to only 125 mm (Blacker, 1980), and there seems little hope of obtaining increases to 160 mm. Fishing mortality has never dropped below 0.6 for this stock; indeed catches as high as 900,000 tons were taken in the late 1970s. High catches caused by the successful year class encouraged heavy fishing pressure, but the stock is now in some danger and is heavily overexploited, just as Garrod and Jones predicted in their model. The stock is now managed by the USSR and Norway, but it must be greatly disappointing to fishery scientists that one of the most important cod stocks in the world is still not managed rationally ten years after clear recommendations were made. Recent surveys (late 1980) have shown that all of the north Atlantic cod stocks are very seriously depleted as a result of overfishing. We will return to this topic in the final chapter, but in the meantime note that despite Garrod and Jones' success in predictions using a full dynamic pool model, they did not include the very large amount of scatter around the Ricker curve. The consequence of this is that fluctuations between years are progressively smoothed out as the simulated year moves away from the last one in which full age class data is available. This effect can be seen in Figure 8.11, and implies that such a model is best employed in predicting the trend of catches over the long term, rather than the actual catch in the coming year. Nevertheless, it does provide a significant advance over the Beverton and Holt analysis without recruitment.

A surprisingly early example of work covering both of these last points was published by Carl Walters from the Seattle College of Fisheries in 1969. Walters used a Ricker curve in a simulation of the same Arctic cod stock providing a dramatic and instructive comparison with the then standard Beverton and Holt predictions. His results, in the form of a yield contour map, are given in Figure 8.12. The Beverton and Holt analysis indicates optimum yields with early and heavy fishing, whereas the simulation shows a clear peak of yield at low fishing mortality rates. The Ricker curve used by Walters was not as sophisticated as the version used by Garrod and Jones six years later,

Figure 8.12: Yield per Recruit Contours for the Artic Cod Stock from Two Contrasting Models. (a) Top diagram, according to the classic Beverton and Holt model. (b) Lower diagram, according to Walters (1969) simulation model which included recruitment.

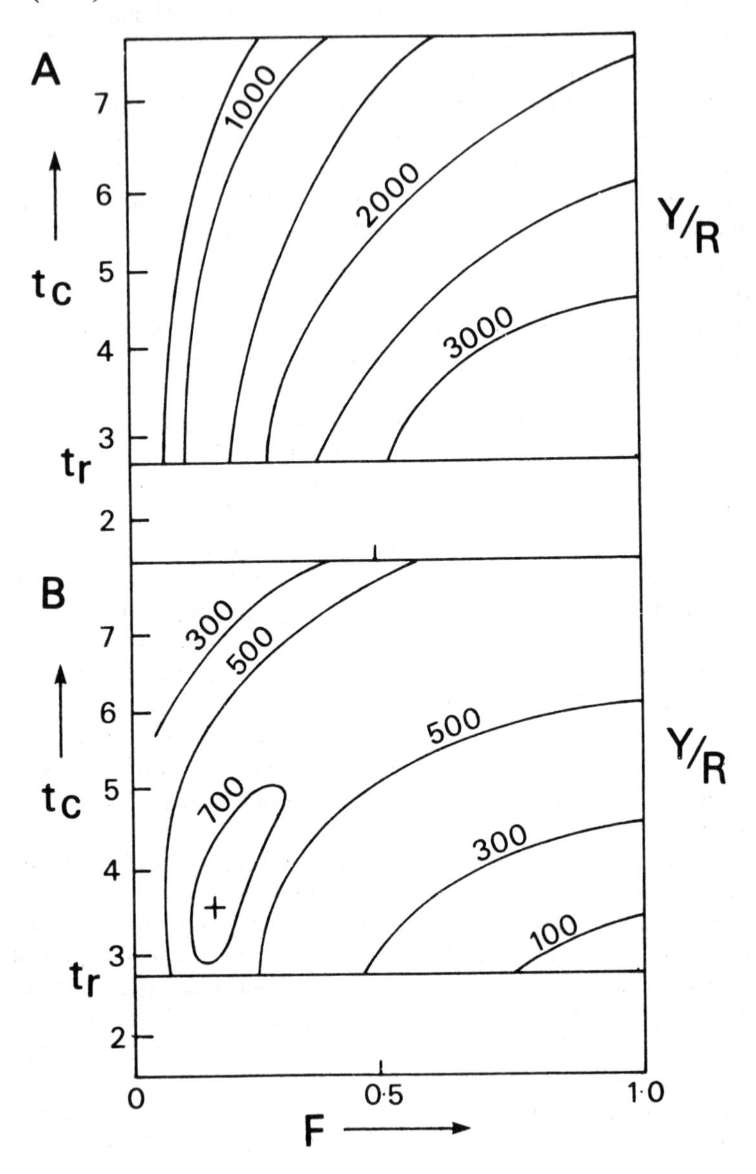

and his gear-selection values and natural mortality rates were not as accurate, but nevertheless his analysis provided a remarkably similar general management recommendation for the same stock.

Two important points can be made about Walters' work. First, on a practical level, comparison with the simulation shows that management of the Arcto-Norwegian stock along classical Beverton and Holt lines would have been disastrous. Secondly, Walters' main purpose was not to present management recommendations for any particular stock (despite three examples being included in his paper), but was to present a completely general dynamic pool fishery model, very similar to Equation 8.9 in this chapter. In his model, the separate sub-models of growth, natural mortality, fishing mortality and recruitment could be of any desired form, just as we have indicated is desirable. Walters even made a computer program available to any who requested it in 1969. The program included an effective routine to search for optimal yields, and the effect of various alterations in fishing could be simulated. In our experience this general program works very well. It really is amazing that this seminal paper has hardly ever been cited in the fisheries literature, especially on the European side of the Atlantic, perhaps illustrating the danger of being too far ahead of one's time! It is unfortunate that Garrod and Jones did not refer to Walters, considering that the similarity of their recommendations would have inspired greater confidence in their results, even though they were working from somewhat different premises.

As we explained in Chapter 6, the main problem with a realistic model of recruitment in many stocks is the wide scatter of values. Including white noise in the model provides a satisfying theoretical solution to the problem, but this approach has important consequences in the way in which predictions from the model may be used. Once we incorporate the randomness, predictions can no longer apply to any particular year, but only to one of many possible years which may ensue. At first sight, this may not seem very useful, but it is in fact of great value and provides a rather different kind of management prediction to the long-term optima derived by Walters in the analysis just discussed. If we run the model and make a number of predictions for each subsequent year, we can adopt a sound statistical approach, which has been in use in other ecological fields for a very long time (e.g. Bartlett, 1960). If we examine repeated simulated predictions for the year to come, under whatever management plan we are hoping to adopt, we can examine the actual distribution of the predictions. The probabilities of each particular outcome can then be estimated very simply by counting the number of times they turn up in 100 random trials. This neat technique is termed 'Monte Carlo' modelling. In plain language, it gives us a way of assessing the likelihood of any particular consequence of our simulated management option. This could be the likelihood of obtaining a particular level of yield in the coming year, or it could even encompass the risk of causing our fishery to collapse, as we will see shortly.

One of the few examples of the use of a randomised recruitment model to date is that of Wilson (1979) on a freshwater seine and trawl fishery for the pollan, a whitefish found in Lough Neagh in Northern Ireland (see Chapter 1). In this work, predictions were made from a dynamic pool model which incorporated Ricker recruitment randomised with normally distributed 'white noise'. The predictions were compared with others from a similarly randomised Beverton and Holt recruitment curve (see Chapter 6). An example of the results for $F = 1.0$ is shown in Figure 8.13b, which gives the pollan population sizes under the two different recruitment regimes over 50 simulated years. In both cases the stock persists over the long term, fluctuating within rather narrower limits in the case of Beverton and Holt recruitment. Figure 8.13c demonstrates that yield contours for the pollan fishery are different for the two types of recruitment. In fact, recruitment is not understood in detail for these fish because the fishery has been studied for only a few years, although there is some evidence that the Ricker model is more appropriate. In this fishery the type of recruitment turned out to be crucial to the management plan adopted, and so the study indicates that all further research effort should be directed to investigating the fry and juvenile stages. At the same time, it suggests that management is best employed in surveying the pre-recruit stages of each year class before they leave the shallow rocky areas and join the fishable stock of pollan over the mud and sand bottom regions of the lake (Wilson and Pitcher, in prep.).

Where data are extensive, as a simple and effective alternative to employing a randomised recruitment model, we can put an actual sequence of recruitment figures into the dynamic pool simulation, chosen at random from recent historical records of the fishery. Yield predictions for the North Sea plaice stock based on four such randomly-chosen sequences are given in Figure 8.14 (Blacker, 1980). The results predict a general slow diminution of plaice yields under current fishing pressure even if yields should in the near future be enhanced by an outstandingly abundant year class, such as occurred in one of the four random sequences. In this example it would have been helpful to have

Figure 8.13: Two Alternative Scenarios for the Lough Neagh Pollan Fishery Using the Ricker and the Beverton and Holt Recruitment Sub-models. (a) Recruitment curves. (b) Example of values obtained over a 50-year simulation run at F = 1.0. (c) Yield per recruit contours plotted against t_c and F for the two alternatives.

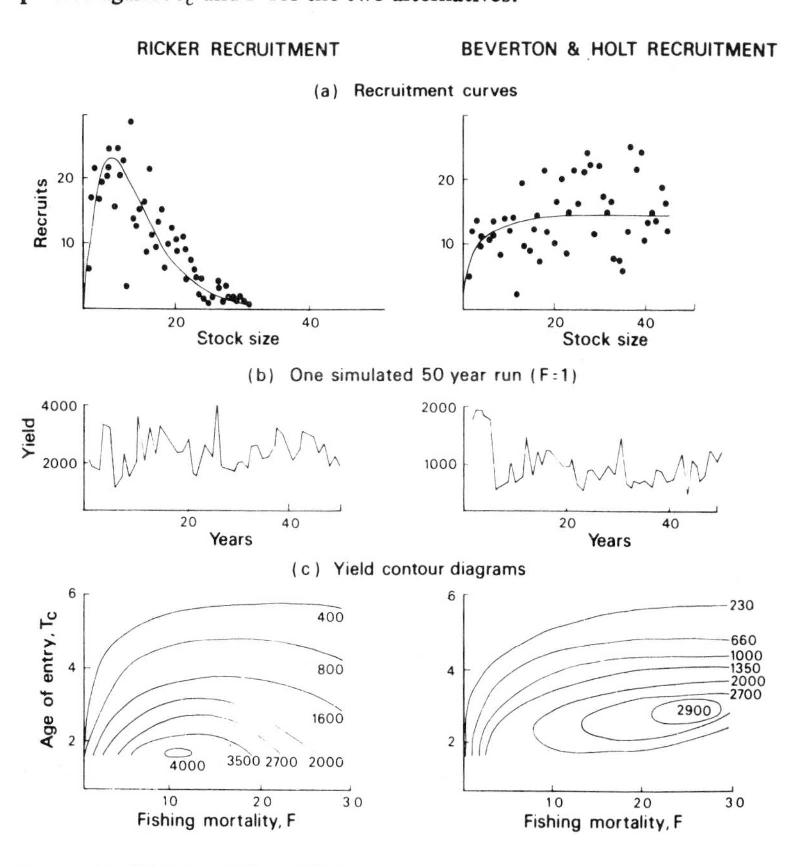

Source: Modified from Wilson (1979).

run a randomised recruitment model as well, so that the chances of such an exceptional year turning up could also be assessed. (A casual glance at the figure creates the incorrect illusion that this is about 25 per cent.) Even so, the exercise clearly shows that the beneficial effects of a good year class will not persist for many years in this plaice fishery. The actual management recommendations for this stock are to maintain the present level of total allowable catch (TAC) of 112,000 tonnes

Figure 8.14: Predicted Yields for the North Sea Plaice Fishery from Four Randomly Chosen Sequences of Recruitment Selected from Historical Records of the Fishery.

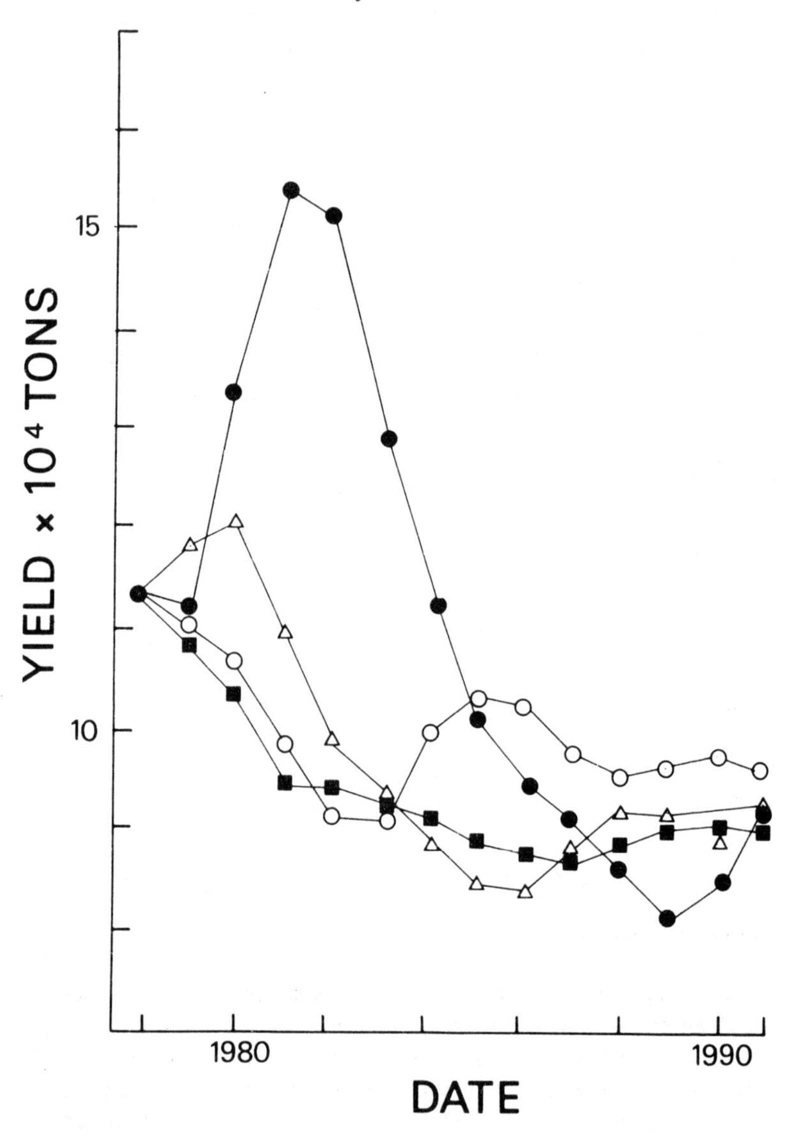

Source: Modified from Blacker (1980).

unless poor recruitment should occur, and this does seem to be a relatively safe option, although higher yields might be obtained by enforcing higher minimum size limits. The choice of real recruitment figures overcomes the problems of model fitting, but this approach too has its drawbacks; concern about the adequacy of fit when using a recruitment model is replaced by worry that any historical sequence of figures may not mimic current conditions.

With these examples in mind, it should now be clear how the dynamic pool model may be used in the optimal control of a fishery. In the ideal situation, summarised in Figure 8.14, there is good information on each of our four sub-models. In particular, recruitment is known through a good pre-recruit survey for the coming fishing season and in addition 'Monte Carlo' simulation has given us the likelihood of each category of outcomes from our management policies. The weakest links in the process are the poor information on natural mortality in most stocks (see Chapter 5), inadequate details of recruitment (Chapter 6), and inaccurate estimates of numbers and fishing mortality rates.

Knowledge of fishing mortality is usually obtained by using a technique known as cohort analysis if figures are not available directly. The method produces figures for the numbers of fish in each age class in the stock together with the fishing mortality to which it was subjected, given certain approximations and assumptions about natural mortality and fishing. The method is fully described in Appendix 1, where we show that the largest errors from the cohort analysis are likely to be in the most recent years, since the method works back into the history of each cohort now present in the stock, and is based on guessed values for F in the current year. Unfortunately this is precisely the region where we require the highest accuracy if our predictions for the coming year are to be viable ones. There seems to be no simple way around this problem although much current research is devoted to it (see Appendix 1). It is of course preferable to have actual figures for the age class composition, but it is much more difficult to get similar values for fishing mortality as these require a great deal of research effort and just would not be feasible for many fisheries outside a few developed nations.

Optimal Control Using the Dynamic Pool Approach

We have seen that the weakest links in the optimal control of fisheries using the dynamic pool model are all due to imperfect information, the most important of these being errors from cohort analysis, recruitment and natural mortality. However, at the very least, alternative versions of the model can be used to see whether alternative plausible cases make an appreciable difference to the results and so may be used to direct research effort accordingly, as in the pollan example described above.

Once predictions can be made with confidence and the optimal fishing mortality and age of entry to the fishery determined, the fishery manager has, in ideal conditions, the option of deciding upon the desired fishing pattern. In practice of course, this is not so easy. One of the first problems is the translation of the desired fishing mortality into fishing effort; if he is lucky this may already have been done during the model building stage. Beyond this point, three major constraints are put upon the fishery manager, which may not derive from the biology with which he has been dealing up to now. First, the method he chooses has to work quickly; the system of TACs may not be the best here, as we will see. Secondly, the economics of the fishery may indicate that the biological optimum is not the most profitable. Thirdly, the politics both within and between the nations fishing the stock may restrict the choice as to what is feasible. These three factors may also interact with each other and it is to these complexities of contemporary fishery management that we will return in the final chapter of this book.

To conclude this chapter, we will use a hypothetical dynamic pool simulation model to illustrate how management of fishing mortality and age of entry can deal with the range of biological situations likely to be encountered in fish populations, and how risks may be assessed in practice.

Assessment of Risk Using a Fishery Simulation Model

Yield contours plotted in F/t_c space for a realistic fishery simulation model are shown in Figure 8.15a. The peak yield (shaded) occurs on an elongated promontory which swings rapidly to older ages of entry as fishing mortality increases. For F greater than 1, the yield for t_c less than 4 is very low, so management of this fishery should clearly aim for t_c about 3 and F of about 0.5 (39 per cent per annum). The simulation model from which this yield contour plot was produced included submodels for growth, recruitment, natural and fishing mortality acting on a population composed of ten age classes.

The growth sub-model was empirical and based on a set of mean weights for each age as illustrated in Figure 8.15c. Growth type I exhibited a smooth increase in weight whereas type II simulated a more realistic phased growth curve, such as might be obtained by cubing the von Bertalanffy model. Basic growth obtained from these curves was modified to be mildly density-dependent: the multiplier shown in Figure 8.15d was applied to the mean weights according to the total population numbers. In practice, the multiplier was near to 1 for most of the simulation, except in collapsing stocks where it increased to about 3 and in dense lightly-fished stocks where growth was effectively reduced to about a half.

The natural mortality sub-model was equally simple, M being set to 0.2 for all ages except the first, which was given a value of 0.3. These values of M were modified by a mildly density-dependent multiplier similar to that for weight (Figure 8.15e).

Recruitment was simulated with a relatively high-peaked Ricker curve with normally distributed 'white noise'. The recruitment curve is illustrated in Figure 8.15b, with a set of stock/recruitment points generated by a typical 30-year run of the model superimposed. Compared to the plots for fisheries given in Figure 6.1, this model appears quite realistic. Spawning adults were all those older than the age of maturity at the end of a simulated year. Maturity was set at 5 for the yields shown in Figure 8.15.

'Knife-edge' recruitment was assumed for all the runs reported here, fishing mortality acting equally on all fished ages.

Each of the four sub-models in the simulation could be altered independently so that the model could be used, for instance, to examine the effect of density-dependent growth on the yield contours. Here, however, we have confined ourselves first to comparing the two growth curves in Figures 8.15c, and secondly to examining the effect of the age of maturity relative to the ages in the fishery.

Each run of the model consisted of a 30-year simulation of the fishery during which average yields, mean weights of fish caught, recruitment failures and population profiles were extracted as required. The model was run separately for a range of combinations of t_c and F to produce values for which contours could be plotted.

Figure 8.16a gives the yield contours for growth type II, all other factors being held the same; the stippled line represents the age of maturity. The effect of the phased growth curve is to reduce the high yields from early capture at low F. Peak yields are now confined to a

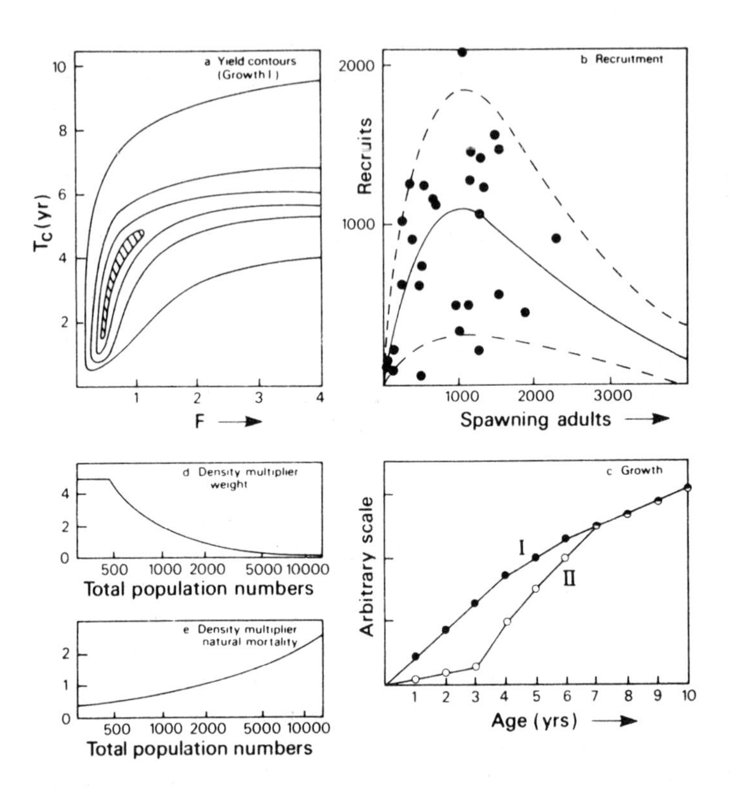

smaller well-defined area just beyond the age of maturity. The peak is approached as the growth rate increased beyond three years old.

The effect of a lower age of maturity is seen in Figure 8.16c. Since most fishing now takes place on mature fish, the peak is shifted towards much higher fishing mortality rates (F = 3.0, or 95 per cent per year), although the best t_c is still at about five years old. Although absolute yield is similar, the whole yield surface is now broader and flatter as the area in which yields are reduced by fishing immature stock is reduced. In both cases in Figure 8.16, management should aim for an age of entry of 5, but heavier fishing can be sustained by the earlier maturing stock. Even so, the increased yield given by increasing F from 1 to 3 in Figure 8.16c might not be worth the extra effort and cost. Therefore it might be best to manage both at t_c = 5 and F = 1.0. This could be decided by plotting the marginal yield as contours to show the

Figure 8.15: Details of the Fishery Simulation Model and Some Sample Results. (a) Yield Contours for Growth Type 1. Contours represent total annual yield averaged over 100 simulations. Each run covered all combinations of age of entry, t_c, and fishing mortality, F. The contours delineate a narrow promontory of yield; the area of maximum average yield is shaded. (Absolute contours are not given since they depend on the arbitrary choice of unit for growth.) Each run was a 30-year simulation of the ten age classes in the fishery population, with growth, mortality and recruitment being generated by separate sub-models as described below. (b) Recruitment Sub-model. Recruitment employed in the simulation was a Ricker curve ($\alpha = 3, \beta = 0.001$) with normally-distributed randomness superimposed (SD = 0.7). Solid line is the Ricker curve, broken lines represent the confidence limits. Points plotted on the graph are a typical set of spawning stock/recruitment figures for one of the 30-year simulations. (c) Growth in Weight. Curve I represents smooth increase in weight, curve II a more realistic phased increase in weight curve. Seasonal growth, differential growth in the two sexes or between individuals was not modelled. (d) Density-dependent Multiplier for Mean Weight. Weights from graph (c) were multiplied by this factor according to population numbers. Relationship was chosen to represent a fairly mild density-dependent factor. (e) Density-dependent Multiplier for Natural Mortality. Natural mortality was set at 0.3 for age 1 and 0.2 for all other ages and then multiplied by this factor according to population numbers, to simulate a mild degree of density dependence.

improvement of yield which would result from a 0.1 increase in F at each point on the map.

The two graphs at the bottom of Figure 8.16 each derive from 100 complete runs of the model shown above them. Variability in yield, expressed as the coefficient of variation (COV) over the 30-year run is plotted as contour values in Figure 8.16b and d. It can be seen that peak yields in this model have an associated penalty of 50–60 per cent variability from year to year. Overfishing beyond peak yield, or at ages too young, are both associated with greatly increased variability in yield, which would of course make the economic planning of the fishery very difficult. These graphs represent a similar demonstration of increased variability with heavier fishing to that published for the surplus yield model by Beddington and May, described in Chapter 7. Most of the variation in this model derives from the recruitment sub-model; random variation incorporated into growth and natural mortality

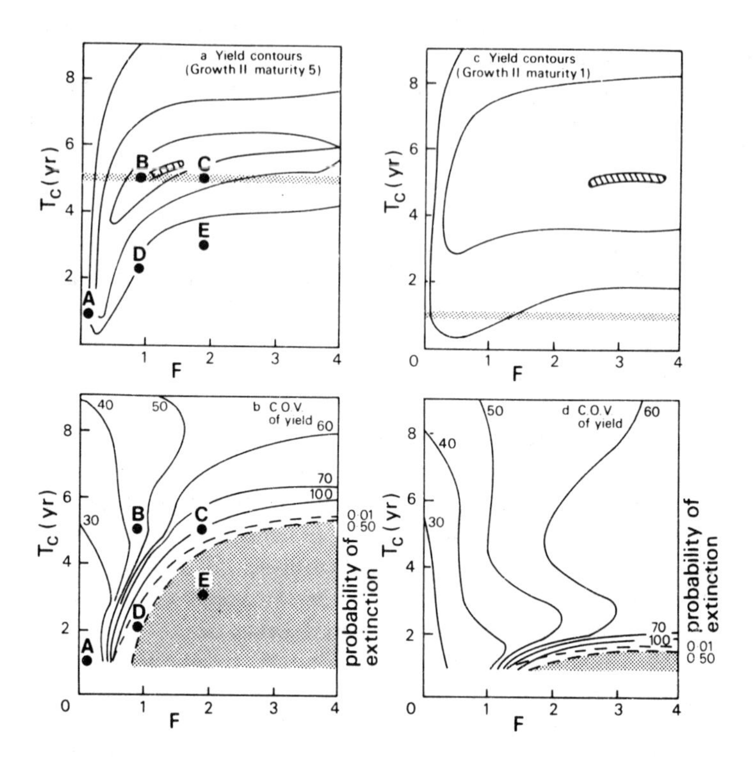

sub-models does not make much difference to this overall picture. We think that this is quite realistic in the light of our discussions in Chapters 6 and 7. For any real fishery, recruitment based on an actual set of figures might be substituted for a Ricker-based model. The important point is that although the absolute validity of any recruitment model can be questioned, the comparison of the relative levels of variability at different points on the yield surface will remain a useful management exercise.

Also shown in Figures 8.16b and d are contours of risk of complete stock extermination through recruitment failure (broken lines). Similar but wider contours can map the occurrence of one or more years of recruitment failure within the 30-year simulation period. The same caveat applies to these risk contours as to the COV contours discussed above, but again the general pattern will be useful in planning the fishery so that risky areas can be avoided. Comparison of Figures 8.16b and d shows that, unlike yield itself, age of maturity has an

Figure 8.16: Results of the Fishery Simulation Model – Yield Contours for Phased Growth, Curve II in Figure 8.15c. (a) and (b) refer to simulation with age of maturity at 5 years old; in (c) and (d) maturity is at 1 year old. (a) and (c) are yield contour plots with the area of maximum yield shaded as in Figure 8.15. The age of maturity is indicated by the shaded line. Points A to E in (a) refer to particular combinations of t_c and F where population profiles were sampled – see Figure 8.17 below. (b) and (d) are contour plots for variation in yield plotted in the same t_c/f space as (a) and (c). Two sets of contours are shown. Solid lines are the contours of the coefficient of variation of yield calculated over the same 100 runs of the 30-year simulation model as in (a) and (b). Contours for COVs of 30, 40, 50, 60, 70 and 100 per cent are drawn. Other values follow the contours drawn smoothly. Broken lines represent contours of equal probability of completely exterminating the simulated population within the 30-year period. The probabilities were obtained by counting the number of times this happened in the 100 runs of the model. Contours for probabilities of 0.01 and 0.05 are drawn; other values follow the lines drawn smoothly. The area within which the simulated population had a more than 0.5 chance of being wiped out is shaded.

enormous effect on both risk of collapse and variability of yield. In general, these results indicate that the penalties of fishing before maturity are large, so that this policy could only be sustained with very careful restriction of F to low values. As explained in Chapter 11, this can be very difficult.

Population profiles obtained at points A to E for the model results given in Figure 8.16a and b are drawn in Figure 8.17. Point A shows a wide population profile, with many age groups present when F is low. Year-class dominance can be seen in this plot, chosen at random from the simulation. Successful year classes can in fact be followed realistically through the fishery in a series of such plots. Despite the low age of entry and year-class dominance, variability in yield is low at this point since the many age groups present in the fishery are each fished moderately. Point B represents the population at the peak yield. Absolute numbers in the population are smaller than at A, but there is a similar pattern of year classes up to the age of entry to the fishery, when numbers are sharply reduced by fishing at F = 1.0 (63 per cent per annum). Point C represents growth overfishing, where the yield at an F twice the optimum is reduced and becomes much more variable. Since the fish do not mature until 5, heavy fishing of spawners introduces

Figure 8.17: Population Profiles from the Simulation Model Run at Points Labelled A to E in Figure 8.16, Representing Values of t_c and F Given Below. Heights of the bars in the histograms show the numbers in each of the ten age classes in the simulated population. Arrows indicate the ages of entry to the fishery, t_c. In each case the population profile has been taken from one randomly-chosen run of the model. For further discussion see text.

	t_c	F
A	1	0.1
B	5	0.9
C	5	1.9
D	2	0.9
E	3	1.9

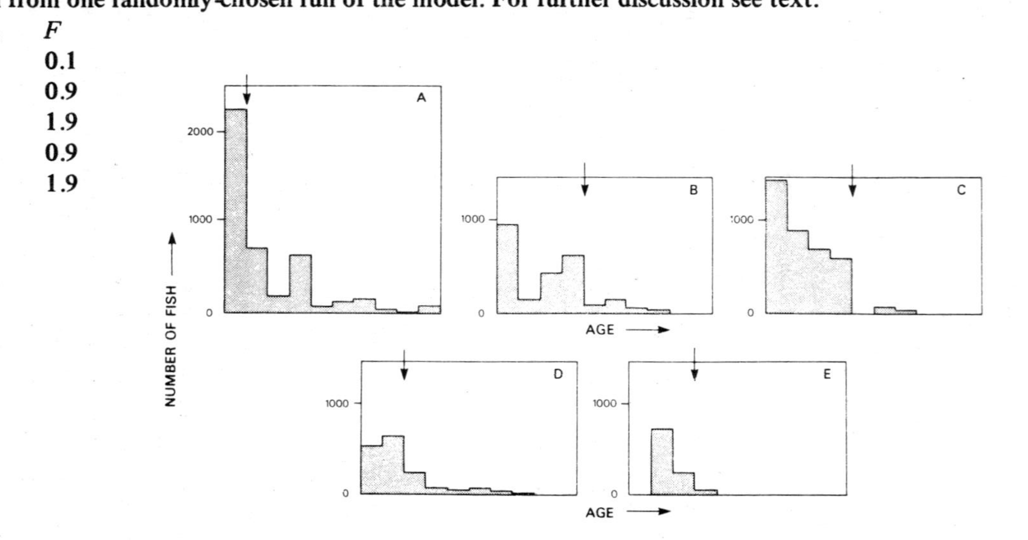

the possibility of occasional recruitment failure. One such failed year class happened to occur five years before the sample C was taken. Point D is a population profile from the correct F, but too low an age of entry. Yield here is similar to that at A, because F is low enough to allow cropping of the peak cohort biomass, but yield is five times as variable and there is a finite risk of recruitment failures wiping out the population (about 0.15 at some time during the 30 years for point D). Point E illustrates the dangers of both recruitment and growth overfishing. Here the population is sharply reduced at three years old before the fish mature and can contribute to recruits. So few fish reach maturity that even the high survival and growth of the small number of recruits cannot compensate so that failed year classes are common; one has just occurred in the population sampled to give point E. This stock stands a greater than 50 per cent chance of being wiped out during the 30-year simulation of the fishery.

Detailed simulation models such as this example enable the fishery manager to weigh clearly the consequences of various fishing strategies, and are increasingly coming into use in the management of world fisheries.

A New Departure?

Deriso (1980) has recently introduced a simulation model which apparently combines some of the features of both surplus yield and dynamic pool models. Its major advantage is that although it includes sub-models for recruitment, natural and fishing mortality, age structure and growth in the population, its seven parameters may be fitted with only catch and effort data like the surplus yield models. In general terms, Deriso's model is similar to Walter's (1973) delay differential model described in Chapter 7; the major new feature is the inclusion of age structure.

Recruitment is given by a generalised three-parameter model, which includes the Ricker and Beverton and Holt curves as special cases, so that recruitment can have a hump or not as appropriate. Recruitment of a particular cohort can be delayed by the appropriate number of years and once recruitment age is attained, the fish join either a catchable or an uncatchable portion of the stock. Uncatchable fish progressively join the catchable stock at the end of each season. This has been termed 'platoon' recruitment by Ricker (1975).

Growth is simply generated by a von Bertalanffy curve in Deriso's paper, but the model apparently can accommodate more realistic growth figures.

Economic values as well as yield can be included and the model can be used to solve for Maximum Present Value (MPV, see Chapter 10) when interest rates are in the region of 2–4 per cent per annum. Deriso does not state what happens to the model's predictions if more realistic financial interest rates are incorporated (see Chapter 9).

Deriso shows how to fit the model to data on catch and effort for a fishery over a number of years. As examples he uses data from the US yellowtail flounder fishery, the Georges Bank haddock stock (see Chapter 2), and the north Pacific halibut fishery. The predicted catches are impressively close to the real values over long periods for all three examples, although Deriso admits that the accuracy of the model is reduced if one attempts to extract more detailed information from it.

This new departure has been greeted with some acclaim by fishery scientists (see Walters, 1980), but it remains to be seen if the rather difficult non-linear estimation procedures which have to be used to solve the basic model equation for the seven parameters will be robust for further examples. If this proves to be the case, the Deriso model may turn out to be of wide application in the management of those fisheries which have previously been treated by surplus yield models. The lack of detailed predictions for age of entry, growth, etc., means, however, that for well-documented fisheries the conventional dynamic pool model will continue to be the most appropriate for management.

We will return to the problem of managing fisheries in the final chapter of the book, but for the next two chapters will digress slightly. We will introduce some simple aspects of fishery economics, and then discuss one way in which some have hoped to see all the intricate problems raised by the population dynamics of natural populations bypassed, i.e. fish farming.

9 FISHERY ECONOMICS

In this chapter we will be concerned with ways in which economic factors influence fishery management. The fishery biologist may strive to provide sound advice on attainable yields and appropriate fishing efforts, but advice based purely on biological analysis will, at best, be modified by economic factors. At worst, biological advice can be ignored totally in favour of economic policies which appear superficially to a biologist to be little more than national or company self-interest. But these actions are on deeper analysis the result of a web of economic factors, not all of which are under control by governments, let alone fishery managers, and may be complicated by international politics. In the past, economics has not always been taken into account explicitly, but what has happened is that fisheries have been nudged (some would say 'dragged') in certain directions by the economic facts of life. Economics has a direct interaction with the biological state of the exploited fish stock, and so, although it might be claimed that this field is properly the concern of economists, we think that biologists should know about the problem. The economic analysis of fisheries is in fact a new and growing discipline; all we hope to provide in this book is a simple introduction for the biologist.

In general terms, an economic analysis subtracts the costs of going fishing and adds the benefits of the fish harvest onto the purely biological parts of the fishery system we have considered in previous chapters. These costs and benefits are usually counted in money, but they could equally well be taken in units of energy and the same sorts of analysis applied. The fundamental point is that fishing is not free; it always costs something (in money, fossil fuel energy, human manpower), and so these costs must be offset against the benefits of the value of the catch (money, or nutritionally useful energy). In some cases what appear to be costs are benefits too; for example, it may be considered a social benefit to provide employment. We will return to the idea of accounting in energy (in Chapters 10 and 11) and for the rest of this chapter we will make the assumption that fisheries will be run to make a financial profit. The exception will be subsistence fisheries, which at first sight appear to escape the strictures of economic analysis, but we will find that even here economic factors can become very significant. Accounting in money does have one

Figure 9.1: Maximum Economic Rent and Maximum Sustainable Yield in the Steady State.

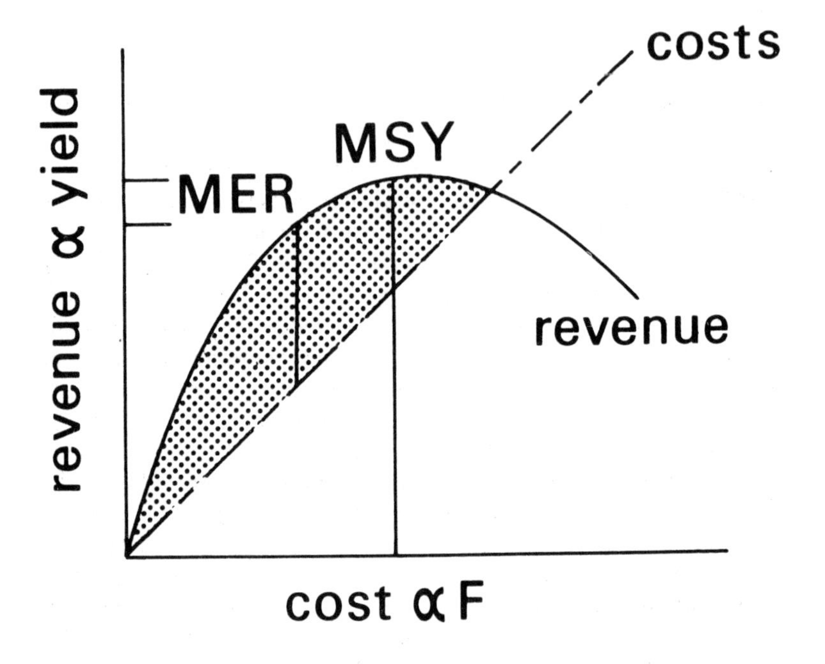

important difference compared to energy: because money can itself be bought and sold as a commodity, interest rates have to be included. As we will see, this is a most important and far-reaching consequence.

The simplest economic analysis, illustrated in Figure 9.1, serves to show that when costs are included the straightforward biological MSY is not the catch around which the fishery should stabilise. We assume here that total costs, represented by the broken line, are directly proportional to fishing effort. So a plot of total costs against fishing effort will form a straight line whose slope is the unit cost of harvest. We next assume that the revenue gained by selling the fish catch is directly proportional to the yield, i.e. that the unit price of fish does not alter with harvest size, a fairly unrealistic assumption which we will relax later. If this is the case, a plot of revenue against fishing effort will have the same type of humped shape as the general surplus yield model; the exact shape does not matter as we are not relying on the surplus yield model for this analysis. When these two plots are superimposed on the same scale, as in Figure 9.1, we can see that the fishery makes a profit only in the shaded region. In economic jargon this profit is termed the economic rent from the fishery. The MSY is at the peak of the yield and revenue curve, but the maximum economic rent (MER) is at the peak of the shaded area. In Figure 9.1, whatever the shape of the broken line representing costs, the MER must always be less than the MSY, and will therefore be taken with a smaller effort than f_{opt}. So this graph just expresses the commonsense idea that including the costs of going fishing reduces the optimum amount of fishing which should be done for maximum returns.

The management strategy favoured by this simple steady-state analysis is apparently a safe one, since a fishery below MSY encounters less variation in yields (see Chapters 7 and 8) and has shorter response times in recovering from changes. Therefore if MER were the only economic pressure which applied to fisheries we would have no biological worries. However, both the cost and revenue curves in Figure 9.1 are likely to vary in practice. For example, the assumption that unit harvest costs are constant whatever the fishing effort is unlikely to hold when one considers a real fleet of trawlers setting out from a fishing port. The number of boats, their size and relative power, how they compete to locate and catch the fish, the density of the stock, price of fuel purchased in bulk, and many other factors, will all have an influence on the cost per unit of effort. A more realistic total cost curve is likely to be S-shaped as unit costs decrease in larger fisheries. Similarly, the total revenue curve is likely to be a plateau as the proceeds from the sale of the catch of fish are not directly proportional to the catch size, because fish will command higher unit prices when scarce. When these more realistic factors are taken into account, MER may lie almost anywhere in relation to MSY, as in the two examples sketched in Figure 9.2.

The simple MER is important for two reasons. First, it demonstrates that the inclusion of costs does not alter the kind of management which must be applied to a fishery. We still work in terms of selecting an optimum fishing effort. Secondly, MER can be taken as an index of performance of the fishery and a baseline to which the more complex analyses may be related.

In the rest of this chapter we are going to examine four aspects of moderately-realistic economic analyses. These include: factors which tend to produce overfishing in a commonly-owned resource; the effects of interest rates; the consequences of overcapacity generated during periods of overexploitation; and the effect of some economic factors included in a model applied to an Australian prawn fishery.

Figure 9.2: Alternative Cost and Revenue Curves.

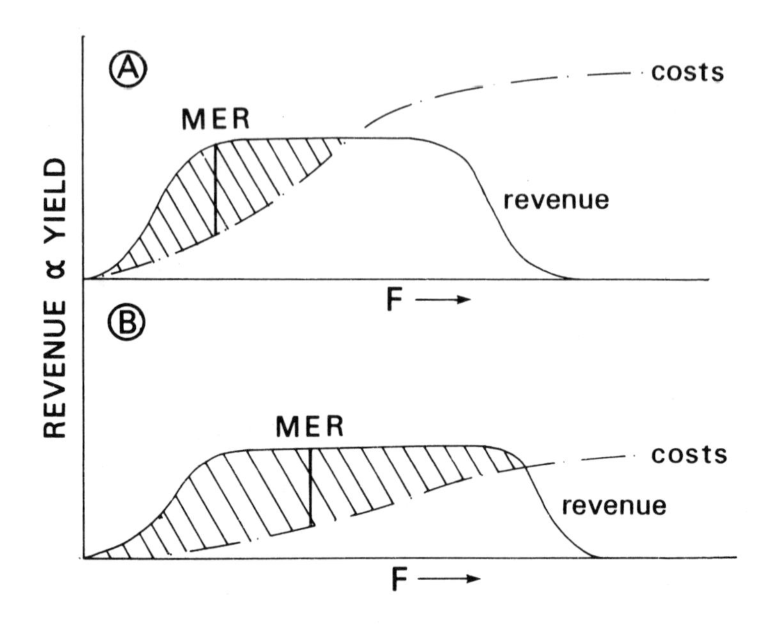

Overfishing in the Free-access Fishery

Overfishing of a commonly-owned fishery to which anyone has access can be seen to be virtually inevitable by a simple economic analysis. At the early stages of expansion of a fishery and over a wide range of conditions, the individual fisherman expects a return roughly proportional to the effort he puts in. Hard work producing handsome profits suits the small entrepreneur, an ethic popularised by Samuel Smiles in the Victorian era. This is represented by the solid straight line in Figure 9.3. New fishermen (or investors) are attracted to the profitable fishery in false hopes of similar gains. In fact, the addition of a new boat to the fishery lowers the slope of the line in direction A on Figure 9.3, so that the profits for each fisherman are reduced slightly. The effect occurs in part because the resource is now fished more heavily, producing smaller returns per unit effort, and partly because the unit price of the fish is reduced by the larger catch. In the early stages of exploitation this reduction in profitability may be barely discernible; in later stages it may be crucial. The process of attraction of new fishermen

Figure 9.3: Common Property Rent Dissipation and the Effects of Improving Fishing Technology in an Open Access Factory. For further explanation, see text.

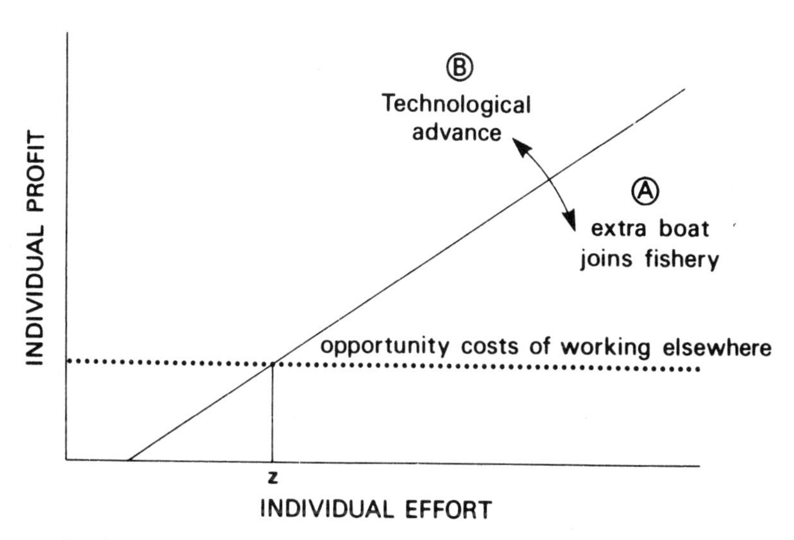

to the fishery can be expected to continue as long as the rewards are greater than can be gained from working elsewhere (or placing capital elsewhere). This is termed the 'opportunity cost' of the resource, and is shown as the broken line in Figure 9.3.

As more and more fishermen join the fishery, the slope of the line decreases until fishing is no longer profitable to the individual. By this stage, however, the stock may be so heavily fished that it is severely depleted or even in danger. Of course this may not be so, but the main point is that this process of common property rent dissipation automatically proceeds without any reference to the biological status of the stock. The process has been termed the 'tragedy of the commons' (Hardin, 1968). As can be inferred from Figure 9.3, fish stocks are in greater danger from rent dissipation if opportunity costs are low, as is likely to be the case in remote areas with subsistence economies.

There appear to be only two salvations for the commonly-owned resource which can preserve it from rent dissipation, and neither of these provides great comfort! First, if the fish stock is in superabundance compared to the greatest possible fishing pressure, it will be safe even though commonly owned. This may have been the case with the North American salmon stocks before European settlement. Secondly, and partially linked to the first, many subsistence-level fisheries may be preserved by the sheer hazard and hardship of fishing in a small boat in treacherous seas, such that expected gain for the potential fisherman is counterbalanced by his assessment of the risks and dangers he may encounter. In terms of Figure 9.3, even though opportunity costs are low, the slope of the 'profits' line is also low and unpredictable. This situation will remain stable only so long as the level of technology in the fishery remains undeveloped. Once advanced technology, such as echosounders, radar navigation and larger powerful boats become available, the slope of the line increases in direction B. However, the improvement to profitability is evanescent, if not illusory, since without limitation of access the process of attraction of new fishermen now begins in earnest. The new gear provides much larger catches and the stock is put in further danger. In addition a pernicious competition for the new technological aids begins between the fishermen. The declining CPUE of the fish stock on the way to depletion means that an individual can temporarily do better than average by employing the latest improvement to his gear. In the initial stages of expansion of the 'new' fishery the large catches may bring about reduced prices, so that this and competition with his fellows may force the individual fisherman to overcapitalise. He commissions gear and boats on a scale greater than the maximum sustainable yield of the resource can support. The technological race is a usual if not a necessary accompaniment to the process of rent dissipation, and explains why many previously stable subsistence fisheries can be rapidly depleted once new technology becomes available. Raising the local opportunity costs in the early stages of the process, perhaps by resiting suitable industries or the fish processing industry itself, may be the only viable policy if restriction of access is not possible. Fisheries can be difficult, costly and even dangerous to police.

The economic analysis of rent dissipation puts the blame for the depletion of fish stocks fairly and squarely on the competition that occurs between individual fishermen. One might therefore anticipate that rescue might come from a takeover of the fishery by a single large company, a fishermen's co-operative, or by a government monopoly. It might reasonably be expected that such a monopoly would be unlikely to endanger its own future by overexploiting the resource on which it depends, and should therefore manage the resource for the MER, as we have seen. However, fisheries which have been severely depleted have been under monopoly management and have not been run as free-access resources, for example the Peruvian anchovy fishery described in Chapter 7. So the 'tragedy of the commons' cannot be the whole story.

Overfishing by Maximising Present Values

Working from the University of British Columbia in Vancouver, Professor Colin Clark has pioneered a fresh approach to the economics of renewable natural resources. He has demonstrated several ways in which the assumption that a monopoly will not overfish is invalid. One simple way in which this can occur is for costs to be low at small stock sizes and fishing effort, coupled with high prices for small landings of fish in demand. Under these circumstances the fishery can make a profit even from catching the last fish ever (e.g. lobster, halibut) rather like the curves drawn in Figure 9.2b. One would not expect this situation to be common since the management should have the foresight to ensure the future of the fishery as an asset even though present catches become very valuable. The recent ban on herring fishing in the north Atlantic agreed by European governments is an example. The dangers of this particular situation then are not as great as those of common property rent dissipation because they do not occur automatically.

A much more dangerous and plausible mechanism by which even a monopoly can be forced by economic factors to deplete the fishery has been analysed by Clark (1973). The analysis is based on the consequences of maximising the present value (MPV) of income gained from fishing which in turn is influenced by the interest rates applying to lent money.

Calculation of Present Value

The concept of present value derives from the way invested money increases according to the current rate of interest. If £100 is invested at 5 per cent compound interest being calculated once a year, the sum invested will be worth £100$(1 + 0.05)^n$ after n years. If we now call P the sum first invested and i the rate of interest after n years, the amount to be expected, A, is given by the general equation

$$A = P(1 + i)^n \tag{9.1}$$

We could now ask what sum of money invested now would give £A after n years by rearranging Equation 9.1:

$$P = \frac{A}{(1 + i)^n} \tag{9.2}$$

P is said to be the present value of the future amount A. In the jargon of economics A has been discounted at the rate of interest i to give P.

The process of discounting future sums is used to investigate the worth of alternative investment strategies. The process can be examined if we take as an example whether or not a manufacturing company with a sum £S should buy a new machine costing this amount. The machine will have a life of n years during which it will yield a stream of yearly incomes (or rents), a_1, a_2, a_3. Annual earnings cannot be compared one with another since they occur at different times, and are therefore subject to a different series of interest rates. To assess the total income expected, each annual sum must be discounted, which means that the present value (PV_m) of the stream of rents will be calculated as

$$PV_m = \frac{a_1}{(1+i)} + \frac{a_2}{(1+i)^2} + \ldots + \frac{a_m}{(1+i)^m}$$

$$PV_m = \sum_{n=1}^{m} \left[\frac{a_n}{(1+i)^n}\right] = \sum_{n=1}^{m} \frac{a_n}{(1+i)^n} \qquad (9.3)$$

The manager now compares the machine's PV_m from Equation 9.3 with the PV_B of £S invested in a bank. If PV_m is greater than PV_B then the manager will buy the machine and conversely if PV_m is less than PV_B. So the decision is based on comparison of the present values of the two strategies. If PV_B is greater than PV_m then the manager could opt to invest a smaller amount in the bank to get the same return (PV_m) as he would have got from buying the machine. This would save some money which he could use for other purposes!

Present Value and Fisheries

How do these ideas apply to fisheries? A fish stock is a renewable resource such that a part of it can be removed at regular intervals and sold to provide a stream of rentals. The remaining part of the stock is left to provide, through growth and recruitment, next year's yield. The fishery manager has to ask whether it is worth leaving a portion of the stock to provide an annual income or whether it would be better to catch all the fish now, sell the catch, and invest the money elsewhere? This is where the concept of present value enters, because by using it in the way we described for the machine example, the manager can compare the expected returns from the two strategies. As we will see the crucial point is the relationship between the rate of interest i in the financial markets and the growth rate r of the fish population.

In real life, decisions are never as simple as described so far. The expected stream of rentals is never secure, particularly in a fishery, and must be estimated. The interest rate does not remain constant so that present value fluctuates. Calculations should also take estimated inflation into account. All of these factors are difficult to estimate.

Clark shows how the present value of the fish stock can be split up into two portions, the expendable surplus (ES) and the conservable flow (CF). The conservable flow represents the part of the total fish production whose present value is worth more left where it is than if sold because it will generate a stream of future catches. The expendable surplus is the part whose present value is worth more when sold and the proceeds invested. Maximum present value is defined as:

$$MPV = \text{maximum}\,(ES + CF) \qquad (9.4)$$

and theory shows that this is the case when the rates of change with fish population size of expendable surplus and conservable flow are equal, i.e.

$$\text{marginal}\,ES = \text{marginal}\,CF \qquad (9.5)$$

Marginal conservable flow represents the value of an annuity which could be purchased on the strength of leaving one unit in the fish population. Marginal expendable surplus stands for the value gained by harvesting a unit from the population and investing the proceeds. Note that marginal values are the equivalent of rates.

Clark has developed functions relating marginal expendable surplus and marginal conservable flow to fish population size, using a surplus yield model to give the catches. He has shown that there are two cases depending on the relationship between the unit harvesting cost, b, and the unit price of the catch, p. The two cases are illustrated in Figure 9.4 where values are plotted against population size. One case leads to relative safety; the other can be dangerous for the fishery. In both graphs the marginal conservable flow is a straight line increasing with greater population size; the slope of this line is the rate at which future values are discounted. If this discount rate increases, the slope increases and this means that a given marginal conservable flow is attained at smaller stock sizes. The conservable flow changes more rapidly with increasing population size as discount rates increase; in Clark's model this change is linear. If the discount rate was zero, then the *mgCF* line in Figure 9.4 would be horizontal because Clark assumes that conservable flow would not change with stock size.

If $b < p$, Clark shows that the marginal ES forms a negative

Figure 9.4: Two Alternative Consequences of Maximisation of Present Value (MPV): b = unit harvest cost, p = unit price of fish, d = market discount rate, *mgES* = marginal expendable surplus, *mgCF* = marginal conservable flow, v = minimum viable population size. Further explanation in text.

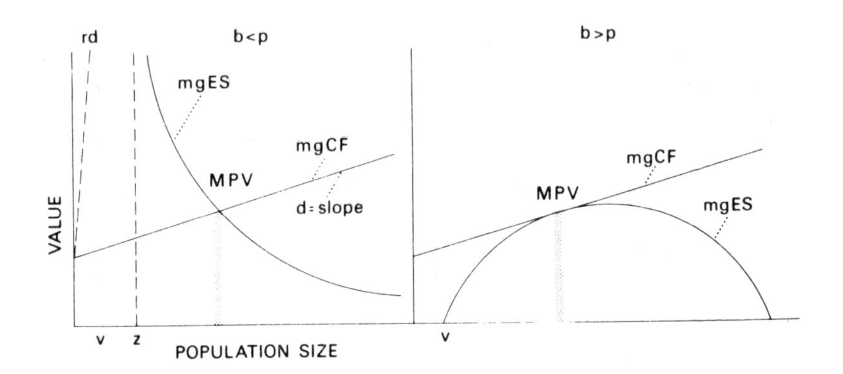

Source: From Clark (1973).

exponential curve declining with increasing stock size. The point labelled z represents the asymptote of this curve at the minimum stock size capable of producing any expendable surplus. Unlike the *mgES* curve drawn on the graph, the *ES* itself would be a curve rising to a plateau. The maximum present value (MPV) is the point where the two marginal values are equal (Equation 9.5 above) and where the lines cross on the graph, as indicated on Figure 9.4. If the discount rate is low, MPV tends towards MER, which would be represented by the intersection we would find if the marginal CF line was horizontal. As discount rates rise, the slope of the marginal CF line increases, and MPV occurs at smaller stock sizes. Provided that z is greater than v, which seems likely, the situation is safe and the stock will not be destroyed by economic factors unless rent dissipation occurs. Interestingly, Clark has pointed out that the individual fisherman in rent dissipation is effectively adopting an infinitely large discount rate. In other words he is 'living entirely for today', and evaluating future catches as worthless or nearly so. This is represented by the dotted line *rd* on Figure 9.4.

In the second type of situation, when $b > p$, the marginal ES curve becomes parabolic with the same shape as the surplus yield curve itself. When this occurs, the outcome depends entirely on the discount rate, as shown in Figure 9.4. If the discount rate is low, marginal CF has a low slope and MPV tends towards the MER as before. However, as the discount rate gets higher, there is nothing to stop the depletion of the stock to very low levels. For the Schaefer surplus yield model, MPV will favour the complete extermination of the stock once the discount rate is equal to twice the maximum reproductive rate of the species. At this point MPV is achieved by catching every last fish and investing the proceeds! Species with relatively few offspring are particularly vulnerable to this process. For example, blue whales have a reproductive rate of about 4 per cent per annum, and so an industry operating on MPV will tend to exterminate all the whales when interest rates rise to 8 per cent per annum, a relatively modest value in Western nations in recent years!

It should be emphasised that the MPV takes virtually no account of the biological properties of the stock, except in as far as predicted yields are used to calculate values. Since this particular analysis is based on the surplus yield model, it too must suffer from all the deficiencies we have described in Chapter 7. Clark (1976) has considered more realistic analyses based on the dynamic pool approach, but the mathematics involved goes beyond the scope of this book.

Although MPV and inflation accounting are elaborate tools of modern economics, they lead to a situation which is clearly absurd. No one can claim that a policy which deprives their grandchildren of the fishery is a rational one. Unfortunately, in a free economy, the MPV process seems to happen virtually automatically without it necessarily being a conscious aim of management. At this stage of the book we have reached an apparent impasse; at first sight neither the biological MSY nor sophisticated economics can provide an easily quantifiable goal towards which management of the fishery can be directed. Clearly, pure capitalist market forces cannot be allowed to destroy renewable natural food resources, but how far can management intervene in economics, and how much should political decisions be taken and legislation enacted to protect the stock when we have no quantitative tools with which to make predictions? We will return to this problem, of identifying the actual goal of management, in the final chapter – the problem is a general one affecting all natural resource management.

Economic Model of a Prawn Fishery

We conclude this chapter with an account of Clark and Kirkwood's (1979) analysis of the Gulf of Carpentaria (north Australia) prawn fishery as an example of a model which includes economic factors. The coastal stocks of banana prawns (*Penaeus marguiensis*) were discovered by a government resource survey in 1961 and have since developed into one of the largest Australian prawn fisheries, yielding between 3 and 9 kilotons per year, most of which is exported to Japan. The government can manage the stock by limiting access to the fishery through the licensing of boats, and by altering the length of the fishing season. Two different kinds of boats, 30 m long freezer trawlers and smaller 15 m brine trawlers, operate in the fishery for the prawns, which are easily caught because of their habit of schooling. The aim of the work was to determine the optimum number and combination of vessels in the fishing fleet which would give the maximum sustained economic rent. The model which was eventually used contains a number of simplifying assumptions which are worth describing in detail because they illustrate the kinds of problems which arise.

The freezer trawlers produce a quick-frozen valuable catch, can stay away at sea for long periods and travel long distances, whereas the brine boats' range and endurance is limited by capacity and the need to unload the prawns stored in their cooled brine vats. The brine vessels need to unload at a port equipped with processing factories, which represent a considerable capital investment (or a rental cost) for the owners, although shore-based plant is cheaper, more flexible and more secure than ship-based plant. However, freezer trawlers cost four times as much as brine trawlers and so increase the level of capitalisation of the fishery devoted to these stocks on which the owners/investors expect a return. Nevertheless, freezer trawlers have the option of travelling a long distance from their base to the Gulf fishery or to alternative fishing grounds, whereas the brine boats must remain close to the processing facilities during the fishing season. In this fishery the brine vessels were built for the East Queensland fishery and have to undertake the journey to the Gulf for the banana prawn season. Clark and Kirkwood estimate that owners will set out to do this if they expect a profit of about $8,000 (Australian). This represents the 'relocation opportunity cost' of this type of vessel in the fishery and, together with the costs of getting to the Gulf, is included in the fixed operating costs of brine trawlers in the model. In fact this value is not very precisely estimated and unfortunately makes a large difference to

Figure 9.5: Biomass and Revenue for the Gulf of Carpenteria Banana Prawn Fishery, Showing Start (t_{start}) and End (t_{end}) of the Fishing Season in Relation to the Unfished Biomass of the Stock and the Time at Which the Peak Biomass Would be Achieved.

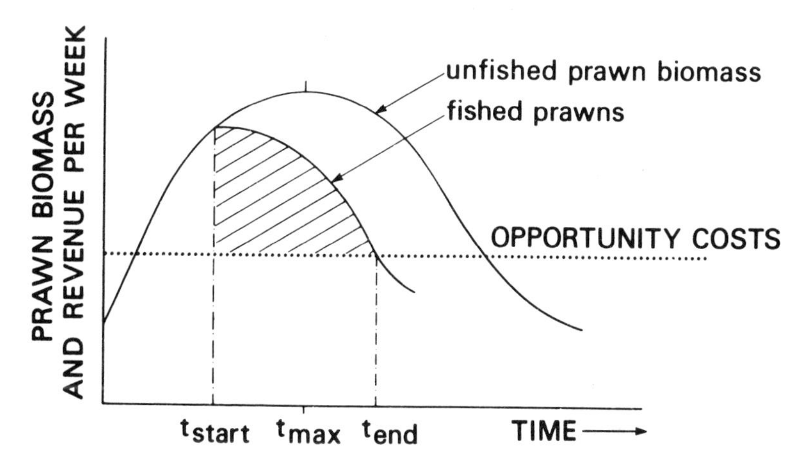

Source: After Clark and Kirkwood (1979).

the predictions of the model. The fixed costs of operating the freezer trawlers are much higher, and include a 10 per cent return on capital invested for the owners, again a figure which is difficult to estimate precisely.

The prawn stock reaches a maximum biomass in the summer of each year; biological parameters of the stock, including von Bertalanffy growth, are built into the model, although the geographically distinct regions in the fishery are subsumed into one variable characterising prawn population size. The authors show that the end of the fishing season should occur automatically once the returns from catching banana prawns drop below the opportunity costs of fishing on other stocks elsewhere, especially tiger and endeavour prawns to the north of the Gulf. On the other hand, the start of the fishing season must be set by legislation to gain maximum revenue, because otherwise fishing will begin once opportunity costs are exceeded and this will deplete the prawn stock too soon for maximum growth to be taken advantage of. The situation is illustrated in Figure 9.5 where the relationship of the start of the fishing season to the peak prawn biomass can be clearly seen. Clark shows that for a larger fleet the start of the fishing season should be delayed even more to gain the maximum yield. In theory,

Table 9.1: Variables Used in the Gulf of Carpentaria Banana Prawn Fishery Model of Clark and Kirkwood (1979). See text and Equation 9.6 for further explanation.

Variable	Freezer Trawlers		Brine Trawlers		
Economic Parameters	Symbol	Value	Symbol	Value	Units
Average catch rate per unit of effort, alt. stocks	y_f	1,750	y_b	875	kg/boat/week
Catchability coefficient	q_f	1.8×10^{-3}	q_b	1.4×10^{-3}	boat/week
Boat serviceability	s_f	0.5	s_b	0.85	–
Landed price of banana prawns	p_f	1.1	p_b	0.9	$/kg*
Landed price, alt. stocks	h_f	1.5	h_b	1.1	$/kg*
Operating costs	c_f	1,600	c_b	700	$/week*
Fixed costs	K_f	101,000	K_b	10,000	$/year*
Number of boats	Z_f	–	Z_b	–	–

Biological Parameters	Symbol	Value
Number of prawns in stock at time *t*	x_t	from fishery model
Average wt of a prawn, time *t*	W_t	from von Bertalanffy
Natural mortality rate of prawns	M	0.05/week
Average recruitment prawns	R	3.7×10^{8}/year
Von Bertalanffy parameters	k	0.08/week
	t_o	17.3 weeks
	W_∞	0.05 kg

Note: * Dollars are Australian.

the optimum solution could give different opening dates for the two types of ships, but this was not considered a politically viable option, and so was excluded from the model.

The model includes: biological parameters; technological parameters such as fishing effort, catchability and boat serviceability; and economic parameters such as the landed price of banana prawns and alternative stocks, as well as fixed and operating costs of the boats. The main variables used in the model are listed in Table 9.1 and the basic equation for net economic revenue of the prawn fishery, simplified from Clark and Kirkwood, is given below together with the meaning of each part of the equation:

$$V = \int_{t_{\text{start}}}^{t_{\text{end}}} [\underbrace{(p_f q_f s_f Z_f + p_b q_b s_b Z_b) x_t W_t}_{\text{(total income from banana prawns)}} \underbrace{- c_f s_f Z_f - c_b s_b Z_b}_{\text{(operating costs)}}$$

$$+ \underbrace{(h_f y_f - c_f) s_f Z_f + (h_b y_b - c_b) s_b Z_b}_{\text{(income from alternative stocks)}}] \underbrace{- K_f Z_f - K_b Z_b}_{\text{(fixed costs)}} \qquad (9.6)$$

This model is analysed using advanced optimisation mathematics to calculate the best fleet of fishing vessels. One surprising conclusion is that the optimal fleet will consist exclusively of either brine trawlers or freezer trawlers. The type of vessel in the optimal fleet depends on the value assumed for the relocation opportunity cost of the brine trawlers. When this cost exceeds $3,500, freezer trawlers are preferred over brine vessels. The results of the model when a value of $8,000 is used are illustrated in Figure 9.6, which shows the net income from the fishery for each combination of vessels in the fleet. The optimal fleet consists of just 24 freezer trawlers producing over $2 million revenue per year. This prediction represents a considerable reduction on the present fleet which contains 58 freezer trawlers and 87 brine boats. If the relocation opportunity cost is totally disregarded however, the model predicts that the optimal fleet would be composed of 350 brine trawlers, producing $2.5 million a year.

It is interesting to note on Figure 9.6 the situation predicted for an unlimited access fishery with its economic rent dissipated to zero by a fleet of 90 brine boats and 50 freezer trawlers, although this specific prediction applies to the fishery when its season is regulated for maximum prawn yield as explained above.

Clark and Kirkwood do not place great reliance on the exact predictions of the model. This is partly because of the high sensitivity to

Figure 9.6: Net Annual Economic Revenue Predicted by the Model of the Gulf of Carpenteria Banana Prawn Industry. The two axes refer to the two kinds of vessel in operating the fishery. For further discussion, see text.

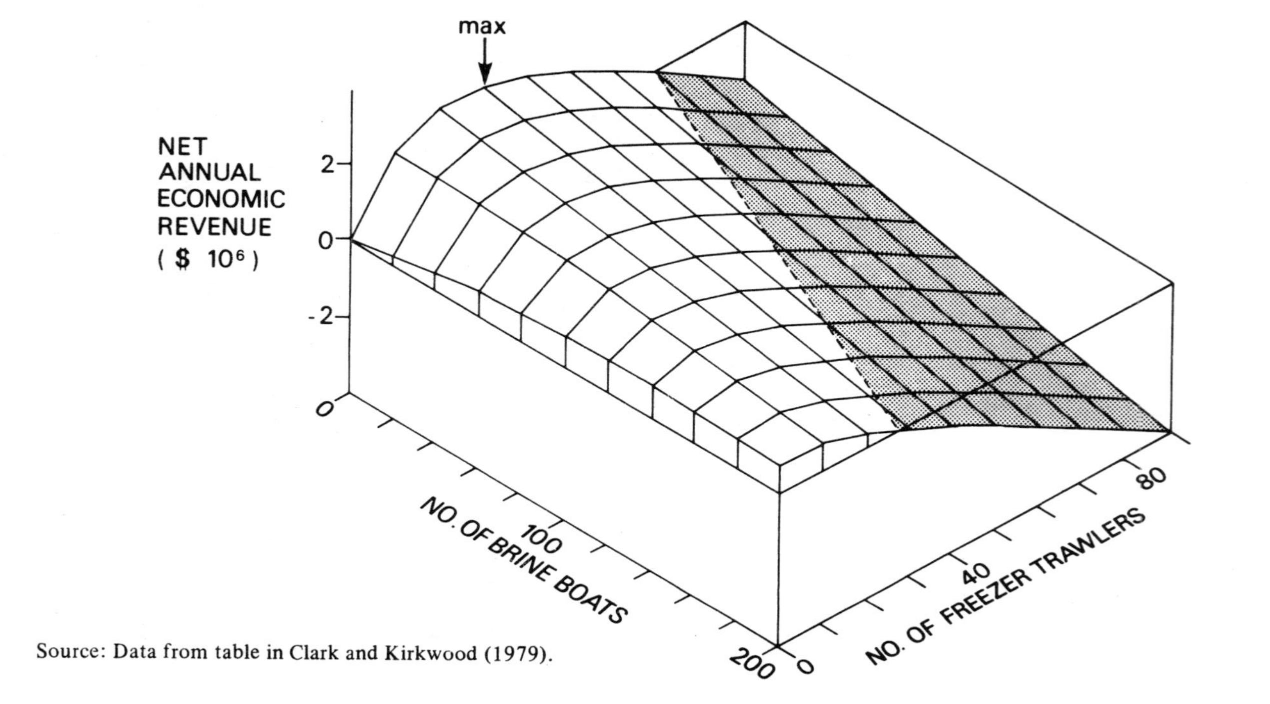

Source: Data from table in Clark and Kirkwood (1979).

the imprecisely estimated relocation costs, and partly because of the simplifying assumptions which had to be made. More modern vessels, smaller locally-based freezer trawlers, and brine boats which unload to freezer ships have altered the picture recently. The main point, however, is that it is a relatively easy task to modify the model to take account of these new factors and begin to use it in the actual management of the fishery. This would have been virtually impossible with a purely biological model.

One problem with the biological portion of the model was that only three years' data on recruitment were available. This was a fairly serious deficiency because the average value of 3.7×10^8 recruits had a coefficient of variation of 81 per cent, reflecting the high variability in the actual prawn recruitment from year to year. Although more accurate predictions are not possible until a longer run of data is available, one advantage of models like this is that they can be run repeatedly to see how sensitive each prediction is to changes in a particular variable, in this case the recruitment value. A 10 per cent error in recruitment brought about a roughly equivalent change in the predicted optimal catch of prawns and in the optimal fleet size, but generated a much larger 25 per cent change in the total net revenue. Since this is the parameter of most direct interest to the fishing industry, further data on prawn recruitment would be most valuable, and in fact this exercise sets an exact value on the further knowledge needed. Since prawn stocks are so variable, the most likely modification would be to make recruitment a normally-random variable, such as the white noise discussed in Chapters 7 and 8.

A number of factors were not built into the model at all. Some of these probably made little difference, such as the lack of a link between prawn stocks in successive years, the lack of geographical separation of the prawn sub-stocks, and ignoring the effects of fishing on the alternative stocks. Other economic factors missed out could alter things considerably though, perhaps especially the prediction that the entire fleet is always optimally allocated either to the banana prawns or to the alternative tiger and endeavour prawn stocks. This might make the number of ships of each type, Z_f and Z_b in Equation 9.6, a more complex variable. Some more realistic factors might be the crowding of vessels on the banana prawn grounds, the saturation of processing plant by brine trawlers unloading at the port, and the competitive incentive for individual boat owners to acquire even more sophisticated ships to capture a larger share of the catch for themselves. This latter factor increases the dangers of stock depletion and the dangers of overcapitalisation even in a regulated fishery, and is a strong argument for legislative control measures beyond merely limiting access to the resource, perhaps through the licensing of only certain types of vessel and equipment. Other ways of regulating the gear and boats used are discussed in Chapter 11. If we attempted to apply Clark and Kirkwood's model to fish stocks where the catch is sold on the open market at the time of landing, more sophisticated models of the relation of price to supply would have to be included. The interested reader is referred to Clark (1976, 1980), Anderson (1977) and Hannesson (1975) for an extensive discussion of these more advanced economic models.

In summary, in this chapter, we have shown that the attractive safety of the steady-state maximum economic rent is too naive to apply to real fisheries. Rent from the free-access fishery can be dissipated and the fish stock depleted through a seemingly inexorable process, which is itself exacerbated by evanescent gains to the individual fisherman through improvements in fishing technology. The expansion of fisheries and the harvesting of new stocks has often led to overcapitalisation in relation to the sustained yields of which the fishery is capable. Subsequent depletion of the stocks can lead to severe social and economic problems as boats and gear become unemployed. Such socioeconomic factors have rarely been incorporated in fishery management models. The theory of maximisation of present values demonstrates how even monopolies can be led to exterminate their natural resources, creating a dilemma as to what the exact aim of management should be. Despite the problems associated with some economic factors, contemporary models of real fisheries can easily include an economic dimension. Even the simplest economic model is useful in pinpointing the areas in which new knowledge would be valuable, and can often set a value on that knowledge. Mere limitation of access is not an adequate control with which to stem the economic tides; only a detailed and precise regulation appropriate to each resource is likely to ensure its future. It is to this network of control measures, based on the biology of the stock, that we will return in the final chapter.

10 FISH FARMING

People start fish farms to make money. The activity is actually rather a perilous way to do this and in this chapter we want to examine some of the biological constraints that influence the path to profit. Fish farming is no more than the commonsense application of biological principles to the business of rearing fish. As a result our narrative will mostly concentrate on general principles and the discussion of areas of fish biology where advances could enhance production. For detailed information on particular species and on methods see Bardach *et al.* (1972), Reay (1979) and Pillay and Dill (1979).

The biggest advantage that fish farming offers to the producer and consumer is security of supply in space and time. As we have emphasised in several chapters, capture fisheries can be influenced unpredictably by environmental events which make it hard for processors and retailers to guarantee supplies to the consumer at a steady price. Ideally, a fish farm run for profit should have a continuous stream of production unaffected by environmental events.

For the fish farmer to make money, the growth rate of the stock must be as fast and efficient as possible, producing maximum flesh from minimum inputs of food. The mortality rate must also be kept as low as possible, achieved mostly through the provision of the correct environment with absence of predators and organisms causing disease. How these aims are achieved varies with the species of fish being farmed, with the climate in which the farm is placed and the social system within which the farmer is living.

In the developed economies most fish farming has aimed at controlling as much as possible of the fish's life history. Ripe adults are spawned in captivity or stripped by hand, the eggs and larvae are reared in indoor tanks, fish are grown to market size in special containers and are fed artificial diets. Such a pattern is shown by salmon and trout farming in Europe and the United States. In many of the developing tropical countries, however, fish farming is part of a subsistence economy where much of the food produced is for home consumption rather than being sold as a cash crop. The farm might have a pond which will be stocked with small carp, perhaps gathered from a local river. The productivity of the pond might be boosted by adding pig or chicken manure, but nothing else is done to increase growth or minimise mortality. These two types of system, the technically-advanced and the simple, are termed *intensive* and *extensive* farming respectively and are at the opposite ends of a continuous scale. A second dimension can be added in the polyculture systems common in the tropics which attempt to farm more than one species in a pond. We will now briefly look at salmon farming in Scotland as an example of intensive farming, at milkfish farming in the Phillipines as an example of extensive farming, and at polyculture of carps in India.

Salmon Farming

Atlantic and Pacific salmon are now farmed under varying degrees of control in several European countries, in North America, the USSR and in Japan. Some details of freshwater rainbow trout farming are given later in this chapter, but marine farming of Atlantic salmon by Marine Harvest Ltd (Unilever) in Scotland will be described as an example.

The main features of the Scottish system are shown in Figure 10.1 (Glen, 1974; Webb, 1975). Very similar systems have been developed by Dømsea Farms Inc., Washington, USA (Lindbergh, 1979), and by Government Laboratories in Canada (Brett *et al.*, 1978) for farming coho and chinook salmon. The freshwater hatchery is at Invergarry about 70 km inland from Loch Ailort where the marine phase of the life history is spent. Many of the procedures for hatching eggs and raising alevins have been well known for some time. Care is taken to maintain strict hygiene and to reduce handling to a minimum; a good source of clean water is of course essential. The mortality rate from fertilisation to the point where the alevins begin to hatch is about 10 per cent in the indoor hatchery.

As for the larvae discussed in Chapter 6, the start of feeding is a critical phase for the alevins as the yolk sac is lost. The time at which food is first presented and the type of food are vital to achieve high survival. Once feeding is established, the formulated feed containing a careful balance of protein, fats, carbohydrates and minerals (Chapter 4) is presented frequently during daylight hours. During this final indoor stage mortality rates are between 10 and 25 per cent. After transfer to larger tanks outdoors, feeding is continued, but with a change in size of pellet to allow for the increase in the mouth size of the fish. At the beginning of the first summer at about nine months old the fish begin to change to the characteristically marked parr form. Mortality during the first outdoor phase is between 1 and 2 per cent.

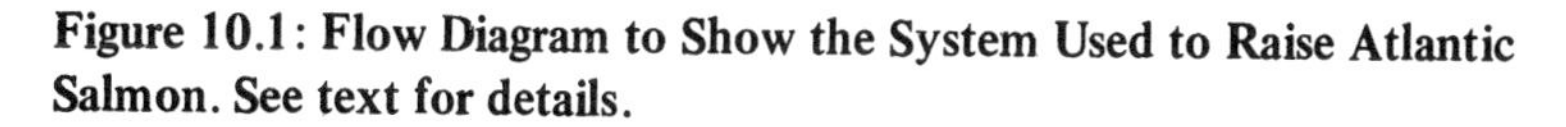

Figure 10.1: Flow Diagram to Show the System Used to Raise Atlantic Salmon. See text for details.

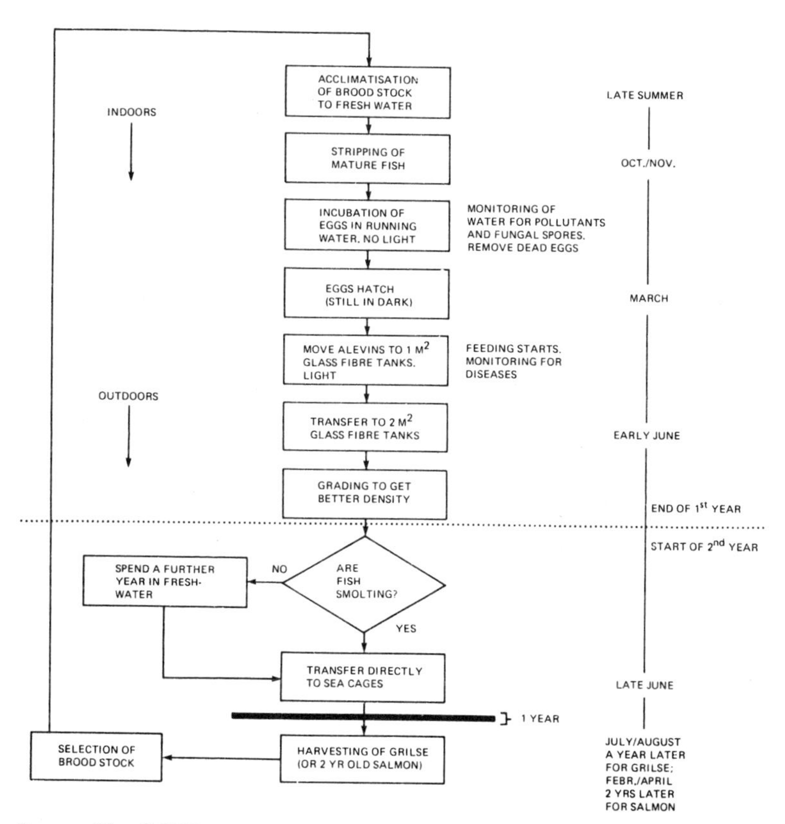

Source: Glen (1974).

Careful husbandry is required at all times to ensure that the water remains clean and plentiful and that hygiene is maintained.

During June, 18 months after hatching, some of the bigger parr begin to become silvery smolts with a higher metabolic rate and a general readiness to move into the sea. The rest remain as parr until the following spring. At the freshwater hatchery the first-year smolts are put into a separate tank and a loss of about 5 per cent is not abnormal at this stage. It was once thought necessary to acclimatise the smolts gradually to sea water, but direct transfer once fish are large enough produces satisfactory results.

The smolts are taken by road to Loch Ailort and put into floating pens at a density of about 2,000 fish per pen. The pens have a wooden frame with an expanded metal mesh netting and are grouped together to form a pontoon with walkways between; the latter are built on top of the steel floats which support the pens. Each pen is surrounded by a raised portion of netting to stop the salmon from jumping out. Protection from bird predation is given by wires stretched across the top of the pen and against seals by an outer protective net. The whole structure is about 4 m deep and is anchored firmly to the bottom.

Once in the sea pens, the smolts begin to grow fast, being fed with a pellet diet. During the first winter in the marine environment growth stops, but starts again in spring so that by late July, when the fish are harvested as grilse, they are between 2 and 3 kg. About 50 per cent of the fish are harvested at this stage, so that density in the pens is reduced to about 1,000. Continuous monitoring of the stocks in the pens is extremely important. Growth must be measured so that the correct amount of food can be given and fish density in the cages is regulated in accordance with measured oxygen concentration. Fish destined to become brood stock are individually tagged and husbanded.

Following a five-year study on the suitability of the five species of Pacific salmon for sea-pen culture in British Colombia, Brett *et al.* (1978) concluded that coho salmon was the best for marine farming. chinook salmon was ranked second, equal with rainbow trout, mainly because wild stocks infected with IHN (infectious haemopoietic necrosis) on the west coast could not be transported elsewhere, although Dømsea workers had some other problems with chinook. Pink and chum salmon, although exhibiting good growth, were unduly susceptible to a range of bacterial and viral diseases aggravated by gill damage from marine diatom spines. Paradoxically, the best species for commercial exploitation of wild stocks, the sockeye salmon, was the worst for farming. It showed poor growth in sea-pen culture and its gills were easily damaged by the barbed spines of the planktonic diatom, *Chaetoceros convolutus*.

A common problem with salmonid culture is the precocious sexual maturity of the males (see later in Chapter).

One attractive feature of salmonids is the pink flesh developed in many wild stocks when they feed on organisms such as crustaceans containing high levels of carotenes. Some wild stocks of salmon and most freshwater trout have white flesh. In farmed fish, pink flesh commands a premium price and can easily be developed by adding carotenoids to the diet for a few months prior to marketing.

Milkfish Culture in the Phillipines

The milk fish is a herbivorous silvery fish found in coastal waters throughout the Indo-Pacific region and is related to the carps. It enters estuaries, rivers and lakes mainly to feed using its gillrakers, and spawns in the sea. Individuals can grow to 180 cm but are commonly found up to 100 cm. The flesh, said to taste like herring, is high in protein (19–20 per cent) but is fatty and is best processed by canning or smoking rather than by freezing.

Milkfish are farmed mainly in the Phillipines, Indonesia and Taiwan. Methods of cultivation vary from the intensive to the extensive. In the latter approach the fish produced are mainly eaten on the farm, so providing a valuable protein supplement to the diet.

The milkfish fry are caught in estuaries, bays and off sandy beaches from March to June. The fry are caught using push nets, small trawls or traps. They are counted and stored in fresh or brackish water in wide-mouthed earthenware jars containing about 20 1 of water. The fry are then distributed, via a broker or wholesaler, to the milkfish farmer who will raise the fry to fingerlings or to marketable size.

On a milkfish farm the holding and rearing ponds vary from 10–40 ha and are usually 1.5–2.0 m deep with an earth bottom. The ponds need a good supply of fresh water or should be close to the sea if they are to contain brackish water. It should be possible to drain the ponds completely and the water table needs to be low enough to allow the emptied pond to dry out completely. Clay, clay loam or sandy clay are considered to be the best type of soil. The ponds are drained after use, the bottom being cultivated to control the pH of the water when the pond is refilled and to promote the release of nutrients into the water. The main application at this stage is of calcium carbonate, but the bottom of the pond can be fertilised to stimulate growth of the main food of the milkfish which is phytoplankton, microbenthic algae (lab-lab) and filamentous green algae (lumut). Both organic and inorganic fertiliser can be used depending on the nutrient supplement required. The various types of food need different water depths and salinity (Tang, 1979).

The pond can also be treated to kill disease-bearing organisms or species of fish which will compete with milkfish for food. Fish predators or competitors can be killed by using camellia (*Camellia sinensis*) seed meal, or the crushed ripe fruit of kanomary (*Biosyros multiflora*). Various types of pesticides are used to kill polychaete worms and snails which are parasite vectors.

Figure 10.2: Flow Diagram to Show the System Used in the Phillipines to Collect and Raise Milkfish.

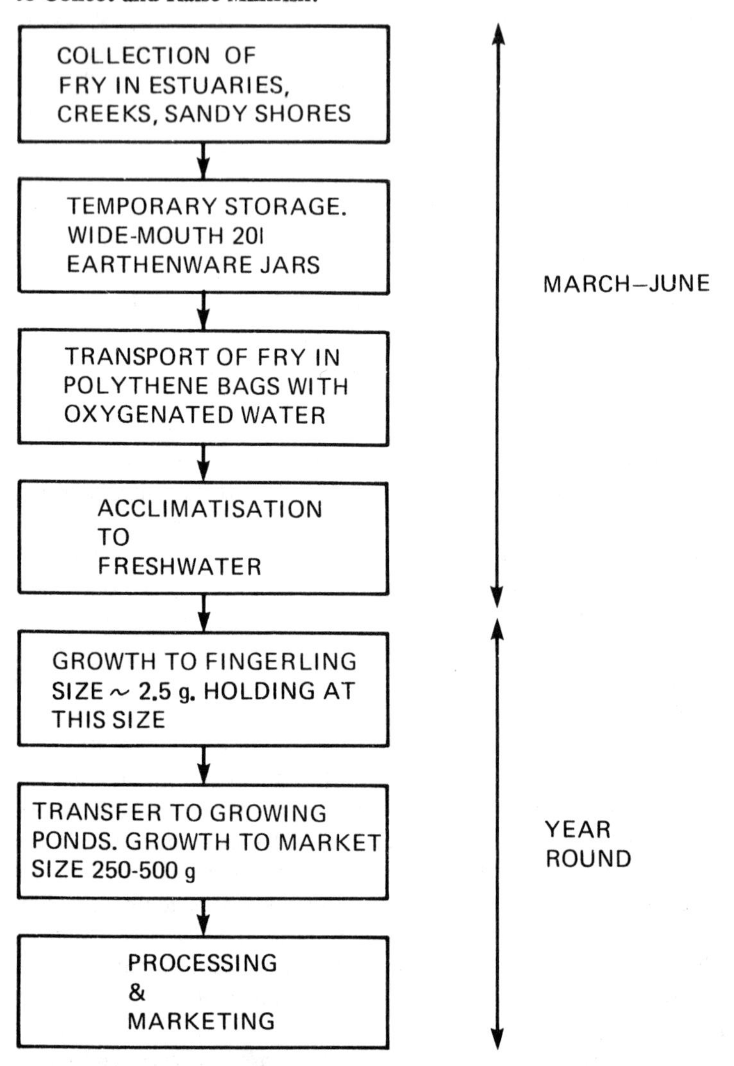

Milkfish farms can be divided into three types (Tang, 1979). Some just raise fry and grow them to the fingerling stage (about 2.5 g), while other farmers buy fingerlings and grow them on to market size which is 250–500 g. The remainder have an integrated system which raises market-sized fish from the fry stage as illustrated in Figure 10.2. When fry are received by the farm they are acclimatised for 24–28 hr to the salinity used in the fingerling and fry ponds; this is done in a small enclosed section of the fingerling rearing pond. Once acclimatised, the fry are grown in the fingerling pond until they are the right size for transferring to the growing-on ponds. Very often a farmer will operate a batch system, holding fingerlings back on a maintenance diet. At intervals, batches of fingerlings are removed and put into the growing-on ponds. By doing this the farm can produce up to eight crops a year, even though fry supply is seasonal. The fish feed on algal material produced in the pond and no supplementary foods are added. It is becoming increasingly common for supplementary crops of prawns, such as *Penaeus monodon*, to be grown in the milkfish ponds, usually in the most intensive farms.

Polyculture of Carps in India

Indian major carps form the basis of aquaculture in India (Chakrabarty *et al*., 1979). The principal species used are the catla, rohu and mrigal. Rearing these fish in association makes better use of the different niches in the pond and raises the production from each hectare. Recently three species of carp from China have been introduced into the system, trial results showing that yield can almost be doubled (Sinha, 1979). The Chinese carps introduced are the grass carp, the silver carp and the Chinese race of the common carp.

Management techniques have also been improved to increase the yields (Sinha, 1979). Amongst the improvements are an application of organic and inorganic fertiliser, the control of aquatic vegetation, eradication of pests and predators, the use of correct stocking densities, the provision of supplementary feed, and development of methods for the transport of larvae. A diagram showing the general approach to farming is shown in Figure 10.3.

The system shown in the figure assumes the most sophisticated arrangement possible. Under more primitive conditions the fry might be obtained from the natural environment, the traditional method (Chaudhuri and Tripathi, 1979). However, under advanced management, brood stock will be selected and injected with a carp pituitary extract when the fish have been brought close to spawning by environmental conditions such as changes in oxygen concentration and pH.

Dry weather can mean poor development of the gonads and hence such hypophysation is wasted. Assuming that the breeding fish are

Figure 10.3: Flow Diagram to Show the System Used in India to Raise Carp.

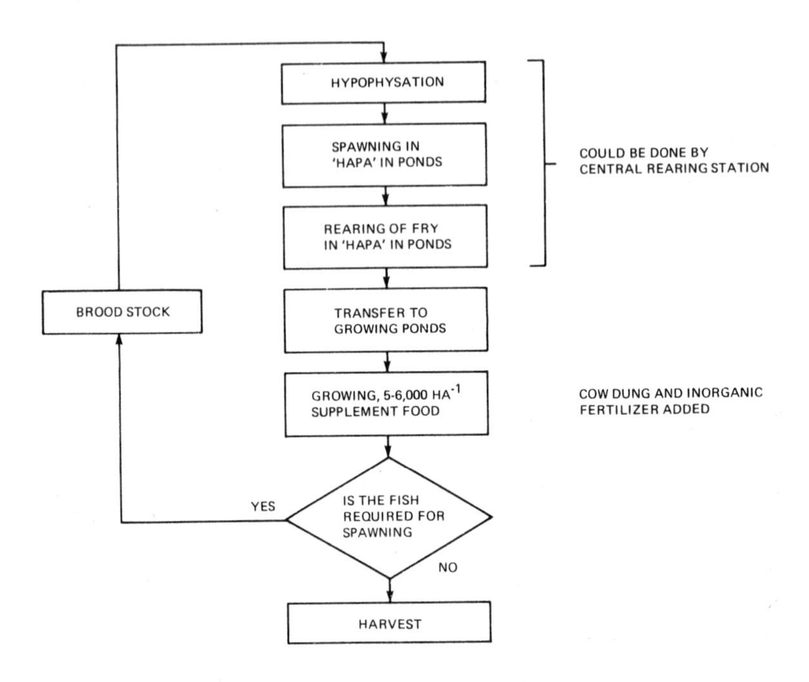

Source: Mainly from Chakrabarty *et al.* (1979).

ripe, they are injected with pituitary extract and put in pairs into a special enclosure called a hapa which is set up in the breeding pond. Once spawning has taken place, the brood fish are removed and the eggs transferred to a larger hapa where they can develop and hatch. This procedure is only successful with the Indian species of carp, as Chinese carp eggs must be in running water for successful development. Consequently it has been necessary to adopt a technique once used in salmonid culture whereby the eggs are reared in jars with a continuous flow of water. This makes it harder for smaller farmers to use the Chinese carps in their polyculture system. However, as indicated in Figure 10.3, it could be possible for a central rearing station to produce the fish fry which could then be bought by the individual farmer.

The young fish are put into earth ponds for further rearing. The ponds will have had all their unwanted fish removed using makua oil cake derived from the plant *Madhuka latufolia*. This is applied in the form of flakes and after a while it is naturally broken down and becomes harmless. Likewise predatory air-breathing aquatic insects are killed with a detergent mixture which is stirred into the pond water.

Once in the pond the fish are given supplementary feed. Chakrabarty *et al.* (1979) describe an experiment done to examine the productivity of a system with only the Indian major carps as farmed species. The fish were stocked at a density of 6,000 ha^{-1} at a ratio of 3 catla to 4 rohu to 3 mrigal. At stocking, the fish were between 102 and 262 mm long. The supplementary feed given was groundnut oil cake and de-oiled rice polish applied at a rate of between 8,900 and 9,900 kg $ha^{-1} yr^{-1}$. When Chinese grass carps are included in the culture, weed must be provided as food and the types used are *Hydrilla*, *Spirodella*, *Lemna*, *Azolla*, *Ceratophyllum* and *Najas* (Jhingran, 1976). The natural productivity of the ponds can also be increased by adding both organic and inorganic fertilisers.

The yields obtained from the Indian system of carp polyculture vary considerably. The experiments of Chakrabarty *et al.* (1979), where only the three Indian carps were cultured, produced gross yields of 3–4 tons per hectare per year. Jhingran (1976) reports gross annual yields of well over 6 tons per hectare for polyculture with Indian carps, silver carp, grass carp and common carp. However, most work in India with varying combinations of carps has achieved a mean gross annual yield of 4.4 ± 1.9 tons per hectare. Culture of three Indian major carps together with common carp yielded a gross mean annual yield of 3 ± 0.6 tons per hectare (Jhingran, 1976).

Areas of Fish Biology of Special Interest to Aquaculture

These three examples serve to illustrate the range of fish farming techniques that can be found. The critical activities in any farmed system are the cheap and continuous production of young, prevention of disease, and the maximisation of both the rate and efficiency of growth which are both influenced by the size and composition of the diet. Diet and growth dynamics have been discussed in Chapter 4. Most of the work on diet is aimed at providing the knowledge necessary for the formulation of an artificial dry diet. A diet in this form is easy to store and dispense and can be made from standard ingredients which can be combined using the principle of least cost. Much experimental work, on flatfish for example, is done using a diet composed of ground-up trash fish (small fish that are unsuitable for direct human consumption) or molluscs. This can be suitable for trials, but is usually uneconomic for full-scale farming. One exception is the production of the Japanese yellowtail which is fed ground-up fish such as sandeel. Since to produce 90,000–100,000 tons of yellowtail it is necessary to catch and process 500,000 tons of trash fish, this is an expensive activity and only feasible because yellowtail fetch a high price. Artificial diets aim to give growth which is as good as that on natural diets, but the farmer must also know the critical relationship between diet size and growth rate. Brett *et al.* (1969) and Elliott (1979) have shown that salmonids require a certain combination of temperature and ration size to produce an optimum growth rate, and this is probably true of other species too (see Chapter 4).

Diseases of Farmed Fish

Disease is an enemy the fish farmer must always be watchful for. The conditions in intensive aquaculture are often favourable to the spread of diseases. Fish are in close proximity and this can stress the fish in itself, but it can also cause low oxygen tension, can change the pH and can increase organic material which in turn can compete with the fish for oxygen.

Other factors such as a poorly-balanced diet and excessive handling can also induce stress which makes the fish much more susceptible to infection. Over the past 15–20 years research has produced knowledge about the characteristics of the important fish diseases, about the causative agents and about methods of treatment. Present intensive methods of aquaculture would be impossible without this knowledge, but there is still a great deal to learn. The fish farmer needs to know how to diagnose specific diseases, what methods he can use to cure infected fish and, most important, what techniques to use to prevent the onset and spread of disease. Diseased fish cause direct and obvious loss through death, but great loss also occurs when disease causes reduced growth, fecundity and stamina, and increases the vulnerability of the fish to predation.

Diseases are caused by five main groups of organisms (Sarig, 1979): ectoparasites and fungi, endoparasites, bacteria, viruses and organisms which produce toxins in the water which subsequently cause fish death. Treatment, where possible, takes two forms: either the pond water is treated with a pesticide or the fish are given food which contains a drug. Viral diseases cannot be cured and the only measures possible are preventive. We will here only give a short outline of some of the major diseases and the reader is referred to Roberts (1978) for further discussion.

Parasites and fungi cause the largest number of fish diseases. Most such organisms exist in equilibrium with wild fish populations but artificially-dense populations in a fish farm can introduce factors which upset the equilibrium and cause the parasite or fungus to become pathogenic. A common disorder in salmonids caused by a parasite is whirling disease caused by the sporozoan *Myxosoma cerebralis*. The protozoan infects young fish only and can cause high mortality. The spores of the parasite live in the pond mud and can last there for 10–15 years. Outbreaks of the disease can be prevented by keeping young fish in concrete or fibreglass tanks and ensuring that spore-free water, obtained by irradiating the water with ultraviolet light, is used. Many other parasites are treated by pesticides added to the water.

Bacterial diseases are more common in cold-water species of fish. They are most often cured or prevented by using medicated food. Of the eight groups of bacteria identified by Sarig (1979), three are common causes of loss in fish farms. In Europe a serious disease of salmonids and cyprinids is furunculosis caused by *Aeromonas salmonicida*. The disease takes hold as water temperatures rise and is transmitted by diseased individuals, by eggs and on contaminated equipment. The treatment is to add sulphonamide antibiotics or nitrofurans to the food. Good hygiene is the best preventative.

Bacterial kidney disease is possibly caused by Corynbacteria although the diagnosis and method of transmission is not yet clea-. There is no effective control or treatment.

The myxobacteria cause a wide range of diseases such as fin rot, tail rot, columnaris, cold water disease, peduncle disease and gill disease. These ailments are aggravated by crowding, sub-optimal temperatures and poor diet, and treatment is by adding sulphonamides to the diet. In many European countries it is not permitted to sell for slaughter livestock that have recently been treated with antibiotics. It is usual practice to stop the treatment some weeks before slaughter.

There are eight important fish diseases caused by viruses. Some of the more important diseases are: infectious haematopoietic necrosis (IHN) attacking salmonids; infectious pancreatic necrosis (IPN); a very contagious disease of salmonids; viral haemorrhagic septicaemia (VHS); ulcerative dermal necrosis (UDN) which attacks adult salmon and sea trout; and finally spring virema of carp (SVC) caused by *Rhabdovirus carpio*. The incidence of these diseases must be kept low by preventive measures, such as strict handling practices and the use of only disease-free stock.

Advances in aquaculture must be accompanied by better knowledge of fish diseases. This must include suitable treatments so that methods specific to fish can be developed. An example is the treatment of fish which have caught vibrio disease, caused by the bacterium *Vibrio anguillarum*. The best method of control is to treat the fish with an antigen, but it is too expensive to give each fish an injection. The method of hyperosmotic infiltration overcomes this problem (Antipa and Amend, 1977). The fish are first dehydrated in a concentrated salt bath; they are then transferred to a second bath with ordinary water which contains the antigen in solution. The compound is then absorbed across the gills as the fish replace lost water. Specially-designed techniques of treatment will become more important, but the need for good husbandry will never be replaced. As in conventional farming, it is the key to healthy fish and a profitable farm.

Control of Reproduction

At several points we have emphasised the benefits of obtaining continuous production from a fish farm. One of the requirements for this is a continuous supply of eggs or young fish. All temperate fish species and also many from tropical waters reproduce seasonally. The hormonally-mediated processes of vitellogenesis, gonad maturation and spawning are triggered and perhaps controlled by environmental and behavioural cues. We need to understand this to provide a year-round supply of eggs.

The teleost reproductive cycle often reaches a peak of activity for a limited time each year. This peak is the spawning period when the eggs and sperm are released and fertilisation occurs. The spawning period is usually of fixed duration and occurs at a definite time each year (see Chapter 5). The timing is adaptive in that it ensures that young are produced when conditions are most conducive to survival. To the casual observer, the peak of reproductive activity seems to be the critical event, but the attention it attracts obscures the fact that all phases of the cycle are synchronised with environmental events and must occur in the correct order (Scott, 1979). These earlier phases, such as vitellogenesis, are also influenced by environmental effects. In aquaculture, fish are often left to complete vitellogenesis under the influence of natural stimuli. Hormone injections (hypophysation) are then used to bring on maturation and ovulation.

The Indian major carps mentioned above were injected with pituitary extracts to bring them to the point where they could be stripped. It is not always easy to estimate the optimal time when the injections should be given; mistakes can lead to the production of poor-quality eggs having a low survival rate. Consequently a study of the phenomena that occur in the ovarian follicles after stimulation by a gonadotropic hormone (GTH) can lead to better methods of intervention (Jalabert, 1976). It is also important to study which environmental factors stimulate gonadal maturation. Once the environmental cues and the internal control system are understood, the fish farmer of the future will be able to achieve a steady egg production all year round. Before a consideration of some of these factors, we must take a brief look at the reproductive process as it effects the gonad.

The ovaries of fish undergo a cyclic and generally annual change. The first stage in the production of eggs is oogenesis, during which the oogonia multiply to produce the primary oocytes. These undergo two phases of growth; the first is non-vitellogenic involving no yolk production and the second is vitellogenic, meaning that yolk is synthesised. Primary oocytes which are in the first meiotic phase can remain quiescent for some time. Maturation occurs when the primary oocyte completes the first meiotic division to become a secondary oocyte, which then passes through the second meiotic division to become an ovum. Two polar bodies are produced in this process. At ovulation, the egg is released from the follicle. Vitellogenic oocytes are sensitive to hormones and changes in these can cause the egg to be reabsorbed, a stage called atresia

In males, 11-keto-testosterone triggers primary spermatogonia in the seminiferous tubules to divide, producing secondary spermatogonia which are found in clusters enclosed in a cyst. This male hormone is characteristic of teleosts. The secondary spermatogonia divide synchronously and mitotically to produce large numbers of primary spermatocytes. These then undergo meiotic division to give secondary spermatocytes. A final meiotic division yields haploid spermatids. The cyst containing these then bursts and the spermatids metamorphose into spermatozoa in the lumen of the testis tubules. Maturation of males is more reliably achieved and so the rest of this discussion will concentrate on the control of the ovary, as this is a process more critical to the success of a fish farm.

Jalabert (1976) summarises some of the results that have been obtained from *in vitro* studies of maturation. These have shown that gonadotropic action is probably mediated by direct steroid action on oocyte maturation in the following species: *Misgurnus fossilis* (a loach), the sturgeon, the Indian catfish, the medaka, the rainbow trout, the

Figure 10.4: A Flow Diagram to Show the Interrelationships Between Endocrine Organs in the Control of Ovarian Development and Ovulation in Fish. The arrows show the point at which the hormones act. (+) indicates a stimulatory response while (–) indicates inhibition.

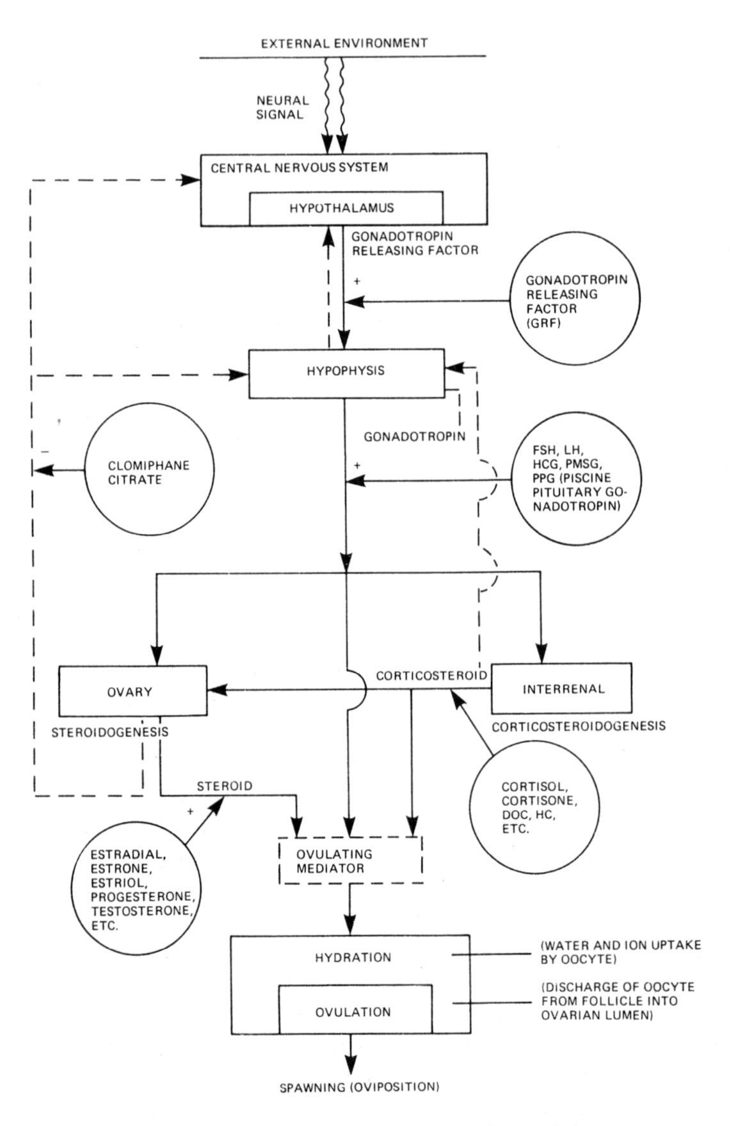

Source: Redrawn from Kuo and Nash (1975), with permission.

goldfish and the pike. However, there are contradictions as to the nature of the active steroids, their place of production and the link between *in vitro* maturation and ovulation. In the Indian catfish it has been shown that the route from the pituitary to the gonad is indirect, via the interrenal gland. In other species, pituitary hormones produce a direct stimulation of the gonad.

Jalabert (1976) gives details of experiments done with oocytes of rainbow trout, goldfish and pike to determine how oocyte maturation is stimulated and how ovulation is instigated. This and other work (e.g. Kuo and Nash, 1975), has shown that the control system for oocyte maturation could look as shown in Figure 10.4. Some sort of environmental event triggers the pituitary to produce gonadotropin; this binds to receptors on the follicle, stimulating mRNA production. The mRNA then synthesises steroids in two phases; specific enzymes for steroid synthesis are produced first followed by steroid synthesis itself, the most active one being 17α-hydroxy-20β-dihydroprogesterone (abbreviated to 17α-20βPg). This steroid binds to steroid receptors on the oocyte which cause protein synthesis and egg maturation. As shown in the figure, a second subsidiary pathway could operate with ACTH stimulating the interrenal gland to produce steroids. The sequence of events in this system has been built up from *in vitro* studies and it clearly needs to be established in the whole organism. This type of detailed knowledge makes it possible to intervene more specifically in the system. The injection of gonadotropin triggers the system from the top down; injecting other hormones and steroids could give a much more controlled and specific response.

Experiments in which gonadotropin was injected together with other steroids (Jalabert, 1977) have good results, whereas injections of one substance or the other alone had a weak effect. Kuo and Nash (1975) discuss the use of various mammalian and fish hormones and steroids to induce ovulation in grey mullet. They concluded that although mammalian hormones induce ovulation and eventual spawning, the best results were achieved with partially-purified salmon pituitary gonadotropin. Often, this was needed to prime the fish when other hormones were being used. Steroids used alone were not very effective either *in vitro* or *in vivo*. Therefore for practical application Kuo and Nash recommend the use of salmon gonadotropin.

We have seen that the process leading to the maturation of oocytes is probably initiated by environmental cues. Rather than use hormone injections to trigger spawning as is now practised, it might be more economic and biologically suitable to initiate maturation and spawning by changing the light or temperature required to initiate maturation.

Whether photoperiod or temperature is most important in maturation varies from one species to another (Kuo *et al*., 1974; Scott, 1979). Photoperiod has been shown to be dominant in the banded sunfish, the shiner (*Notropis bifrenatus*), the killifish, stickleback, brook trout and medaka. However, other experiments have demonstrated that temperature is more important in killifish, the European minnow, the four-spined stickleback, mosquito fish and the lake chub. As some species occur in both lists, it must be concluded that at least in these, both factors act together and it has been shown, with brook trout for example, that photoperiod has a different effect at different states of gonad development.

Kuo *et al*. (1974) studied the influence of temperature and photoperiod on the maturation of oocytes in the grey mullet. In Hawaii, where the work was done, the grey mullet normally spawn in the winter months of November to March when days are shorter than nights and water temperatures are cooler. A photoperiod of 6-hours light to 18-hours dark was chosen to simulate conditions in the spawning period. The series of experiments showed that with the chosen dark-light regime vitellogenic oocytes appeared approximately eight weeks after the cycle began. Once maturation has been triggered, its speed was unevenly proportional to temperature with more rapid development at 17 and 21°C than at 26°C. Examination of the mature oocytes showed that only at 21°C were normal viable oocytes produced so that this appeared to be the optimum temperature. Even with this knowledge it was still necessary to inject gonadotropin to bring about egg release.

We have emphasised the problems concerned with the initiation of oocyte maturation and spawning in captive fish, but it should not be forgotten that several species spawn spontaneously in captivity and at most require manipulation of the photoperiod. Turbot spawn in rearing units and photoperiod manipulation now makes it possible to obtain continuous egg production by exploiting stocks in different localities.

These examples demonstrate the depth of physiological knowledge that is necessary if reproduction in farmed fish is to be controlled. Ultimately, it is most likely that a combination of hormone injection and environmental manipulation will be used to achieve continuous egg production.

Breeding of New Strains

A fish farmer wants his stock to reach marketable size quickly and to eat as little food as possible in the process. It is also desirable to get as many eggs as possible from a small breeding stock. At present some farming operations use wild fish and consequently the farmer can only control performance in growth, food conversion and egg production by manipulating diet and environment in an empirical way. The most satisfactory approach would be to breed special stocks of each farmed species so that optimum performance could be obtained, combined of course with the correct diet and environmental conditions. As reported by Purdom (1974a) and Moav (1979), breeding in fish has been common practice for many years, but the aims have usually been to produce ornamental types, particularly of goldfish and carp. The process yielding these types has been to select for particular visible characters such as reduced scale number in carp. The features of interest to fish farmers are usually quantitative characters (see Chapter 5) which can only be assessed by measurement of large numbers of individuals. Examples are growth rates and food conversion efficiency. These characters are controlled by many genes working together. Consequently selection for growth rate is often not as straightforward as selection for fin colour or scale number, which can be controlled by just one gene.

The methods open to the geneticist to produce suitable species for farming are selection, inbreeding and heterosis, hybridisation, sex control, parthenogenesis and induced polyploidy. We will briefly discuss each in turn.

Selection can be done in two ways. The first is called mass selection, by which the experimenter chooses individuals for breeding by virtue of their possession of the desired characteristic. This is done irrespective of the family history of the individuals. The second method is to choose individuals who come from a family possessing the desired characteristics. The mass selection method is good when the character is easy to measure and has high heritability (see Chapter 5). Where heritability is lower it is often better to use family selection and to control closely environmental variability.

The fish breeder can select for resistance to disease, improved growth rate, better food conversion, behavioural characteristics that make the fish more amenable to farming, ratio of flesh to bone, reduction of the number of intramuscular bones, maturity times and seasons, egg size and egg viability. The results of work to select for certain features have not always produced consistent results; for example, Moav and Wohlfarth (see Moav, 1979) carried out a mass-selection programme on common carp to improve growth rate. No response was achieved after five generations even though a parallel programme, selecting for slower growth, had had a strong result. Further work with the line taken for fast growth showed that the stock still contained considerable variability for growth rate and when a programme of family selection was carried out, improved growth resulted (see also Chapter 5).

Family breeding can lead to pure-bred lines which show great uniformity for certain characters. This means that one can be sure that the desired characters are always present after each turn of the breeding cycle. An undesirable side effect is that the inbred pure line often loses vigour and fitness. For example, in rainbow trout a recessive allele affecting the pituitary which results in stunted growth, tends to accumulate in inbred lines. The affected fish, called 'cobalts', can be recognised by their lack of spots and irridescent blue colour. One can overcome this by using a technique much used in chicken rearing and the production of new vegetable varieties! The breeder maintains two lines, pure for the same characters; the lines are then crossed and the result is F_1 offspring with the same characters as before but with increased vigour. This technique is known as reciprocal recurrence selection, and since the selection of the pure-bred lines is on the basis of the desirability of their F_1 hybrids and not their own phenotypes, it is an example of selection for variance resulting from interactions between genes rather than from additive effects (see Chapter 5). Purdom (1974a) considers this technique as being one of the most promising ways of producing changes in fish in characters such as growth rate. F_1 hybrids have to be bred anew for each stock and so specialist fish breeders would have to supply F_1 hybrids for sale to fish farmers.

The principle displayed by the method of inbreeding two pure-bred lines can be used to produce hybrids between two closely related species. This process is less successful than is inbreeding, as many interspecific hybrids are weak animals. Usually interspecific hybrids have characters which are intermediate between those of their parents, and the phenotypes are uniform. When hybridisation works, the resulting offspring can have greater vigour than either of the parent species. An example is the hybrid between *Tilapia nilotica* and *T. mossambica* which shows faster growth and better food conversion. The breeder can also use hybridisation for other purposes. By introducing into a farmed species certain desirable genes from another, one might improve undesirable or unwanted characters. An example might be transference of a gene for resistance to a certain disease from wild stock to the farmed stock. A more specific promise of possible advantages from hybridisation comes from the Pacific salmon – the sockeye salmon and the pink salmon in particular. Sockeye salmon are good for canning but they mature late. Conversely the pink salmon is poor for processing and spawns early. It was proposed by Foerster (1968) that crossing of these two species could produce a variety which would have good flesh and early maturity. In the USSR several hybrids have been produced which have been successfully employed in fish farming. A cross between the mighty migratory Caspian sturgeon and the small freshwater sterlet has provided a fast-growing hybrid which can tolerate farm conditions; pituitary injections are used to hasten maturation for caviar production. Hybrid carp tolerant of colder northern conditions have also been produced. In the UK plaice (*f*)/flounder (*m*) hybrids and turbot (*m*)/brill (*f*) hybrids have shown good growth rates and other features of promise for flatfish farming. The flounder/plaice hybrid is naturally common, especially in the Baltic. In these interspecific hybrids care has to be taken to make sure that the sexes used are the right way around as the reciprocal hybrids are often not viable.

A further use of a hybridising programme is to increase the variability available to the breeder for further selection work. As mentioned, the first generation of hybrids will be uniform but the second will show greatly increased variability. This can be used as the raw material for continued selection for desired traits.

In many species of fish the sexes grow at different rates. Very often the mean size of females at a given age is greater than an equivalent mean size for males and they can have a better condition factor. Consequently it could be an advantage to have a stock of single-sex fish for farming. Single-sex fish would also prevent unwanted breeding. The sex of fish is under rather loose genetic control in teleosts (Purdom, 1979); many species can have their functional sex altered by hormone treatment, especially during the early part of their lives. Fish treated in this way still possess the genetic make-up corresponding to their original sex; this means that they can be used in a breeding programme aimed at producing offspring all of one sex. It is the homogametic sex which is of interest in these programmes. If, for example, the female is homogametic then it will be fertilised by females that have been converted to functional males. The offspring will all be homogametic. These methods are currently being used to produce all-female lines of trout and all-male lines of tilapia.

The methods discussed so far have used normal breeding methods to achieve the required results. Induced parthenogenesis and polyploidy are ways of altering the genetic make-up of a species by experimental treatment during development. This occurs frequently in nature in several animal groups including some species of fish. In the Amazon molly most populations are nearly all females with an insignificant proportion of males. Egg development is stimulated when sperm from related species try unsuccessfully to fertilise the eggs. Parthenogenesis, or gynogenesis as it is also called, induced artificially, results in fish with a haploid genotype unless treated further. These haploid individuals do not develop normally so that it is necessary to give further treatment to produce a diploid condition. The sequence of events is usually as follows. Sperm of the desired species are de-activated with ionising radiation. These sperm will stimulate the eggs to divide, but contribute no genetic material. To bring about a diploid offspring the developing egg is subjected to a cold shock during a specific stage in the cell division process, which causes fusion of the haploid egg with its own second polar body. This results in a diploid offspring which, genetically, contains about 60 per cent of its mother's genotype. In this way it is possible to get pure-bred lines in far fewer generations than the usual 15–20 (Purdom, 1974a). Using this technique the fish breeder can maintain the purity of a successful stock. The technique has been used successfully in plaice, carp, turbot and trout.

By applying a cold shock to normally-fertilised eggs, the number of chromosomes can be increased beyond the normal diploid state. Such fish are often sterile, which is a great advantage to the farmer; food is not wasted in the formation of gonads. Also, the polyploid fish grows faster than normal fish. Triploid plaice/flounder, turbot/brill and carp hybrids have been produced and are being introduced to fish farms. Unfortunately experiments with trout have failed to date. In salmonids sterility would be particularly valuable as fish in this family use much energy defending territories before spawning and virtually cease growing.

The Energy Subsidy Required for Intensive Aquaculture

Although fish farming is started to make money, insight may be gained about the long-term viability and economics of farming by constructing an industrial energy budget. Blaxter (1975) distinguished between biological and industrial energy budgets; biological energetics describes the efficiencies with which solar radiation is converted by the metabolism of animals and plants, whereas industrial energetics estimates and

Table 10.1: Annual Industrial Energy Budget for a 25 ton/yr Rainbow Trout Farm.

		Energy Equivalent GJ/yr
Farm Inputs		
Eggs	500,000 eyed ova from hatchery at 44.97 GJ/yr	44.9
Labour	553 man-days on main farm at 7 MJ per working day	3.8
Maintenance	£100 per year at 162 MJ/£	16.2
Treatment	prophylactics and disease treatments, £20 at 162 MJ/£	3.2
Processes	sorting, grading, etc., small tractor for average ¼ hr per day on 240 days = 60 hrs at 188.7 MJ/hr	11.3
Transport	delivery of food, 50 tons in 5 batches at 290 MJ/ton	14.5
	delivery of fish, 25 tons in 5 batches at 362 MJ/ton	9.0
	Sub-total – non-food inputs	103.1
Fish food	50 tons pellet food at 25.7 MJ/kg	1,291.4
Total Input Energy to Farm		1,394.6
Farm Output		
Gross	25 tons wet-weight rainbow trout at 6.477 MJ/kg	161.9
Gross edible	maximum 60% of wet-weight 15 tons at 8.165 MJ/kg	122.5
Protein	22% of gross edible 3.3 tons protein at 17.376 MJ/kg	57.3

Source: Pitcher (1977).

apportions the energy which has to be expended by man in order to carry out the various stages of the food-producing process. Much of this energy derives from fossil fuels, which have to be paid for and which, in the long term, man must conserve. In the energy budget, the total energy put into the system is compared with nutritionally useful output in the form of energy represented by the food produced, or with the energy equivalent of the protein in the food. The ratio of input to output describes the energetic efficiency and measures the subsidy of energy inherent in a particular kind of food production (Leach, 1975).

As an example we will examine an energy budget for a rainbow trout

Table 10.2: Energy Loading on Food Pellets for Rainbow Trout. Budget for 1 kg of 'Growers No. 1' Pellet Food. Details of the Calculations of the Unit Costs are given in Pitcher (1977).

Constituent	Unit Cost (MJ/kg)	Proportion in Pellet	Energy Cost (MJ)
South African fishmeal	69.0	0.263	18.1
Whole dried milk	80.0	0.050	4.0
Wheatmeal middlings	4.3	0.300	1.0
Salt, mineral mix and vitamins	5.0	0.237	1.0
Meat and bone meal	1.0	0.150	0.1
Processing	1.0	–	1.0
Total		1.000	25.2

Table 10.3: Energetic Efficiencies for a Rainbow Trout Farm, from an Industrial Energy Budget.

Gross output/energy input	0.116
Gross edible output/energy input	0.088
Protein energy output/energy input	0.041
Energy subsidy per kg protein output	422.6 MJ/kg
Energy subsidy per kcal protein output	24.3 MJ input per MJ protein output
Energy subsidy per kg wet-weight fish produced	55.8 MJ/kg

farm marketing 25 tons of fish per year (Pitcher, 1977). The trout are held in running-water ponds and tanks, size-graded, fed on dry pellets containing 40 per cent crude protein, and are brought to sale in 15 months. The overall energy budget for the two-man, 1 ha farm is given in Table 10.1. Full details of the assumptions and calculations in drawing up the budget are discussed in the original paper, together with an energy budget for operating the hatchery which supplies the trout eggs for the farm. The separate budget for the food pellets is given in Table 10.2 and the efficiencies of the farm outputs in Table 10.3.

The results show that there is an enormous energy subsidy involved in trout farming; each kilogram of fresh trout has cost the energy equivalent of an equal weight of petroleum to produce. Only one-tenth of the input is converted into fish. On the other hand, in terms of protein production, the farm's energy subsidy of 423 MJ/kg of edible protein compares favourably with the average for UK marine fisheries of 489 MJ/kg (Leach, 1975), as does the wet-weight subsidy of 55 MJ/kg wet-weight produced. The fish food is responsible for over 90 per cent of the farm's total energy costs, much larger than the 30–50 per cent regarded as typical of terrestrial farming in temperate climates. The next largest single item is the hatchery (at 3 per cent based either on the farm or through purchase of eggs from a large specialist supplier. Transport, maintenance and other essential activities consume negligible amounts of energy compared to the food, a large contrast to the financial budget where labour costs and transport are significant.

The fishmeal incorporated into the food pellets accounts for 70 per cent of the food energy costs (Table 10.2), most of this (54 per cent of the total farm budget) being consumed in the original fishery for fishmeal. At the time that this budget was drawn up, the pilchard fisheries of southern Africa formed the main fishmeal supply for UK feedstuffs, and the figures are based on an estimated fishing cost for the Namibian purse seine fishery (see Chapter 2) of 10 GJ/ton. Until the early 1970s, the Peruvian anchovy fishery was the main world source of fishmeal, and this was extremely energy-efficient (see Chapter 11); should this fishery recover from its collapse, the fishmeal costs would be almost halved. Fishmeals from European sources would produce pellets costing almost three times as much when deriving from trawled whitefish, or rather less from other sources such as mackerel and sandeels.

If foreign fishmeal is used, it is interesting to note that the purely local UK portion of the energy costs give an input/output ratio of 0.34 (similar to milk) and a protein subsidy of 144 MJ/kg (similar to potatoes). However, a global view should be taken, as it is clearly not sensible to ignore the energy debts of the foreign fishery which supplies the 25 ton/year trout farm with fishmeal equivalent to the total annual nutritional needs of 50 persons in the country of origin. In the long term, it is probably unwise to rely on fishmeal industries in countries with their own large populations to feed. Furthermore, as explained in Chapter 8, such fisheries are often exploited in a way which damages the fish stocks (Macer, 1974; Popiel and Josinski, 1973).

With a labour force of two men, the farm can provide the equivalent of the total nutritional needs of about 30 people from its one hectare,

Figure 10.5: Response of Trout Fish Farm Energy Efficiencies to Reduced Energy Costs of Producing Trout Food and Comparison with Production of Other Foods. (a) Response of protein energy subsidy. (b) Response of gross input/output efficiency ratio. Key to diets: 1 = South African fish meal from purse seiners. 2 = European fishmeal from freezer trawlers. 3 = Peruvian fishmeal from drifters and seiners. 4 = 50 per cent vegetable protein in diet, 10 per cent loss of yield. 5 = All vegetable or single-cell protein in diet, 20 per cent loss of yield.

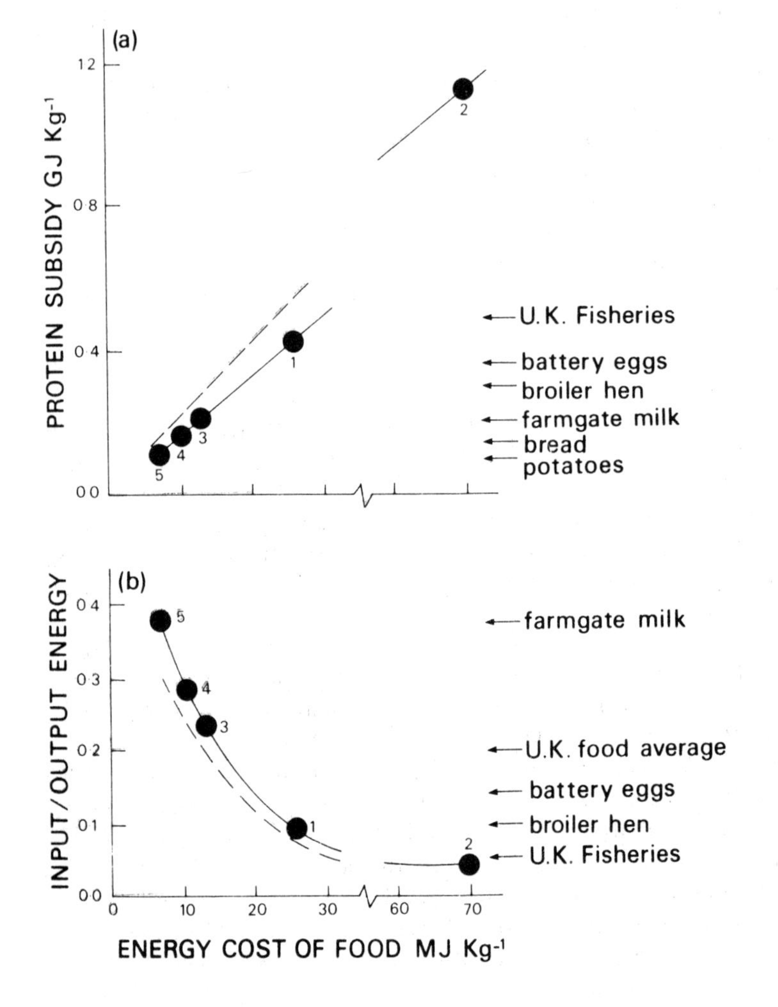

provided adequate clean water is available. The total yield, of 25 tons $ha^{-1} yr^{-1}$, is exceptionally high, and even though it is achieved at the expense of an enormous food energy input, it is much higher than yields obtained with terrestrial systems. However, it is worth noting that the pure water supply requirement constrains other activities which might take place in the same catchment area which contains a fish farm.

If the trout could be fed on a diet composed of suitable vegetable protein, grown at low energy costs in the tropics, the performance of the farm could be improved to an overall efficiency of about one-third and a subsidy of about 160 MJ/kg of protein, a similar improvement to that obtained if the Peruvian fishmeal were restored. The problems here are that fish like trout are carnivores and piscivores in the wild and need the ten essential amino acids in their diet (Chapter 4). Unlike mammals, these fish seem to have no mechanism to conserve the essential amino acids in the diet even when fed a high protein food like that used in fish farming (Cowey, 1975). Unfortunately most plant proteins are low in the sulphur-based amino acids such as lysine and methionine, and fish get diseases and grow very poorly on them. Some plant proteins, such as soya varieties, are better, but even here the best results have been obtained only when the plant protein in the diet is supplemented, either by adding the amino acids which are deficient, or by supplementing with additional fishmeal (see Chapter 4). Single-cell proteins (SCPs) produced by various fermentation processes have amino acid balances which mimic animal tissues much more closely, and can be grown in very efficient industrial conditions. Some preliminary work with trout using SCP from algal, yeast and bacterial proteins has been reported by Matty and Smith (1978), and on a yeast protein obtained from whisky distillery waste by Dabrowski *et al.* (1980). Should these experiments result in successful large-scale production of feeds, we can expect to see energy savings as good as, or better than, those which might be achieved with plant proteins. Some of these potential gains are set out in the graphs in Figure 10.5.

The energy costs of selected fish farming operations on different species are listed in Table 10.4 along with the value from the rainbow trout farm budget, and including examples of most of the systems described in this chapter. Within any one species a wide range of costs can be seen; extensive farming usually has low energy costs while intensive systems with recirculated filtered and aerated water incur very high energy costs. In general, culture in the warmer waters of the tropics costs less, as does farming of species like milkfish which are

Table 10.4: Energy Costs of Selected Fish Farming Operations.

Species and Location	Type of Farm	Energy Costs (GJ/ton harvested)
Carps		
Japan	intensive, recirculation and aeration chambers	300
Germany	sewage grown, chambers	130
Czechoslovakia	intensive, ponds	54
Germany	traditional earth ponds	25
Israel	intensive, ponds	25
Japan	net pens, intensive	18
Philippines	extensive, ponds	2
Tilapia		
Israel	intensive, aeration chambers	90
Liberia	ponds, rice fields	17
Thailand	ponds, intensive	11
Congo	ponds, extensive	0.2
Catfish		
USA	intensive, ponds	92
Thailand	ponds, intensive	56
Milkfish		
Taiwan	intensive, ponds	7
Philippines	intensive, ponds	5
Philippines	extensive, traditional ponds	1
Trout		
UK	intensive, recirculation, aeration chambers	280
UK	intensive, ponds	55
Polyculture		
Sri Lanka	milkfish/prawns, intensive	170
Israel	carps/mullet/tilapia, intensive	65
Israel	carps/tilapia, aeration, intensive	26
Thailand	carps/tilapia, extensive	0.13

Sources: Edwardson (1976) and Pitcher (1977).

further down the food chain, although intensive farming of any fish can be surprisingly costly. Intensive polyculture systems do not appear to improve automatically the energy budget, probably because of the greater amount of care and husbandry they involve. On the other hand extensive polyculture systems have the most favourable energy costs in this survey (Edwardson, 1976).

Prospects

In this chapter we have discussed the factors which influence the biological feasibility of aquaculture. Biological factors are not the only ones to be considered. We have not mentioned the economics of fish farming, which are perhaps more important to an industrial enterprise, but it needs to be pointed out in this context that economic judgements cannot be made until the biological and engineering aspects are successfully solved. The design and construction of enclosures are not reviewed here but there are many papers on the subject (see, for example, Pillay and Dill, 1979). Despite these comments, fish farming is being carried out successfully, but most of the biological problems, such as diet and stock density, are solved empirically. This approach is only good whilst aquaculture is a marginal contributor to food. Aquaculture will only achieve its estimated promise when our biological knowledge about fish has increased far beyond what we know today.

The first part of this chapter could be taken as describing the state of the art. The farming of salmon in Scotland and of carp in India are two systems which solve the same problem in very different ways. The Marine Harvest Ltd system needs lots of capital and not much labour, whilst the system for carp farming is the reverse. Much more is known about salmon nutrition than that of any other fish, so that the farming of salmon must be the least dependent on skilled husbandry. The salmon farm uses formulated manufactured food whilst the carp in India feed on organisms growing naturally in the culture pond. Within their respective social and economic contexts both methods work, but the carp farming method is more likely to yield large crops in the future. With greater knowledge of reproductive physiology it could be possible for a central egg production station to provide the carp farmer with eggs as and when required. The salmon system is at present too dependent on an expensive food input and consequently on the finished product fetching a high price. It is possible that further understanding of nutrition could cheapen the diet, but then the heavy capital costs and overheads could only be recovered by mass sales of the cheaper product. Perhaps with improved strains and cheaper diets this may be possible.

In Israel the most successful polyculture has been in adding *Tilapia aurea*, a detritivore, and silver carp, a planktivore, to fertilised ponds where an omnivore, the common carp, is being cultivated. The silver carp enable higher levels of pond fertilisation to be employed, since by grazing they reduce the shading effect of heavy phytoplankton growth which would otherwise set an upper limit to primary production. Under farm conditions, the silver carp defecate small pellets which still contain appreciable levels of undigested protein, and these are consumed by the other two species. The tilapia acts to clean up the bottom of the pond. Food pellets are added for the common carp. With correct farming practice, the yields for each species obtained in this system are greater than those in monoculture, despite the competition between the species for some of the food available (Figure 10.6). This type of fish farming combines the best features of extensive and intensive systems, partly exploiting the natural ecology of the pond system and partly achieving high yields through the use of added food and fertiliser. This system has the added advantage that it is less dependent on the fluctuations of the animal feeds market than fully-intensive fish farming, because the amount of added nutrients and food can easily be adjusted to the current economic optimum. It is this type of farming which is most likely to be able to contribute significantly to world food resources in the future. In temperate climates, growing carp in sewage treatment settling ponds has made a promising start in the UK, but so far the introduction of herbivores has not been successful (Noble, 1975).

The idea of salmon ranching has recently been proposed as a means by which production of the species could be raised. In many instances ranching involves just raising young and releasing them into their home river in order to migrate out to sea and feed before returning to be caught. Much work has also been done in enhancing the natural spawning areas of Pacific salmon on the west coast of North America. Some projects have proposed transplanting species to new areas before release. These need closer examination.

There are many parts of the world where environmental conditions appear to be suitable for salmon. One could mention in particular the Chile coast where there are rivers which could provide good spawning sites (Joyner, 1975). The krill, *Euphausia superba*, distributed round the Antarctic continent, might be a suitable food for the salmon once it has grown large enough in its marine phase, but one must also assume

Figure 10.6: Polyculture in Israel. Solid circles represent yields in monoculture, overlapping areas represent yield lost due to competition between the species in polyculture. Broken circles represent actual yields in correctly managed polyculture.

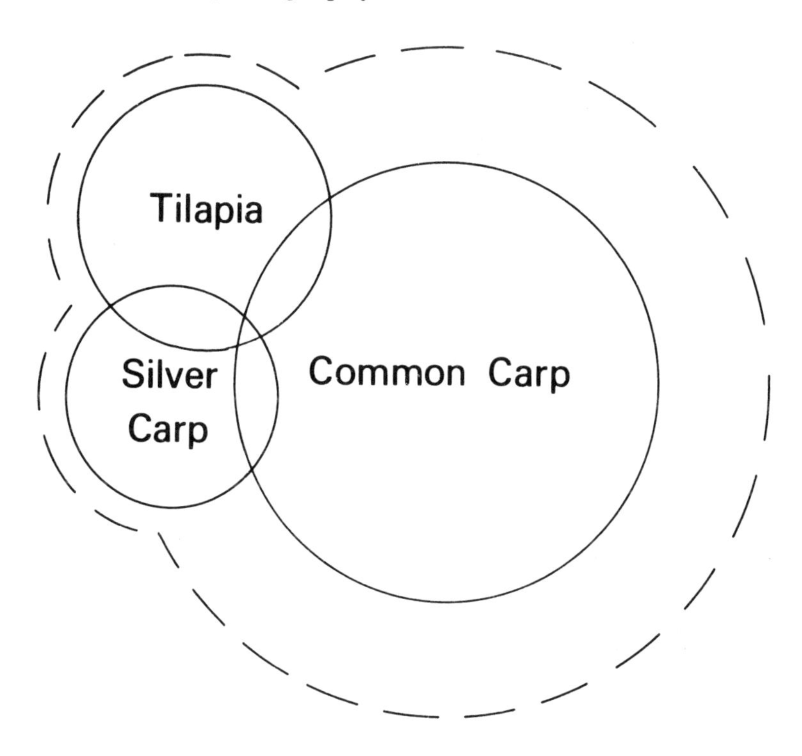

Source: After Reich (1975).

that the oceanographic environment has the right properties to stimulate the salmon in its search for the home stream. This type of speculation has led people to try transferring species of Pacific salmon to many parts of the world (reviewed by McNeill, 1979). The experiments started at the beginning of the century, and by 1930 100 million eggs had been distributed from the American Pacific coast to 17 Atlantic states, to Hawaii and 15 other nations. However, very few of the plantings have succeeded and only New Zealand has an established population of chinook resulting from an egg transplant. A selection of examples from McNeill (1979) is given in Table 10.5 and it illustrates the variety of attempts that have been made and the high

Table 10.5: Transplants of *Onchorhynchus spp*.

Species	Experimental Areas		Time of Transplant	Number of Eggs Used	Result
	Origin	Destination			
O. gorbuscha	Alaska USA	Washington State USA	1922–32	65.5×10^6	no return
	Pacific coast USA	Maine USA	early 1900s	not known	no return
	eastern USSR	north-west USSR	1960	15.3×10^6 (juveniles)	100,000 returned initially, now lower
O. keta	Amur River	Lake Baikal	1929 on	not known	no return
	eastern USSR	north-west USSR	1933–9	9×10^6	no return
	eastern USSR	Caspian Sea	1960s	not known	1,500 initial returns
	USA	Korea	not given	not known	no return
O. kisutch	west coast of USA	Great Lakes	1960s	not known	population established
	west coast of USA	Korea	1960s	not known	no return
	west coast of USA	New England	1970s	not known	no results yet
	west coast of USA	Chile	1970s	not known	no results yet
O. nerka	Skagit River	Lake Washington	1936	not known	population established
O. tshawytscha	west coast of USA	Great Lakes	not given	not known	population established

Source: Extracted from McNeill (1979).

percentage of failures.

All living organisms in their natural states have evolved together with their environment; thus their genotypes have achieved a balance which ensures that the population remains fitted to conditions. With most species we only know the major environmental requirements (Mayr, 1963) so that even with such well-studied groups as Pacific salmon there can be unknown factors that are essential for its success.

Consequently what appears to be a good host environment for an introduced species could well lack a vital factor. It is therefore hardly surprising that most transplants fail. A transplant of pink salmon from British Columbia to North Harbour River, Newfoundland, in Canada, is described and analysed by Lear (1975). The transplant was considered to have failed because of predation by brook trout and eels, unfavourable surface temperatures in the river during the fry run, predation of herring on fry in the sea, year-class failure of the even-year stocks that were introduced, unsuitability of the donor stocks with respect to migration pattern and homing behaviour, and an inadequate number of eggs transplantd to produce populations to maintain runs in sub-optimal environmental conditions. Lear pointed out that the salmon's ability to find its home stream is partly genetically-determined and therefore a transplanted stock must adapt its genotype to the new habitat. This adaptation would be by the usual agent of natural selection; the fish that do not return to the new parent stream do not contribute to the genotypes of the next generation. Because of the high losses in the early years, Lear states that large numbers of eggs must be used over a period of several years. However, it would seem that the process is self-defeating if one continues to bring in new eggs from the original donor stocks, as the adults deriving from these will have the pure donor genotype so that their presence will continually frustrate the process of genotype adaptation.

These attempts to transplant salmon and other species to new areas need to be approached with new methods. Merely transporting eggs has proved to be fruitless in most of the attempts made. Work described by Seneviratne (1977) investigated runs of salmon in rivers in British Colombia where the natural run only occurs every other year. The work starts with the assumption that the oceanic part of homing behaviour in the salmon is genetically determined. A run in the off year can only be started with salmon from the same river. It is only these fish that have the correct homing behaviour specific to the rivers worked on. By using frozen sperm and by speeding up maturity with sex hormones, the Canadian workers are trying to fill the gap. The work is in its early stages and the problems being solved are small compared with those which arise when salmon are transplanted from one continent or hemisphere to another. Success may come when a breeding programme can produce new strains specifically adapted to the new environment. Before such a programme can start, we need to know much more about the genetic control of migration and the process of migration itself.

If reproduction could be controlled then the complete farming system could be freed from the seasonal cycle. New eggs, and consequently new finished product, could be available in a steady stream. This is the ideal sought by most enterprises processing frozen or canned products. The ability to produce eggs continuously would make it necessary to refine the rearing techniques. The Marine Harvest system is mostly outdoors; as a result the fish do not grow through the winter. It would therefore be necessary to either heat the water or to site the farm near a power station with a hot water effluent if continuous growth was to be achieved. In the tropics the problems would be less as climatic conditions are less variable. Nevertheless, there could be environmental factors which would need control if continuous growth is required. For example, it was found for Indian carp that the right environmental conditions are necessary to bring carp gonads to their spawning state even though the fish are living in a fairly constant tropical environment.

Not all farmers will necessarily want to produce their own stock, but they will have to feed their growing fish. The broad nutritional requirements of fish are known, but what is lacking is a detailed knowledge of the nutritional dynamics of species of interest to farmers. The work of Brett *et al.* (1969) has illustrated how vital it is to do experiments linking diet size to growth rate. The farmer needs to know how much food he has to feed to get the maximum conversion into flesh, i.e. the maximum specific growth rate. Brett's approach needs to be applied to all the species we are interested in for farming under intensive conditions.

For fish farmed in ponds on natural foods, such precise control over diet will not be possible, but it will still be necessary to know the food requirements of the fish so that one can be sure that each fish is going to have sufficient to satisfy its needs and allow fast growth. In the open-pond environment it will be more important to know about feeding behaviour and about competition between different species. Ideally, one should have a range of species that each make use of food sources situated in different parts of the habitat. For example, one species would feed on detritus, one on marginal vegation, and a third on the planktonic organisms of the open water. In France and Yugoslavia predatory pike are being stocked together with prey species. Presumably a better overall yield is obtained in this way (Sinha, 1979).

Fish breeding is only just beginning. Its contribution to aquaculture can only be great when we know more about fish reproduction, nutrition and growth. At present the techniques exist for producing new strains but before this can be done we need to know what to select for. Selecting for faster growth is not easy, as discussed above and in Chapter 5. Growth rate is influenced by diet, quantity of food available, temperature, the density of other individuals of the same species and genetic make-up. Consequently it is necessary to be able to evaluate the effects of each before a breeding programme can produce a faster-growing race that will perform as expected in a known environment. At present the salmonids are the best known fish. Salmon and trout have an established market and a high price which is sufficient to cover the high operating costs.

We know quite a lot about fish physiology, but often only in general terms and for only a few species. Consequently for those who want to farm fish intensively, either a species must be selected about which much is already known, or the organisation must be prepared to do a lot of preliminary research. The first way was chosen by Unilever, the second by the White Fish Authority of the UK, when they decided to develop plaice and turbot farming. For those wishing to farm fish less intensively, as with the Indian carps, it is possible to proceed on a more empirical basis. It might even be more useful to have greater ecological and behavioural knowledge than detailed physiological or biochemical information. Because of the reasons reviewed above it seems clear that the extensive method will continue to be the main contributor to world protein supplies.

11 FISHERIES AND THE ECOLOGY OF MAN

Like other species, humans have to live within the limitations of their resources. Technology has enabled our species to expand its niche through the use of inventions and by expenditure of fossil energy, postponing the day when we are brought up against global limits to food. These limits seem not far off, and so the rational exploitation of renewable resources like fisheries will increasingly become a vital aspect of the ecology of man.

The current picture is rather bleak. Over the past 30 years vital protein resources have been squandered by failure to manage stocks properly. Until ten years ago there was perhaps some excuse for ignoring the advice of biologists because classical fishery management models were often misleading and inadequate, as we have seen in Chapters 7 and 8. However, this excuse is no longer tenable; the problem now is that biologists' advice is still ignored even though based on more accurate and sophisticated analyses. It is a sad reflection on our ability to manage resources that the past decade has seen no amelioration of overfishing. A long list of stocks endangered in 1970 would be longer still in 1980, the only removals being those stocks that have collapsed. Despite the wealth of helpful theory, there have been very few success stories of fishery management in practice. Most stocks of herring, cod, hake, sardine, anchovy, pilchard, tuna, mackerel and many flatfish, together with the great whales, are all in a worse state today than ever before. In region after region the familiar pattern of increasing CPUE, stock depletion and collapse, followed by a switch to the next stock, has been brought about by the inexorable increase in effort symptomatic of open-access fisheries (Chapter 9). At present, there are both optimistic and pessimistic factors for the prospect of conservation and optimal management of fishery resources. We are at a turning point, but it is debatable whether this will be for the better or will lead to the edge of a precipice! In this chapter we will examine these factors and look at the causes and possible remedies for this endemic and chronic over-exploitation.

The 200-mile EEZs, declared in the mid-1970s, are factors on the optimistic side. The FAO considers that they have given us a 'window in time' in which to get things right. Resource Adjacent Nations (RANs) now have clear responsibilities to manage and conserve their stocks, and can trade with nations (non-RANs) with no fishery resources of their own. So biological and political pressures for the preservation of the resource now coincide more than under the old system with its ineffectual international advisory bodies. Scientific monitoring of stocks is similar to before, but policing and the imposition of sanctions on those who do not obey regulations are much easier. The main management problem is in choosing control measures which counter the pernicious trend to increasing effort. Failure to manage stocks will be to the detriment of government in both democratic and non-democratic nations, and so it is a fairly safe bet that important food fisheries will be conserved in the long term, provided they are not destroyed before this political maturity occurs. Perhaps we can regard events like the French fishermen's blockade of ports in 1980, the tuna war between Canada and USA in 1979, and similar disputes around the world, as the teething troubles of the new system. The EEZs might be seen to parallel the enclosures of common land in the middle ages in Britain (Coull, 1975), so such problems may take a while to sort out. The FAO estimates that some 95 per cent of global fishery resources are now within EEZs; there is probably not much hope for those outside which are effectively up for grabs.

Icelandic Cod

Seventy per cent of Iceland's wealth depends on fish, and a case history of her industry provides an interesting example of current trends. Despite clear warnings about the consequences of overfishing of northerly cod stocks made in the late 1960s and 1970s (see Chapter 8), Icelandic cod were heavily fished prior to the establishment of the 200-mile EEZs in 1975. These cod recruit to the fishery several years before first spawning, making the risks of collapse greater (Chapter 8). As stocks began to show serious signs of depletion during the early 1970s, the Icelandic government attempted to establish its right to sole exploitation of the cod, in the face of considerable opposition from European countries, whose distant-water trawler fleets had been built on the profits from the Icelandic fishery and who refused to obey Icelandic gear limitations or to follow quotas. A 'cod war' developed, both sides claiming that cod was on their side. The Icelanders' mistake was to try to claim management rights prior to the general acceptance of 200-mile limits. There was clear evidence of overfishing, but the case was obscured by unbecoming jingoism in the UK and British naval vessels were sent when Icelanders attempted to protect their stocks by force, British fishery scientists maintaining a prudent silence. Eventually the Icelanders' rights to manage their stocks were conceded, at the same time as the UK declared its own EEZ, which is unfortunately still entangled within the EEC. It may take some time before widespread observance of the law of the sea renders this kind of dispute as obsolete as cattle raiding!

After the 200-mile limit was introduced at Iceland, mesh limitations and restricted zones were introduced. Foreign fleets were generously allowed to take a reducing quota for a transitional period, but evidence from cohort analysis now suggests that these quotas were greatly exceeded at the time, delaying the recovery of the cod stocks. Even with home ownership, management is not as effective as might be hoped. The 1979 and 1980 quotas of 300,000 tons were exceeded by about 15 per cent, mainly as a result of chronic overcapacity of the Icelandic fleet. The only good year classes recently were in 1973 and 1976, and the situation is critical since the 1976 fish will not spawn until 1983 but are already being caught. Properly regulated, the stock could yield about half a million tons a year on average; the main problem is to enforce existing regulations and hold quotas. Small foreign quotas are still allowed, but are regularly exceeded by individual vessels.

If such teething problems of the new EEZs were the only ones, then we would have cause for optimism, but there are at least two other factors to be considered. The first is biological and concerns the outstanding difficulties and uncertainties in the use of models for yield prediction. The second concerns the economic and social complexities which impinge upon optimal biological management of fish stocks. We will deal with each of these in turn in the rest of this chapter, concluding with a discussion of fishery control devices.

Better Fishery Models – 'Come the Millennium'

According to Robert May (1980), 'come the millennium', with perfect understanding of the ecology of fish stocks, fishery scientists will be able to predict the likely consequences of alternative management strategies with high accuracy, although because of its stochastic nature no biological process can ever be as cut-and-dried as the laws of chemistry and physics. So at this point we will briefly review ways in which the management models outlined in Chapters 7 and 9 can become more effective. We will examine the choice of model, the continuing problem of stock and recruitment, multiple species fisheries, the aims of fishery management and the possibility of active optimal control.

Choice of Fishery Model

Now that there is clear responsibility for most stocks, it should become easier to obtain sufficient data for proper management. In Chapters 7 and 8 we emphasised that neither the short cut of using only catch/effort data, nor blind data gathering could lead to safe and productive management. Carefully-planned research investigations are needed, tailored to the problems found in particular species.

It is evident that, whenever possible, dynamic pool simulations (including recruitment) should be the preferred model, since even the most sophisticated surplus yield models suffer from the problems of estimating parameters outlined at the end of the Chapter 7. Rewarding the large amount of work in measuring growth, mortality, fishing mortality and recruitment, the general dynamic pool model is flexible enough to cover almost any eventuality including seasonal and regional differences. As well as predicting total yield, the dynamic pool model has the advantage of showing the pattern of ages and sizes in the catch. The main danger is in extrapolating one or more of the four component sub-models beyond the region in which the raw data were gathered. The answer to this dilemma is to compare 'scenarios' based on informed guesses about the likely range of relationships outside the measured region, choosing the management strategy which is robust against the maximum number of mistaken assumptions in the model.

Inevitably, for some time to come there will remain fisheries for which only catch/effort data are available, especially in remote areas. In these cases there is no choice but to use a simple fishery model. The difficulties discussed in Chapter 7 make it necessary to explore several different versions of the model. Uhler (1980) and Hilborne (1979) have shown that all Schaefer-type methods can produce poor estimates, f_{opt} falling as far as 50 per cent from the true value. This criticism does not appear to apply to the Walter (1978) equilibrium yield method, which is therefore the method of choice provided some recruitment data are available.

Unlike their classical antecedents, all contemporary fishery models must include recruitment because of the dangers of recruitment over-fishing or collapse. For effective predictions, it seems that there is no substitute for empirical data on recruitment to construct models tailored to the life history of each species. Pre-recruit surveys are an expensive necessity we feel. If recruitment models are defective for the lack of hard data, fishery quotas might as well be pulled out of a hat. The expense of regular pre-recruit surveys should be a worthwhile investment. At the very least, in the absence of recruitment data, management advice should be robust against having got it wrong. In the long term, detailed and painstaking research on the larval and pre-recruit stages of the fish is needed.

Whichever model is chosen, it is clear that just as there is no short cut to sound data to use in the model, there will be a continuing commitment to monitoring research if the fishery is to be managed 'on-line' so that regulations can be adjusted from year to year, or even season to season.

Multiple Species Fishery Management

All or most of the exploited species in a region should be included in the same fishery model. The trend to planning for multiple species fisheries has arisen not just because tropical fisheries with high species diversity (Chapter 2) are under increasing pressure, but also as a consequence of two other factors. First, fish belonging to overexploited species are taken as a by-catch in other fisheries. Secondly, modern fishing vessels are becoming flexible enough to switch from one stock and species to another. Both factors can nullify controls aimed at conserving stocks in danger.

Catching fish from one type of boat with one kind of fishing gear has been the norm, for example plaice fishing with otter trawls or tuna with long lines. Local sea conditions and distance to fishing grounds have also determined the design of vessels, so that fishery models did not, on the face of it, seem to need to include several species of stocks. Nowadays powerful modern fishing vessels over 30 m long exist which can catch several different species in purse seines, trawl for demersal species, or trawl in pairs for pelagic fish in midwater, even laying drift nets if required. Several different nets can be stored on board. This trend will probably continue, and standard boats capable of dealing with almost any species will evolve.

Surplus yield models described in Chapter 7 can be easily altered to include multiple species, but the problem of accurate parameter estimation becomes even greater, so the exercise is not very useful. Fortunately, dynamic pool models can be extended to cover several species. In the simplest modification, models for the different species just run in parallel, although joint effects on growth, gear and fishing mortality may be difficult to partition accurately without extra research. The switching of fishing effort among the stocks can be simulated in detail with this method, adding economic constraints if desired, in a fashion analogous to the early work of Clayden (1972) on the Atlantic cod stocks (Chapter 8). A similar technique can determine the optimal fleet composition; this has been done for the Lowestoft plaice fishery to be described later in this chapter. When many species and gear types are included, the models are very complex, the optimisation problems encountered having been considered by May *et al.* (1979), based on the ideas of Paulik *et al.* (1967) and Garrod (1973). The general problem is the occurrence of many local optima of yield so that solutions have to be constrained within certain fleet and species boundaries. Decisions about where these boundaries lie are still based on guesswork and intuition. Clark (1980) considers it too difficult to try to find them by including economic factors in such a complex fishery model. In fact the search for a universal single realistic fishery model may be unhelpful.

Management of multiple-species fisheries and the problems of switching of effort between stocks can be dealt with reasonably effectively with simple models (e.g. Huppert, 1979), and such modelling is rapidly becoming commonplace. However, there remains a more subtle longer-term problem, which assumes special importance at the two extremes of fishery development. In both seriously overfished and in virgin fisheries the wider consequences of fishing cannot be predicted. These include changes in production and species composition at trophic levels both above and below the fish in the community. In the 1970s fisheries science did not move towards the whole-ecosystem models which were popular with freshwater and terrestrial ecologists. Although useful in primary production studies, these models do not seem to have been particularly successful for polyphagous carnivores like fish. Fisheries scientists have so far wisely avoided making predictions with such ecosystem models, but the general problem still remains.

Anderson and Ursin (1977) have attempted to overcome this by ignoring primary production and modelling the trophic links between various species of fish and their prey, including competition for food supplies. Their model is essentially an extended dynamic pool type, but incorporates food consumption and prey species for several fish stocks. Data for fish prey are sparse, but 1981 is to be International Year of the Fish Stomach in the North Sea to gather data on fish trophic relationships in order to evaluate Anderson and Ursin's model for multi-species fisheries. In freshwater, Hart and Connellan (1979) show how yield of roach is partitioned between anglers and predatory pike. Even though these approaches represent an advance, it is difficult to see how much models could effectively predict the kind of reciprocal effects between fish and production at other trophic levels which will be needed in the long term.

One successful original and purely empirical approach to multi-species and multiple-gear artesan fisheries has been published by Marten (1979a, b) on the fisheries in the Lake Victoria ecosystem which we described in Chapter 1. Lake Victoria fisheries have developed symptoms of overfishing in the past 20 years. Since the 1950s when catches mainly consisted of the two indigenous *Sarotherodon* species caught in large gill nets, smaller mesh nets, seines and long lines have been introduced. Catches, CPUE and fish sizes have declined and the valuable *S. esculenta* has disappeared from much of the lake, although two introduced tilapias have spread quite well. Catch is now composed of six main groups of species varying in size from 6 cm to 2 m, but almost a quarter consists of small *Haplochromis* caught in beach and 'mosquito' small-mesh seines.

Marten believes that there is an alternative to reducing effort in this scattered artesan fishery and has used two years' detailed catch, gear and effort data from 50 stations around the 1500 km lake shoreline. The catches are regarded as the results of local experiments in gear and effort combinations. A stepwise multiple regression (see Chapter 7) analysed the data to find the optimal combinations which were associated with the maximum local yields. This assumes that (a) all inshore areas of the lake are similar, and (b) catches reflect the long-term response to fishing patterns at each locality. With the exceptions of changes in *S. esculenta*, these assumptions were thought sufficiently valid to encourage faith in at least the qualitative pattern of the results. Encouragingly, there were no differences between the results for the two years, so they were combined. Marten's approach is purely statistical, subsuming all the actual relationships between gear, growth, recruitment and species interactions into the actual catch and effort figures for each station. The method has the advantage of producing clear management recommendations without needing any further interpretation.

Six categories of fishing gear (in standard units of effort) were considered after pilot work. There were four sizes of gill net from 3 to 20 cm mesh, hooks set on longlines, and beach seines. In such a multiple fishery, gear which catches one species effectively may depress or enhance the catch of another species indirectly through competition, predation, juvenile mortality or other feeding relationships, so the method can include any subtle effects relayed through the ecosystem

Table 11.1a: Optimal Gear for the Lake Victoria Fisheries.

	Gear (Standard Units per km)						Average Yields	
	Gill Nets				Hooks	Seines	Weight	Value
	small	medium	large	v. large			kg/km/day	$/km/day
Current average	90	12	100	68	1,240	0.7	270	60
Optimal (weight)	0	0	410	0	2,850	0	460	–
Optimal (value)	190	0	170	170	4,300	0	–	150

Source: Marten (1979).

Table 11.1b: Species Composition of Catch.

	Gear	
Species Group	Current %	Optimal %
Tilapias	15	41
Bagrus	12	16
Clarias	15	10
Protopterus	26	32
Haplochromis	22	0
Synodontis	2	0

even if it cannot state precisely what these are. For example, large numbers of hooks were associated with greater catches of the tilapias and *Haplochromis* even though few of either of these species were caught themselves. This was thought to work through the reduction of large predators. Small mesh nets caught *Haplochromis* but depressed tilapia catches. Optimal gear for either of these species could increase their individual yields five-fold, but the gear combinations were mutually exclusive.

Optimal gear for the total catch, maximising yield from all the major species, is summarised in Table 11.1. The current total yield is 110,000 tons per year, providing an important protein source for East Africa. Marten's analysis suggests that this could be increased by 70 per cent and the value of the catch doubled, gains which would be worth the rather difficult enforcement tasks which would be involved in management. Two species would be eliminated from the catch, small beach seines and gill nets being banned, but greater hook effort would increase catches of large predators like *Bagrus*. Under the optimal plan, human fishing time would increase slightly, an important benefit in an area with endemic unemployment. Some new marketing infrastructure would be needed because certain large fish are considered delicacies in some areas but unfit for human consumption in others. Unfortunately, desired and 'trash' species swap identities from tribe to tribe!

Marten's method could be employed in other artesan multiple gear and species fisheries where suitable data could be gathered. Moderate violation of the assumptions would not render the results completely invalid since they can indicate qualitatively how management should proceed. The method would not work in industrialised fisheries where boats travel over the whole stock area and only one type of gear is used.

Models and the Aims of Fishery Management

Once appropriate models for the fishery have been chosen, the next problem is to decide what the aims of management are before optimising to achieve them. Sometimes this is not as clear-cut a decision as might appear at first sight (May *et al.*, 1978). There are two levels of complexity in the problem, the first including only biological problems, the second also considering economic and social factors.

In its classical form, MSY excludes the effects of competition, symbiotic or commensal relationships with other species, trophic relationships, or changes in carrying capacity due to pollution or other human influences. Holt and Talbot (1978) point out that MSY became 'institutionalised in a more absolute and precise role than intended by the biologists who were responsible for its original formulation ... as the sole conceptual basis for management'. Even within a single stock, as Larkin (1977) and Gulland (1978) have pointed out, MSY is no longer the aim of fishery management for four main reasons: (1) classical surplus yield models produced MSY values which were too high, because data did not come from fisheries in equilibrium (see Chapter 7); (2) equilibrium effort is a better regulator than catch (see below); (3) MSY itself does not indicate how to partition the catch or effort (Rothschild, 1973); (4) long-term MSY is an average which has indeterminate usefulness in the face of year-to-year variation in stocks. Larkin aptly summarised the situation in the poem given in the preface to this book.

In Chapter 8 we showed that risks of known magnitude can be assessed for various combinations of age of entry and fishing mortality in a fishery. Similar exercises can compare different management strategies. Strategies robust against the wrong information would be the most conservative and could be chosen more often when stocks were severely overfished. Alternatively, riskier but more rewarding plans might be adopted for fisheries where this was not the case. Both actions are a far cry from the automatic application of MSY based on shaky data.

The model can be run for the optimal 30- or 50-year strategy. This is not necessarily the one which produces the highest aggregate yield depending on the risks involved. A further decision has to be taken: how many years should be included in the risk assessment? In other words, to what extent should the fishery be buffered against the 10-, 50- or even 100-year catastrophe? Bridges, sea walls, tall buildings and other civil engineering structures have to withstand the '50-year wind or wave', but much greater risks have typically been taken with natural resources.

One corollary of both simple Schaefer models and simple dynamic pool models is called the 'bang-bang' policy. This shows that the maximum yield is obtained when stocks above MSY are brought back down to it rapidly in as large a step as possible. This is based on the principle of keeping the stock at the level of maximum biomass production indicated on Figure 7.1. Unfortunately, if this level (MSY) has been estimated incorrectly (as it probably has been) the 'bang-bang' fishing will be more dangerous than gentler adjustments. In Canadian fisheries evidence has shown that large adjustments to MSY delayed recovery of stocks from overfishing; much smaller quotas are now set (Needler, 1979). At the very least, where information is less than perfect, strategies such as those discussed with respect to Walter's AEY fishing in Chapter 7 will be safer in maintaining viable levels of recruitment. There are new quantitative ways of determining how far below MSY fishing effort should be set. For example, the $F_{0.1}$ method is a means of estimating the fishing effort which would result in an improvement of yield of 10 per cent of the rate of increase of yield at very low fishing mortality (Gulland and Robinson, 1973). This method enables fishery managers to trade off increased effort against returns, and is in use in Canadian fisheries (Needler, 1979, and see the end of Chapter 8).

So management with MSY as a central concept has gradually been replaced over the past decade by the more subtle aim of optimal control – in any particular year yield can be taken which gives the most satisfactory combination of as many factors as are necessary to the long-term survival of the fishery.

In the near future, the management of many of the world's depleted fisheries will (if all goes well) include as a major factor the recovery to former yield levels bearing in mind the need to avoid a repetition of overexploitation. Peruvian anchovy, North Sea herring and Icelandic cod are among stocks being currently managed along these lines.

Chapter 9 described how simple economic factors can be included in fishery models. Analysis of our 50-year model to provide an 'optimal economic rent' (OER) is only a little more difficult than for the purely biological model. The economic model could also be used to show how rapidly the gains diminished for increased effort. Again, OER might be below MER just as OSY is below MSY. Work by Hannesson (1975) has shown that 'pulse fishing' in a pattern of alternating periods of heavy and light effort gives the MER in an economic model of cod stocks. The problem with such a management pattern is that it ignores the surplus capacity of the industry in years of low effort. Just as in fleets with the chronic overcapacity which we will turn to later in this chapter, effort may be switched to another stock or there may be pressure leading to illegal fishing.

More realistic economic fishery models suffer from a number of disadvantages when interest rates are included. This is true even if rates are low enough in relation to the fish reproductive rate to avoid the absurd effect described in Chapter 9, where maximum present value is given by catching all the fish and investing the proceeds. The problem is that the optimal solution tends to include catching all the fish in the last year in the simulation period and investing these proceeds. If constrained to avoid this, the solution still tends to include heavier catches towards the end of the period in order to benefit from the interest rates. One way to avoid the problem is to assume a zero interest rate beyond the end of the 50-year simulation period, but this seems intuitively unsatisfactory. A better way is to put a cash value on actually having a resource in the 51st year, but this is difficult to quantify as we will see in the next section. There is no simple solution to the problem at present. While many fisheries are currently barely making a crude operating profit, and the majority in the world continue only with some form of government subsidy, the more complex economic models can be ignored at present until they provide more sensible recommendations. This is not to deny their heuristic use in demonstrating why fisheries are forced to follow certain trends.

Active Optimal Control

So far we have discussed optimal control, or on-line management of the fish stocks, as a passive process responding to natural or fishery-induced changes in the stock as rapidly as possible. A further, but more risky option is the 'active optimal control' advocated by Silvert (1978). Here effort or gear are deliberately manipulated in one fishing season in order to find out more about desired aspects of the stock's biology, such as growth, recruitment, or local movements. Deliberate management experiments are set up with the stock itself as the laboratory. The aim is to improve progressively the accuracy and realism of the fishery management model more rapidly than would be the case by waiting for natural perturbations. Silvert calls this 'creative risk taking'. Recruitment in the US capelin stock has been investigated in this way. It has also been suggested for use on salmon enhancement schemes to determine to what extent marine survival is density-dependent (Peterman, 1978).

Silvert considers that management plans which play it safe may 'freeze the fishery into a pattern of sub-optimal yields', but a danger which optimal control ignores is inaccuracy in the relationship between fishing mortality and fishing effort. No amount of alteration of effort will improve the data until mortality has been estimated in the following year. Similarly, experiments on recruitment may only produce results with a lag of several years. This means that active optimal control can probably only be applied to fisheries which are already well understood. Where this is the case, it seems an attractive general principle to follow provided the gains in precision are greater than any reduction in yield as part of the experiment. Since the aim is to manage the stock indefinitely, any improvement in knowledge presumably has infinitely large value, and the loss in any one year can be set to some reasonable small percentage of the yield.

Fishery management then is an exercise in optimal control using appropriately selected models. The continuous commitment to adjustment in the light of new circumstances or to improve the model implies that Robert May's millennium for fishery scientists in the sense that they can all retire, will never arrive. There will always be need for monitoring research and management of these naturally renewable resources.

Fisheries and Man – Contemporary Fishery Problems

A far greater complexity than we have considered so far underlies fishery management problems. EEZs have eased former problems of lack of ownership, but other social, economic and human factors must be considered. Not all of these assume easily quantified forms, and so some of our discussion will impinge upon political factors. First, we discuss the value of access to fishery resources. Then we look at some aspects of human behaviour in artesan fisheries, why overcapacity occurs widely and finally the general question of how far such factors can or need to be included in fishery management.

The Value of Access to Resources

The Lowestoft fishing fleet, from a small traditional port on the east coast of the UK, is currently afflicted with problems similar to those in other parts of the world where stocks are still overexploited by vessels from several sources. We will consider it in detail as an example.

Favourably-situated close to herring and plaice stocks, Lowestoft expanded as a fishing port in the mid-nineteenth century when trawling and drifting techniques arrived at the same time as the railway. Seventy per cent of the catch is now plaice, caught at some distance out in the southern North Sea on trips lasting about ten days. The fleet is a rather heterogeneous mix of 25 m vessels most of which are specially adapted for the local variety of otter trawling. A few boats are multipurpose stern trawlers.

North Sea fisheries are in a parlous state; restrictions came too late to prevent collapse of herring and mackerel stocks (e.g. see Burd, 1974). Distant-water grounds are now closed except for a favourable agreement with Norway which means that cod can still be caught. Failure to police regulations has led to the sole fishery in the south of the North Sea being overfished and it is in any case more efficiently caught by the Dutch beam trawls rather than by the otter trawls, which are best for plaice. Thanks to recruitment from a run of good year classes, the plaice fishery is fully- rather than overfished, although higher long-term yields could be obtained from regulations aimed to do more than just preserve the status quo. A run of poor year classes would also alter the picture for the worse. All these factors mean that Lowestoft plaice fishermen have stayed in business, but cannot increase revenue by taking a larger catch or switching to other stocks. Rising fuel costs in 1979 brought the fishery into an operating deficit. Optimisation of a simple bioeconomic model (similar to that in Chapter 9) indicated that the best solution was to manage the plaice stock so that CPUE increased, improving efficiency by removing the smallest and largest boats in the fleet.

Government scientists, however, considered that such improvements would be 'merely cosmetic' if a further problem was not tackled. Dutch beam trawlers land many plaice as a by-catch of sole and their operating costs are more than covered by this up-market species sold on French and German markets. The plaice represent a bonus sold at virtually any price; moreover the Dutch fish are eminently marketable since they are accurately graded with modern machinery. Cheap imported Dutch plaice have consequently taken the market in the UK. Fishery scientists conclude that the only way to preserve the Lowestoft fleet is by government subsidy out of taxes, a solution which generates all kinds of political problems which we need not discuss here.

On a global scale such problems appear parochial; surely it matters little who catches the plaice so long as they are not overexploited? There are, of course, further levels to the problem. First, the collapse of a local fishery has direct and quantifiable local costs, the loss of cash revenues and the costs of unemployment. Short-term preservation by subsidy can be compared with this sum. Beyond this timescale we enter into the political realm, which does not make a decision any easier. In a perfect world, no one would mind who did the fishing, but both nations and large companies inevitably act out a game of self-interest. This means that there is political value in preserving access to the renewable natural resource. If the fishery is in another EEZ, we can bid or trade for it, but within a common multiple-ownership such competition is bound to result in changes in fishing fleets. The value of continuing access is a political decision based on an assessment of the disadvantages and risks of no longer having a fishing fleet based at Lowestoft. At a later date it may be possible to modernise the fleet, switch to a recovered stock, or trade for access to another. Current political alliances might falter. Confidence in the relative ability of the two fleets to conserve the stock might also be assessed. Such political decisions are out of the hands of fishery scientists; all we can do is to indicate the likely impact of the various consequences on the fish stocks.

Non-RANs have severe problems since the EEZs were introduced. Russian vessels in particular have become the scavengers of the seas, buying up fish of all types (and conditions!) trans-shipped from local vessels, and even trading excess fish between different parts of the world (e.g. sprats from the UK traded to Portugal). Such nations clearly put a very high value on access to protein resources.

Overcapacity

Overcapacity is a serious and endemic problem in modern fisheries (it is also termed overcapitalisation, overindustrialisation, or over-expansion), and means that the fishing fleet is capable of effort beyond that needed to exploit the stock optimally. In extreme cases the fish could be virtually wiped out by the combined fishing power of modern vessels. In open-access fishing this is the process of rent dissipation we examined in Chapter 9. However, even in regulated fisheries, the problem grows worse since at root it is another but more subtle symptom of rent dissipation beyond that found with open-access fishery (Wilen, 1979, and Chapter 9). In the most detailed economic model of a fishery to date, Clark (1980) has demonstrated that catch quotas alone do not improve economic performance over the open-access fishery. Clark's model illustrates the pressures which generate overcapacity in licensed and quota-controlled fisheries through the competitive introduction of newer ships capable of greater effort, even though the total number of vessels may not increase and even if the TAC has to be adhered to. The symptom of this form of over-capacity is a progressive shortening of the fishing season needed for the TAC. Individual fishermen compete because a vessel with more fishing power can take a larger share of the quota. Even the introduction of vessel quotas does not cure the problem, since a more powerful boat benefits from catching its quota in the shortest possible time. It may be able to offload catch in excess of its quota to another boat, still at some gain to itself, or it may have free time to travel to another fishery. Finally, the most modern and efficient vessels gain by being able to be more selective about the catch they retain and return to port. Less valuable sizes or species caught in the net can be discarded at sea without anyone knowing, fishing being continued until the desired type or size is caught. These factors generate a pernicious trend even in apparently well-regulated fisheries.

North European fisheries will serve to illustrate the problem. Many of the new ships built for distant waters in the early 1970s were made virtually redundant overnight when the EEZs were declared, and switched to local stocks to keep going. North Sea herring and mackerel were overexploited and attention switched to west British mackerel stocks, the relatively small local markets for these fish being suddenly enhanced by the new process of trans-shipment to scavenger ships from non-RANs. Successful switches to new stocks like this are soon reflected in new vessels more flexible than the older generation. For example, the heavily overexploited North Sea sole fishery has financed new Dutch multi-purpose ships capable of pair trawling for herring or sprats as well as conventional demersal fishing. The mackerel bonanza has stimulated the production of ultra-modern purse seiners which, working at night with a battery of sonar aids, can position themselves accurately over shoals even in rough seas. Such boats working out of Ullapool, West Scotland, in 1980, regularly took their week's quota of 70 tons of mackerel per ship in just one night.

Where overcapacity occurs in unregulated fisheries it soon proves disastrous and the loss of employment costs governments money – hence the temptation to subsidise before complete collapse. But the insidious consequences of overcapacity in apparently fully-regulated fisheries are more difficult to manage.

Having caught their quota, the temptation is for idle boats to try their luck with something else or make illegal catches, and this has to be countered by stricter licences and penalties. Regulation will be made all the more complex by new boats capable of different kinds of fishing. In the long run, a small and efficient fleet able to switch its effort between different stocks is no bad thing, but strict control will be crucial.

The control of selectivity of catch is more difficult. Trawled fish discarded at sea all die, so if they are undersized and immature the effect on the stock can be serious. Mackerel discarded from purse seines by 'slippage' are stressed and physically damaged by overcrowding and suffer at least 70 per cent mortality (Pawson and Lockwood, 1980).

The only direct way to reduce this is by control and monitoring of effort, to which we return at the end of this chapter.

In general terms, the capacity of a fishery in catch and effort can be estimated from gear and boat design and compared with the maximum which should be allowed (given by the long-term MSY), so that in principle it should be easy to regulate. At present, most fisheries limit only catch and numbers of vessels, but since overcapacity can still arise, further regulation is required.

Human Behaviour and Artisan Fisheries

Historically, fisheries have been a powerful factor in shaping many human societies. In the middle ages, Amsterdam was said to have been 'built on the bones of the herring' and the Hanseatic League of North German cities derived wealth from the Baltic herring stocks. In the UK seasonal migrations of itinerant fish gutters, picklers and smokers followed herring migrations up and down the east coast. Traditional whaling communities existed in Newfoundland and elsewhere. Although many such communities are now lost or diluted, in democratic societies those who sense strong ethnic ties to traditional ways of life will tend to vote to preserve them. Such artisan fisheries can never produce the spectacular yields of the modern industrialised fisheries, and often lack efficient processing and marketing links. However, in addition to political pressure for their preservation, artisan fisheries exhibit desirable features of human ecology from which we could learn.

Intimately linked with the fishery, complex social behaviours have to some extent acted as a buffer to change and the introduction of modern technology, which invariably increases effort and depletes the resource. Geographically (or ethnically) isolated fishing communities still exist in many parts of the world. They tend to have a limited impact on the resource partly because of the rigours and dangers of traditional fishing outlined in Chapter 9, but also partly for other reasons. Predation on fish by low-technology man exhibits many of the general features of predation which has evolved in other species. We will give two examples, one from native Indian fisheries in British Columbia, and the other from traditional lobster fishermen in Maine, USA.

The first example shows that fisheries which have evolved over a long time in a locality do not deplete the resource at normal stock densities. Peterman (1980) describes how six native Indian fisheries for sockeye salmon in British Columbia act like many top predators. Their impact on the resource is governed by a 'type II' functional response (Holling, 1965) in which the number of salmon taken per predator reaches an asymptote with increasing salmon density. Indians in tribal areas are licensed to operate traditional gear for food, but are prohibited from selling their catch. Considerable ritual is associated with the operation of the fishery, although some of this culture has sadly been lost. Regulation of the fisheries generates much local political conflict in British Columbia.

The functional response of the salmon fishery is governed in two ways, by handling time and search efficiency, as with other predators. There is an upper limit on the area swept by traditional gear, and catch is also limited by time taken to empty the salmon trap. More modern gear would have higher limits. Although the Indian fisheries take less than one-twentieth of the total BC salmon catch, more than a quarter of the fish approaching spawning grounds can be taken. The type II response causes the fishery to be depensatory, or negatively density-dependent. This means that there is higher fishing mortality at low salmon densities than at high. So the Indian fishery cannot have a significant effect on abundant salmon runs, but the higher percentage mortality caused at low stock densities can be destabilising and increase the time taken for stock recovery. Critical depensation, reduction to a density from which numbers cannot recover (Clark, 1976), could occur in some stocks, but these dangers are only likely to occur when fish have already been much reduced by the industrial fisheries for salmon at the river mouths, in the estuary and along the coast. Peterman concludes that the Indian fisheries must be included in salmon management plans and models for the modern fishery because of the depensation effect.

Secondly, there are fisheries where even traditional gear would deplete the stock, but social behaviour among the fishermen restricts effort and conserves the resource, perhaps in an analogous way to other top carnivores.

Legally there are few restrictions on the Maine lobster fishery (Acheson, 1975). There is an easily obtained government licence, protection for breeding females, a minimum size limit, and rules which ensure that lost pots rot to free trapped lobsters. About 6,000 lobstermen from the communities along this 2,500 km coastline form a number of 'harbour gangs', each defending a lobster fishing territory of approximately 100 km^2. Sonar- and radar-equipped 12 m boats are used, each man operating about 600 traps, setting one-third of them in a day. Entry into the fishery is difficult, unofficial apprenticeships being operated by the harbour gangs, whose activities are however only loosely organised and governed more by social custom than by a military-style command structure. Sanctions imposed on interlopers and transgressors of local rules start with verbal abuse, include ritualised warnings (knots tied on buoys), and culminate in destruction of gear, but usually stop short of personal violence.

Traditional perimeter-defended territories tend to be found on islands where ethnic ties are stronger and/or where there has been less pressure to buy larger boats whose loans have to be paid off by increased fishing. Such pressures are encountered most strongly in river-mouth communities where, without long-distance boats, lobsters can be caught only in a few months. Here, and in other mainland areas, there has been a trend to nuclear-defended territories, where deep incursions are defended but mixed fishing is tolerated further out to sea.

Both types of territory restrict effort, increase efficiency by allowing only experienced men to fish, and act to conserve the lobster resource. Acheson demonstrated however that the traditional perimeter-defended territories were more effective as they had a higher CPUE, larger lobsters, and higher per capita income than the nuclear territories. Some of the 'lobster fiefs' had brought in closed seasons and individual pot limits and Acheson deplored the swing from traditional territories which accompanied the expansion of capacity.

This self-organised collective fishery evolved in a closely-knit society with relatively low mobility. Each catch has high value and the dangers of going fishing are not so great that maximum effort could not wipe out the resource. Many artisan fisheries may have been similarly organised to conserve their resources in parts of the world where feudal ownership was absent or lax, and it could be interesting to look for other historical examples. Some features might profitably be adopted by modern fishermen's co-operatives. It may be beneficial to encourage artisan fisheries alongside modern industrialised ones, perhaps by regulating the zones in which each may operate. Different types of regulation might apply to the two kinds of fishery. The artisan fishery would be encouraged to run itself and supply local markets, giving the additional benefit of employment. Artisan fisheries can often continue to operate efficiently in politically difficult areas; for example a co-operative of 200 fishermen now runs Europe's largest eel fishery on Lough Neagh, Northern Ireland, having effectively restricted gear and effort, something which the previous private owners were unable to do in the face of problems such as the bombing of the fishery protection vessels.

Sport Fisheries

Fish are not always caught for food; the economic value of fish as part of recreational and tourist industries is very high. For example, it has been calculated that a salmon is worth twelve times as much in the river waiting to be caught by an angler than on a plate as food. Freshwater sport fisheries, principally for salmonids and cyprinids in Europe and for salmonids and perciforms in North America, support huge recreational industries of equipment suppliers and reading material. Marine sport fisheries for sharks, tuna and billfish contribute greatly to tourism in sub-tropical countries. Regulation of sport fishing via size limits, licences and closed seasons for breeding is widespread, but is often on a very poor scientific basis. However, management aims are often rather different from those in food fisheries, in that anglers prefer to catch a few exceptionally large specimens rather than a larger number of small fish. Most anglers would like the impossibility in nature of catching lots of large fish and this dream can be met, along with giving novices a catch, in the commercial 'put and take' fisheries for freshwater salmonids. Here naive hatchery-reared fish are stocked into small lakes and rivers. Good reviews of the main principles of sport fishery management can be found in Bennett (1971) and Everhart *et al*. (1975).

In addition to the direct commercial value of the industry supporting sport fisheries, the recreational opportunity to fish can also be priced so that the value of sport fisheries can be compared with alternative demands on water use. This can be done through surveys asking how much anglers would be willing to pay for their sport (Cauvin, 1980).

Whales

The characteristics of 'fisheries' for whales deserve special mention for three reasons. First, the recent notorious history of whaling has involved the successive destruction of species after species as ineffectually regulated effort has switched from one stock to the next, despite early and emphatic warnings about the depleted state of stocks. Secondly, these mammals with their low reproductive rate, social behaviour of travelling in family groups, long juvenile period, and migrations beyond the confines of EEZs, have been rendered particularly susceptible to recruitment overfishing and possibly to the deleterious efforts of selection (Estes, 1979, and Chapter 5). Thirdly, whales find themselves naturally the object of international conservation concern because of their high intelligence and complex social behaviours. Some people would wish to prohibit the killing of whales completely on these grounds alone, while others would at least wish to see controlled slaughter carried out more humanely than at present. With decimated stocks, species of blue whales and the larger fin whales are protected from capture at present, although illegal catches are still reported, and there are virtually no sanctions which can be imposed. Other species, notably sperm whales, smaller species of fin whales and humpbacks, are still caught in large numbers and are overexploited despite ostensible reductions in quotas. Statistics on whale populations are difficult to gather (Beddington, 1980), depending to a large extent on the whaling industry, and can hardly be expected to be unbiased. It is now realised that extended searching by whaling vessels for these social animals can mask the symptoms of stock depletion for some time. Humpbacks migrate from the central Atlantic to Newfoundland to feed on spawning capelin, grey whales migrate from the arctic to spawn off California, and such vast migrations mean that many whale stocks pass through a whole series of EEZs. In some of these they are protected (e.g. USA), but in others they will still be caught by small local whaling industries (e.g. Spain).

Large whaling fleets have until recently been maintained by Japan and Russia, who have tended not to follow the recommendations of the IWC, the world-wide advisory body on management. Despite recent improvements in its scientific advice, the IWC has little power, no inspectors independent of the member states, and no effective sanctions which it can impose on nations which do not comply with its quotas. Scientific advice has often been abandoned for political gain and even in the face of the collapse of stocks, catches have only been reduced from 45,000 whale units in 1972 to 14,000 in 1980. Bonner (1980) states: 'Now the damage is done. It remains to be seen if whale stocks will ever recover'.

Whale fisheries can be inhibited by stopping trade in whale products, and by moral pressure of public opinion, neither of which is especially effective for countries like Japan which have large national whale-product industries of their own. One aspect of international trade which has recently been reduced is in sperm-whale oil used in tanning leather, for which synthetic substitutes are now available. The Greenpeace foundation has hindered whale catching at sea, creating effective pro-whale propaganda such that it seems likely that whaling by Western nations will probably cease in the near future. In 1980, Russia disbanded her north Pacific whaling fleet. One aspect of the demise of the great fin whales of the southern ocean is the effect on sub-antarctic ecosystem, and the huge krill populations in particular.

Whether the krill formerly eaten by the whales is now available for other species or ourselves to crop or whether there will be changes in the balance of energy flow through these ecosystems is unknown.

Optimal Management

How far can the social and economic factors we have discussed be included in the management plan for the fishery? Attempts to include as many as possible seem to be made in current fishery models; as we have seen some aspects of access to the resource and overcapacity can be quantified. Anderson (1980) has, for example, incorporated the value of a behavioural factor, the 'worker satisfaction bonus' in an economic fishery model. This represents the aesthetic benefits which are experienced by commercial and sport fishermen alike. Silvert (1978) has set a monetary value on the improved yields which should follow active optimal control management.

Attempts to include everything possible have led to the concept of 'optimal sustainable yield' (OSY), which has been defined by Roedel (1975) 'a deliberate melding of biological, economic, social, and political factors designed to produce the maximum benefit to society'. Larkin (1977) pithily comments that 'your optimum is unlikely to be the same as mine' and questions 'what is a deliberate melding as opposed to an inadvertent one? Is this like being a virgin by intent?'. Indeed it is probably unwise to hope for a single model to include all these relevant factors, just as the search for the universal ecosystem model in the 1970s turned out equally to be a search for the philosopher's stone. Canadians like Sinclair (1978) and Mitchell (1979) have argued strongly for reducing biological restrictions to a minimum in fishery management, apparently on the grounds that these distort the economics of the free market. Their regulations are based largely on economic factors aiming for what appears to be a rather nebulously defined 'best use' of fishery resources. These writers seem to have ignored the problems which arise when fully realistic economic factors (such as MPV) are included in fishery models, considering them only to represent 'fine tuning' imposed on the simpler kinds of economic analysis.

It has also been suggested that fishery biologists can provide management advice for whatever goals others, such as politicians, may set them. The fishery biologist is expected to show how the stock should be exploited in order to achieve OSY, MSY, MER, OER, or any other goal which might be set. Superficially this sounds plausible, until one realises that certain political or economic goals bring unacceptable biological penalties. Two examples are particularly clear. First, management for MPV is associated with large catches whose size is determined more by the whims of the financial markets and interest rates than any regard for risks to the stock (Chapter 9). Secondly, it has been suggested that biological advice might be provided in order to manage the fish stock at the 'zero net revenue' level (with or without subsidy), with the political aim of maintaining employment in the fishing industry. As we have seen in Chapters 7 and 8, this too would be an unwise goal because it would increase the risks of stock collapse. In addition, it is hardly a sensible strategy on economic grounds alone, since greater variability in yields is associated with catches beyond MSY (and this is true whichever fishery model one uses, as we have seen in Chapters 7 and 8). Greater variability in landings from year to year would not be helpful to a policy attempting to keep people in work.

The best policy seems to us to produce sound purely biological recommendations, decide what regulations might achieve them, and then bring in economic and other considerations as a way of choosing between alternative strategies of gaining the *optimal biology*, which may of course include conservation aspects as well as fish yields. To run things the other way round seems a path fraught with dangers, since only by putting the stock biology first can we be sure of continuing to use the valuable naturally renewable resources of fisheries.

Fishery Control Devices

Decisions made about the control of a fishery have to be implemented. This section reviews the range of fishery control devices listed in Table 11.2. When particular controls are proposed, the impact on other species and stocks should also be considered. Optimal control of one fishery may harm another through the food chain or by switch of fishing effort. Therefore in general terms we feel that control devices should always be seen as part of a wider plan for all the fish resources in a region. Table 11.2 also lists the fishery management groups and the interested parties in fisheries agreements. It is never possible to make simple and foolproof recommendations for the best fishery control method; this will depend on the characteristics of particular stocks and local conditions. However, it is possible to indicate which methods might be most useful.

The first two control measures act on the age at first capture, t_c, and are designed to increase recruitment, allowing fish to grow nearer to

Table 11.2: Summary of Fishery Regulation Devices.

Age at first capture:
- inspection of landed fish
- gear limitation – minimum mesh size

Fishing mortality:
- closed seasons/periods
- sanctuary zones
- boat licences
- gear limitation – type
- energy consumption
- catch quotas
  - total allowed catch (TAC)
  - partially allocated quota (PAQ)
  - vessel allocated quota (VAQ)

Taxation:
- profit
- industrial capacity
- catch
- effort

Enhancement and Conservation:
- spawning grounds
  - preservation
  - construction
- stocking
  - sea ranching

Farming:
- food fish farming
- 'put and take' sport fisheries

Fishery Management Groups
Biological research units
Shore inspectors
Fishery inspection/protection vessels

Interested Parties to Agreements
National governments:
- resource adjacent nations (RANs)
- non-RANs

Multinational processing and fishing companies
Port and vessel owners
Working fishermen and co-operatives

the age of peak cohort biomass (Chapter 8). Most fisheries set a lower limit on the size of fish caught by regulating mesh size and specifying minimum landing sizes. While avoiding excessive dangers of fishing before maturity (Chapter 8), such measures alone do not prevent increasing fishing effort for two reasons. First, the number and power of vessels entering the fishery is not limited; and secondly, regulations increasing mesh size are difficult to introduce because fishermen naturally resist anything which reduces their personal catch. In this case their short-term loss becomes a long-term gain, but awareness of this probable benefit is countered by immediate economic pressures (Chapter 9). Icelandic cod fisheries should have moved to 160 mm mesh a decade ago, but even with single ownership of the EEZ, the present increased mesh is still a compromise.

Mesh regulations are best policed by random inspection of gear at sea accompanied by on-the-spot fines. Inspection of catch on shore is too late to prevent damage and on its own causes small fish to be discarded at sea, but infrequent random shore checks back up the sea gear inspection. Industrial fisheries for meal in the same region as food fisheries make regulation difficult. For example, undersized mackerel can be sold for reduction to fish meal. Sale of small mackerel reduces the incidence of slippage, although it will still happen unless banned, as skippers are always driven to a more profitable catch. The best policy would be to ban the landing of small fish altogether, sending stale large fish to the meal plants. This would still not eliminate slippage which would have to be controlled separately, and might need short-term cash subsidies to meal plants in the change-over period. A similar problem is encountered if small species are fished in the same region (e.g. sprats for canning, or sandeels for meal); in such cases it is almost impossible to stop catches of juveniles of other species. Where the stocks are sympatric, the only hope is to open only small areas to fishing.

The next batch of measurements in Table 11.2 all act to reduce fishing mortality by restricting fishing effort. Seasonal closures and sanctuary zones are helpful auxiliary controls which allow stock to reproduce and grow undisturbed in a pattern of closed areas rather like fallow fields in medieval crop rotation. The famous Pacific halibut stock has been managed in this way. The measures need heavy investment in fishery protection vessels in order to work.

The Chesapeake bay drift net fishery for shad (Foerster and Reagan, 1977) provides an example of a small overfished inshore resource with declining CPUE, in danger of recruitment failure, and with estuarine/spawning habitats damaged by dam building. The management plan includes alternation of sanctuary areas, rest days, and restrictions on size and mesh of nets together with a further measure from Table 11.2, enhancement of the spawning grounds. On the Pacific coast of North America large artificial spawning channels are frequently constructed to enhance salmon stocks.

As a last-ditch measure for severely depleted stocks, a complete prohibition on fishing can be easier to enforce than the above, but problems are still encountered. For example, despite the complete ban on fishing, 10,000 tons of North Sea herring were landed in 1979 as a by-catch (Blacker, 1980). This represents about one-tenth of the likely yield of the recovered stock. Some North Sea herring are also caught in the Skaggerak on migration. Similar problems hinder the recovery of blue and the large fin whales which continue to be taken illegally and by countries not adhering to the International Whaling Convention's quotas.

In the future all fishermen may have to be licensed. The arguments presented in Chapter 9 and above indicate that access to resources must be limited, but many exercises (e.g. Hilborn, 1979; Clark, 1980) show that mere limitation of entry alone is not sufficient to prevent increasing fishing effort. As we have described above, nor is a combination of closed seasons, vessel licences and quotas, since the fishing season just gets shorter as effort increases (May, 1980).

Gear limitation, in addition to mesh regulation, is a sensible way of restricting effort. If the temptation to use modern gear to the full breaks quotas, licensing can be reserved only for certain gear. For example, herring fisheries off Iceland are now limited to the traditional nocturnal drift net method, albeit in very modern vessels, rather than permitting purse seines or pair trawlers. Traditional gear may be preserved in artisan fisheries, and is to a large extent essential for their survival.

The cost of energy acts passively to regulate fishing to some extent, but is not very effective because prices vary greatly between regions and over time, and gradual increases can be passed on to the consumer. Effective effort and gear limitation would be aided by actively favouring vessels which were energy efficient. Vessels might be taxed, or perhaps earn quotas, according to fuel consumed per trip and ton of catch. This would help conserve energy supplies as well as fish! Some values for the energy consumption of fishing vessels are given in Table 11.3. The energy budget for fish pellets in Chapter 10 was based on consumption by a modern purse seiner at about 10 GJ/ton of fish landed. Freezer trawlers can consume as much fuel in weight or energy as they land fish, since 1 ton of oil is equivalent to 43 GJ. Drift netting is very energy-efficient, and this is also true where boats are aided by sail, as in the Peruvian fishery for anchovy. In the long term, low energy consumption might be actively encouraged in fisheries as an end in itself.

Table 11.3: Energy Consumption of Various Fisheries, GJ Per Ton Landed (Dockside Weight).

Fish	Location	Energy Cost (GJ/ton landed)
shrimps	Gulf of Mexico[a]	359
all fish average	Adriatic[a]	170
demersal fish	North Sea trawlers[a]	50
all fish	UK average[a]	35
pelagic	UK average, mainly herring drifters[a]	10
pilchards	high-technology purse seines (Namibia)[b]	10
herring mackerel	Scottish purse seines[a]	4
anchovy	seiners with auxiliary wind power (Peru)[c]	0.6

Sources: a. Leach (1975); b. Pitcher (1977); c. Slessor (pers. comm.) and Leach (1975).

Most fisheries are currently managed by setting quotas at the start of each season on the basis of whatever scientific advice is available. Quotas act to limit the number (or weight) of fish which can be legally caught, and can be set at several levels of sophistication and effectiveness. The TAC covers the maximum catch permitted from the whole stock. Escapement is an equivalent term used mainly in North America meaning the number of fish allowed to escape capture, in other words, the stock size minus the catch. In theory TAC fixes the escapement. However, catch can be measured, but escapement can only be estimated. An additional disadvantage is that escapement cannot be applied meaningfully to anything other than the whole stock.

TACs are subject to a number of difficulties. TACs usually cover wide areas so that catches in excess of quotas regularly occur because they are difficult to detect until they have already happened. Often there are few sanctions that can be applied because the vessels which exceeded the quota cannot be fairly identified. Trans-shipment and illegal or unrecorded catches are difficult to detect. For example, French quotas for hake were greatly exceeded in the Bay of Biscay by unlicensed Spanish boats in 1980 (May, 1980). The TAC for west British mackerel was set in 1979 at 435,000 tons, but the actual catch was over 600,000 tons (Blacker, 1980). Worse consequences can ensue;

profits from catches double the TAC in the North Sea sole fishery have helped finance new flexible vessels which will in the words of a Dutch fishery scientist 'make short shrift of plaice and sole' (Anon, 1980).

To avoid these problems, TACs will increasingly be split up into much smaller units, preferably in many cases being brought right down to vessel allocated quotas (VAQs) acting in concert with gear and licence limitation. May (1980) argues very strongly that this is going to be the only effective control measure to limit effort and Clark's model (1980) shows that only 'allocated transferable catch quotas' (essentially VAQs which can be traded between vessels) work in his model. VAQs have the advantage of flexibility within a season, and can be relaxed or made more stringent depending on the speed with which the TAC is being approached. Unforeseen stock changes can also be responded to, so the measure meets all the requirements of optimal control.

A somewhat different strategy has to be adopted where fish like migratory tuna do not remain in one ownership. Both fish and vessels can travel great distances. Tuna boats operate in all the oceans of the world within any one year and consequently regulations enforced in one ocean can mean a rapid switch of effort to another. The only effective management would have to be on a global scale. A recent approach to such a plan (Joseph and Greenhough, 1979) attempts to reconcile the interests of fishing nations of different levels of economic development. RANs for tuna are generally developing nations, whereas non-RANs are industrialised fishing nations controlling large markets and processing industries. Joseph and Greenhough compare a range of measures, including competitive bidding to lease the resource, various forms of TACs, and complete control by individual RANs as the tuna move through each of their EEZs. The best system was thought to be the 'partially allocated overall quota', or PAQ. Under PAQ, the RANs get an exclusive part of the quota in recognition of their proximity to the stock. Non-RANs get open access to the rest of the quota, but pay a fee or tax on catch to the RAN, and also contract marketing of the catch. This money can be used by RANs to develop their own industry and to police the resource in their own region. Management would be co-ordinated by a central body including research staff, funded by a tax on all revenues. An important feature of the plan is that all species encountered, such as frigate mackerel and billfish in the tropical tuna fisheries, should be covered. Clearly, the PAQ system is an attractive plan for highly migratory species.

A number of workers (mainly Canadian) have suggested that fisheries might be regulated by a system of taxes (Smith, 1969; Scott, 1979; Sinclair, 1978). They argue that large and progressive taxes would ensure compliance with fishery regulations. Taxes on profits would be too late, uncertain and avoidable to be of much use. To be effective, fishing taxes would have to be in a different category to other government taxes, so that in big companies they could not be offset against 'losses' elsewhere. Since MSY sets a limit to the size of a fishery, taxing overcapacity is viable and provides a pleasing long-term solution. Furthermore, a tax on the catch of individual boats which progressively forces a VAQ is also attractive, and has the advantage of keeping the TAC within bounds more easily.

Hilborn's (1979) comparison of fishery control systems showed that managing for equilibrium effort is a safer option at low stock sizes but is difficult to restrict directly. Scott (1979) and Anderson (1977) have made a case for a direct tax on fishing effort as a most effective taxation instrument. (Measurement of effort might not be too hard to make; we have one suggestion below.) Clark (1980) demonstrated that taxes on effort were formally equivalent to VAQs as controls, but he preferred to use quotas as being more immediate. May (1980) and Scott (1979) support this view because they consider quotas less easily avoided. Taxes are thought to need detailed economic knowledge of markets and prices which vary and which can only be imprecisely known. On the other hand, we have shown that effort limitation has the benefits of restricting overcapacity, inhibiting slippage and conserving energy. If we want to limit effort, taxes are probably the best way to do it.

Whatever control devices are introduced, Larkin (1977) has pointed out that 'unfortunately, fishermen have the vote', implying that an unpopular measure may be difficult to introduce and police. Larkin suggests doing nothing to subsidise fishing, but governments faced with unemployment, loss of access to a resource, and industrial lobbies in the processing sector, may find short-term subsidies hard to resist. The advantages of optimal biological control of stocks need to be constantly proselytised in the face of this pressure.

So there is a clear need for tough and effective control measures, chosen as appropriate from the spectrum available, enforced and agreed by the interests listed in Table 11.2. Monitoring and research will remain expensive tasks.

One suggestion in this respect is to develop the TACometer, which we envisage as a microprocessor device mandatorily built into all new fishing vessels. It will work by encouraging compliance with the fishery regulations. The memory of the TACometer should be accessible to interrogation by a master computer for the region (via satellite link) as well as routinely transferring data at the end of a voyage. The TACometer would be built to record periodically the use of fishing gear, fuel consumption and navigational fixes. Catches could be entered by the skipper, or perhaps automatically via the holds. The system could be programmed to feed useful data on weather, fish prices or even satellite surveys to the skipper, so that it was seen to provide him with direct benefits too. Landings at ports would be recorded on shore-based TACometers and checked against vessel-recorded catches so that failure to tally could be investigated. The TACometers would provide instant catch/effort data. Problems like slippage might be detected by monitoring the sonar fishing aids, or could perhaps be detected by comparing catch with net use. Small artesan vessels could be exempted, provided their catch was controlled in other ways. TACometers would be a resource-beneficient 'eye on the bridge', which would back up fishery protection vessels and shore staff. The first TACometers would be fairly expensive to build, but the next million would be extremely cheap. We hope we will not be accused of furthering the illusion that perfect information automatically brings perfect control, but the TACometer could act to great benefit to world fishery resources when used with other devices to control the fishery optimally.

APPENDIX: COHORT ANALYSIS

Cohort analysis is a technique for estimating the magnitude of fishing mortality and the numbers at age in a stock from catch data only, provided M is known. The method is based on two population equations. First, the number in the next age group can be predicted from numbers in the current age group diminished by total mortality, Z_i,

$$\text{i.e.}\ N_{i+1} = N_i \exp(-Z_i)$$

But in fisheries, $Z_i = F_i + M_i$, and assuming for the moment that M_i is the same for all fished age groups, then

$$N_{i+1} = N_i \exp(-(F_i + M)) \tag{A.1}$$

The second equation states that the catch in an age group is given by that proportion of total numbers dying which results from fishing:

$$\text{number dying} = N_i(1 - \exp(-Z_i)) = N_i(1 - \exp(-F_i + M))$$

$$\text{and so, } C_i = N_i\left[\frac{F_i}{F_i + M}\right] (1 - \exp(-(F_i + M))) \tag{A.2}$$

Several workers, including Gulland (1965), realised that these two equations can be elegantly combined to estimate fishing mortality,

$$\frac{N_{i+1}}{C_i} = \frac{(F_i + M)\exp(-(F_i + M))}{F_i\,(1 - \exp(-(F_i + M)))} \tag{A.3}$$

which is just Equation A.1 divided by Equation A.2. Thus if N_{i+1}, C_i and M are known, an estimate of F_i may be obtained.

In practice this is not easy since F appears four times on the right-hand side (RHS), twice in rather awkward expressions. Using a computer it can be solved iteratively (i.e. by successive approximations). Tables are also available.

For some years this iterative method was termed 'Virtual Population Analysis' (VPA), because it bore some resemblance to a technique used elsewhere in ecology. The more descriptive term 'Cohort Analysis' is now preferable for the general procedure (Gulland, 1977).

Once F_i is known, the numbers in that age group, N_i, can be

Figure A.1: Flow Chart for Cohort Analysis.

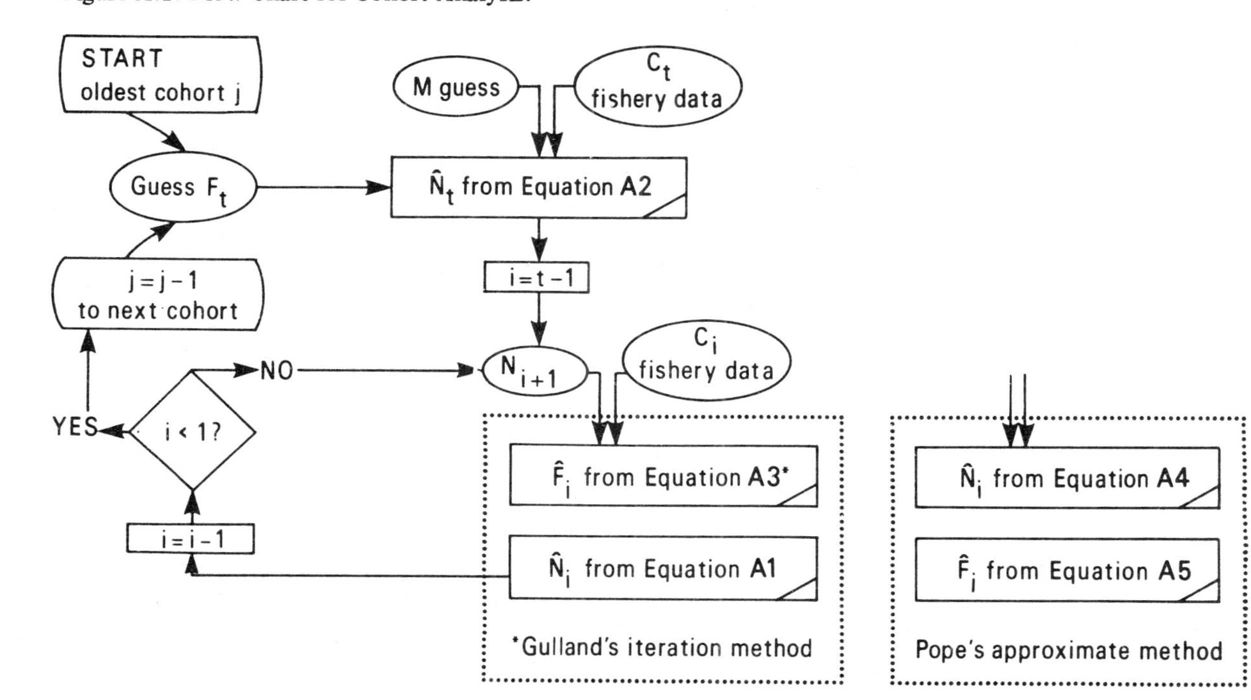

calculated by rearranging Equation A.1. We then set i to $i - 1$ and go to the next younger age group in the cohort, where we solve Equation A.3, again using the N calculated last time as N_{i+1} and plugging in the appropriate catch data. Then we get N_i from Equation A.1 again. We go around this loop until we have estimates of N and F for all the age groups of the cohort present in the fishery. The next cohort is then tackled. The procedure is set out explicitly on the flow chart (Figure A.1).

To start the analysis we have to guess F_t, the fishing rate on the oldest cohort present in the stock in the current year. From this guess we obtain N_t from Equation A.2. Pope has shown that the accuracy of results in the body of the analysis is not very sensitive to an accurate choice of F_t, and errors get progressively smaller in the earlier age classes. In practice the analysis is usually run for several values of F to make sure that any conclusions are robust against this sort of error.

Note that there is a slight difference in procedure depending on whether the data for age class t include only fish of that age, or covers all fish age t and over.

Pope (1972) bypassed the problem of Equation A.3 by substituting an approximation for the RHS which he showed had an accuracy of about ± 5 per cent when $M < 0.3$ and $F < 1.2$. He called his method 'Cohort Analysis' to distinguish it from VPA at the time. However, Pope's method is really an approximate version of general cohort analysis to obtain F. Several other versions were devised in the 1960s (see Gulland, 1977; Ricker, 1975). Pope's method, which is very useful if the calculations have to be carried out by hand, is drawn as an alternative in the flow chart. The two equations are:

$$\hat{N}_i = N_{i+1} \exp(M) + C_i \exp(M/2) \qquad \text{(A.4)}$$

$$F_i = \ln\left[\frac{N_i}{N_{i+1}}\right] - \mathrm{M} \qquad \text{(A.5)}$$

So in Pope's method the two estimates of N_i and F_i are obtained in reverse order to Gulland's method.

In the absence of information of previous year's catch, cohort analysis can be used to estimate the current rate of fishing on the age groups in a single year's catch. When used in this way the method makes the further assumption that the population is in a steady state. This is not likely to be true for any fishery you might be sufficiently worried about to do all this work for.

The standard cohort analysis method also assumes that natural mortality, M, is known and constant. In most fishery work this is more an article of faith than a known fact since M is quite difficult to measure, except in fisheries where F is known accurately anyway ($M = Z - F$). Age-dependent M can be incorporated into the model quite easily, but variations in M in different years, or density-dependent M, would be more difficult to consider.

Cohort analysis is usually carried out separately on data from each cohort in the stock from past years' catch information. It provides a history of F (and N) for each group present in the stock today, stretching back in time to the dates when each age group recruited. It is therefore potentially of great value in showing up large, and possible deleterious, changes in fishing mortality soon after they have happened.

Cohort analysis is currently being used in setting quotas for several important fish stocks (e.g. Bannister, 1977). The estimate of F which is going to be the most valuable in setting TACs is that from the year just past. The ideal system would be one in which man made his fishing perfectly density-dependent.

Unfortunately, this most recent set of F estimates is the least accurate in the analysis. Pope (1977) has attempted to reduce this problem by a modification of his approximate cohort analysis which employs the logs of the ratio of catches in a cohort in successive pairs of years. This has improved the situation somewhat, but he has emphasised that (a) fishery scientists have to learn to live with this problem since there is going to be no easy way around it, and (b) quota adjustments based on this type of analysis should take account of the likely errors involved (see Chapter 11).

Fishing effort can also sometimes be estimated from tagging experiments in which tags are recovered in the commercial fishery. Conventional methods are covered in Ricker (1975). There is some recognition now that none of these approaches to estimating F is sophisticated enough for the needs of modern 'real-time' management of fisheries, but as yet there is little to put in its place (Rothschild, 1977).

REFERENCES

Acheson, J.M. (1975). The lobster fiefs: economic and ecological effects of territoriality in the Maine Lobster industry. *Human Ecology 3*: 183–207

Alderdice, D.F. and C.R. Forrester (1971). Effects of salinity, temperature and dissolved oxygen on the early development of Pacific Cod (*Gadus macrocephalus*). *J. Fish. Res. Board Can. 28*: 883–902

Alexander, R.McN. (1974). *Functional Design in Fishes*, Hutchinson, London

——— (1975). *The Chordates*, Cambridge University Press, Cambridge

Allen, K.R. (1966). A method for fitting growth curves of the von Bertalanffy type to observed data. *J. Fish. Res. Board Can. 23*: 163–75

——— (1971). Relation between production and biomass. *J. Fish. Res. Board Can. 28*: 1573–81

Alverson, D.L., A.R. Longhurst and J.A. Gulland (1970). How much food from the sea? *Science 168*: 503–5

Anderson, K.P. and E. Ursin (1977). The partitioning of natural mortality in a multispecies model of fishing and predation. In: J.H. Steele (ed.), *Fisheries Mathematics*, Academic Press, London, pp. 87–98

Anderson, L.G. (1977). *The Economics of Fishery Management*, The Johns Hopkins University Press, Baltimore

——— (1980). Necessary components of economic surplus in fisheries economics. *Can. J. Fish. Aquat. Sci. 37*: 858–70

Anon (1978). Fishery commodity situation and outlook. FAO Committee on Fisheries 12th session. *COFI/78/inf.* (5 April 1978): 1–15

——— (1980). Eurofish Report. (10 December 1980)

Antipa, R. and D.F. Amend (1977). Immunization of Pacific salmon; a comparison of intraperitoneal injection and hyperosmotic infiltration of *Vibrio anguillarum* and *Aeromonas salmonicida* bacterins. *J. Fish. Res. Board Can. 34*: 203–8

Arthur, D.A. (1976). Food and feeding of three fishes occurring in the Californian current, *Sardinops sagax, Engraulis mordax* and *Trachurus symmetricus*. *Fishery Bulletin, Seattle 74*: 517–30

Backiel, T. and E.D. LeCren (1978). Some density relationships for fish population parameters. In: S.D. Gerkin (ed.), *The Ecology of Freshwater Fish Production*, Blackwell, Oxford, pp. 279–302

Bagenal, T.B. (1971a). The ecological and geographical aspects of the fecundity of the plaice. *J. Mar. Biol. Assocn. UK 46*: 161–86

——— (1971b). The interrelation of the size of fish eggs the date of spawning and the production cycle. *J. Fish. Biol. 3*: 207–19

——— (1973). Fish fecundity and its relationship with stock and recruitment. In: B.B. Parrish (ed.), *Fish Stocks and Recruitment. Rapp. Proc-Verb. Reun. Int. Comm. Explor. Mer. 164*: 186–98

——— (1974). A buoyant net designed to catch freshwater fish larvae quantitatively. *Freshwater Biology 4*: 107–9

——— (1977). Effects of fisheries on Eurasian perch (*Perca fluviatilis*) in Windermere. *J. Fish. Res. Board Can. 34*: 1764–8

——— (1978). *Methods for Assessment of Fish Production in Fresh Waters*, IBP Handbook No. 3, 3rd edn, Blackwell, Oxford

Bailey, N.T.J. (1951). On estimating the size of mobile populations from capture-recapture data. *Biometrika 38*: 293–306

——— (1952). Improvements in the interpretation of recapture data. *J. Anim. Ecol. 21*: 120–7

Bainbridge, V. (1972). The zooplankton of the Gulf of Guinea. *Bull. Mar. Ecol. 8*: 61–97

Balon, E.K. (1972). Possible fish stock size assessment and available production survey as developed on Lake Victoria. *Af. J. Trop. Hydrobiol. Fish 2*: 45–73

——— (1974). Fish production in a tropical ecosystem. In: E.K. Balon and A.G. Coche (eds), *A Man-made Tropical Ecosystem in Central Africa*, Monographiae Biologicae no. 24, Junk, Hague, Holland, pp. 249–87

——— (1975). Terminology of intervals in fish development *J. Fish. Res. Board Can. 32*: 1663–70

Banks, J.A. (1970). Observations on the fish population of Rostherne Mere, Cheshire. *Field Studies 3*: 357–79

Bannister, R.C.A. (1977). North Sea plaice. In: J.A. Gulland (ed.), *Fish Population Dynamics*, Wiley, London, pp. 243–82

Baranov, E.I. (1918). On the question of the biological basis of fisheries. *Nauchn. Issled. Iktologicheskii. Inst. Izv. 1*: 18–218 (cited in Ricker, 1975)

Bardach, J.E., J.H. Ryther and W.D. McLarney (1972). *The Farming and Husbandry of Freshwater and Marine Organisms*, John Wiley & Sons, New York

Bartlett, M.S. (1960). *Stochastic Population Models*, Methuen, London

Beadle, L.C. (1974). *The Inland Waters of Tropical Africa*, Longmans, London

Beamish, F.W.H. (1974). Apparent specific dynamic action of largemouth bass, *Micropterus salmoides*. *J. Fish. Res. Board Can. 31*: 1763–9

Beddington, J. (1980). Counting whales. *New Scientist 87*: 194–6

Beddington, J. and R.M. May (1977). Harvesting populations in a randomly fluctuating environment. *Science 197*: 463–5

Begon, M. (1979). *Investigating Animal Abundance: Capture Recapture for Biologists*, Edward Arnold, London

Behnke, R.J. (1972). The systematics of salmonid fishes of recently glaciated lakes. *J. Fish. Res. Board Can. 29*: 639–71

Bennett, G.W. (1971). *Management of Lakes and Ponds*, 2nd edn, Van Nostrand Reinhold, New York

Berner, L. (1959). The food of the larvae of the northern anchovy, *Engraulis mordax*. *Bull. Inter-Amer. Trop Tuna Commission 4*: 3–22

Berrie, A.D. (1972). Productivity of the River Thames at Reading. *Symp. Zool. Soc. London 29*: 69–86

Beverton, R.J.H. (1962). Long-term dynamics of certain North Sea fish populations. In: E.D. LeCren and M.W. Holdgate (eds), *The Exploitation of Natural Animal Populations*, Blackwell, Oxford, pp. 242–59

Beverton, R.J.H. and S.J. Holt (1957). On the dynamics of exploited fish populations. *Fish. Invest. London, Ser. II, 19*: 1–533

Blacker, R. (ed.) (1980). *Fishing Prospects 1979/1980*, Fisheries Laboratory Lowestoft, UK, 49pp

Blaxter, J.H.S. and G. Hempel (1963). The influence of egg size on herring larvae. *J. Cons. Int. Explor. Mer. 28*: 211–40

Blaxter, J.H.S. and M.E. Staines (1971). Food searching potential in marine fish larvae. In: D.J. Crisp (ed.), *Fourth European Marine Biology Symposium*, Cambridge University Press, pp. 467–85

Blaxter, K.L. (1975). The energetics of British agriculture. *Biologist 22*: 14–18

Boerema, L.K. and J.A. Gulland (1973). Stock assessment of the Peruvian anchovy (*Engraulis ringens*) and management of the fishery. *J. Fish. Res. Board Can. 30*: 2226–35

Bonner, W.N. (1980). *Whales*, Blandford Books, London, 278pp

Braum, E. (1978). Ecological aspects of the survival of fish eggs, embryos and larvae. In: S.D. Gerking (ed.), *The Ecology of Freshwater Fish Production*, Blackwell, Oxford, pp. 102–31

Breder, C.M. and D.E. Rosen (1966). *Modes of Reproduction in Fishes*, TFH Publications, New Jersey

Brett, J.R. (1979). Environmental factors and growth. In: W.S. Hoar, D.J. Randall and J.R. Brett (eds), *Fish Physiology Volume 8: Bioenergetics and Growth*, Academic Press, New York, pp. 599–75

Brett, J.R., J.E. Shelbourn and C.T. Shoop (1969). Growth rate and body composition of fingerling sockeye salmon, *Oncorhynchus nerka*, in relation to temperature and ration size. *J. Fish. Res. Board Can. 26*: 2363–94

Brett, J.R., W. Griffioen and A. Solmie (1978). Suitability of Pacific salmon for seapen culture. *Fish and Marine Service Tech. Rept No. 845*: 1–17

Brett, J.R. and T.D.D. Groves (1979). Physiological energetics. In: W.S. Hoar, D.J. Randall and J.R. Brett (eds), *Fish Physiology Volume 8: Bioenergetics and Growth*, Academic Press, New York, pp. 279–352

Buckley, L.J. (1979). Relationships between RNA-DNA ratio, prey density, and growth rate in Atlantic Cod, (*Gadus morhua*) larvae. *J. Fish. Res. Board Can. 36*: 1497–502

Buhler, D.R. and J.E. Halver (1961). Nutrition of salmonid fishes. 9. Carbohydrate requirements of chinook salmon. *J. Nutri. 74*: 307–18

Burd, A.C. (1974). The North-east Atlantic herring and the failure of an industry. In: F.R. Harden-Jones (ed.), *Sea Fisheries Research*, Elek Brooks, London, pp. 167–91

——— (1978). Long-term changes in the North Sea herring stocks. In: G. Hempel (ed.), *North Sea Fish Stocks – Recent Changes and their Causes. Rapp. Proc-Verb. Reun. Int. Com. Explor. Mer. 172*: 137–53

Burgess, G.H.O. (1980). Response to change: the influence of fish science and technology. In: J.J. Connell (ed.), *Advances in Fish Science and Technology*, Fishing News Books, Farnham, p. 4

Castell, C.H., B. Smith and W.J. Dyer (1973). Effects of formaldehyde on salt-extractable proteins of gadoid muscle. *J. Fish. Res. Board Can. 30*: 1205–13

Cauvin, D. (1980). The valuation of recreational fisheries. *Can. J. Fish. Aquat. Sci. 37*: 1321–7

Cavalli-Sforza, L.L. and W.F. Bodmer (1971). *The Genetics of Human Populations*, W.H. Freeman, San Francisco

Chakrabarty, R.D., N.G.S. Rao and S.R. Ghosh (1979). Intensive culture of Indian major carps. In: T.V.R. Pillay and W.A. Dill (eds),

Advances in Aquaculture, Fishing News Books, Farnham, pp. 153–7
Chaudhuri, H. and S.D. Tripathi (1979). Problems of warmwater fish seed production. In: T.V.R. Pillay and W.A. Dill (eds), *Advances in Aquaculture*, Fishing News Books, Farnham, pp. 127–34
Clark, C.W. (1973). The economics of overexploitation. *Science 181*: 630–4
——— (1976). *Mathematical Bioeconomics*, John Wiley, New York
——— (1980). Towards a predictive model for the economic regulation of commercial fisheries. *Can. J. Fish. Aquat. Sci. 37*: 1111–29
Clark, C.W. and G.P. Kirkwood (1979). Bioeconomic model of the Gulf of Carpentaria prawn fishery. *J. Fish. Res. Board Can. 36*: 1304–12
Clayden, A.D. (1972). Simulation of the changes in abundance of the cod, *Gadus morhua L.*, and the distribution of fishing in the North Atlantic. *Fish Invest., Lond. Ser. II, 27*: 1–57
Colebrook, J.M. and G.A. Robinson (1965). Continuous plankton records; seasonal cycles of phytoplankton and copepods in the North-east Atlantic and North Sea. *Bull. Mar. Ecol. 6*: 123–39
Connell, J.J. (ed.) (1980). *Advances in Fish Science and Technology*, Fishing News Books, Farnham
Coombs, S.H. and R.K. Pipe (1978). The distribution, abundance and seasonal occurrence of the eggs and larvae of the blue whiting, *Micromesistius poutassou* (Risso), in the eastern North Atlantic. *ICES CM 1978/H: 45*: 1–9
Coull, J.R. (1975). The big fish pond: a perspective on the contemporary situation in the world's fisheries. *Area J. Inst. Brit. Geog. 7*: 103–7
Cowey, C.B. (1975). Aspects of protein utilisation by fish. *Proc. Nutri. Soc. 34*: 57–63
Cowey, C.B. and J.R. Sargent (1972). Fish nutrition. In: F.S. Russell and C.M. Yonge (eds), *Advances in Marine Biology, Vol. 10*, Academic Press, New York, pp. 383–492
——— (1979). Nutrition. In: W.S. Hoar, D.J. Randall and J.R. Brett (eds), *Fish Physiology Volume 8: Bioenergetics and Growth*, Academic Press, New York, pp. 1–69
Culley, M. (1971). *The Pilchard*, Pergamon, Oxford
Cushing, D.H. (1967). The grouping of herring populations. *J. Mar. Biol. Assocn. UK 47*: 193–208
——— (1968). *Fisheries Ecology: A Study in Population Dynamics*, University of Wisconsin Press, Madison, Wisconsin
——— (1971). Upwelling and the production of fish. In: F.S. Russell

and C.M. Yonge (eds), *Advances in Marine Biology, Vol. 9*, Academic Press, London, pp. 255–334

——— (1973). *The Detection of Fish*, Pergamon Press, Oxford

——— (1974). The possible density-dependence of larval mortality and adult mortality in fishes. In: J.H.S. Blaxter (ed.), *The Early Life History of Fish*, Springer-Verlag, Berlin, pp. 103–11

——— (1975a). *Marine Ecology and Fisheries*, Cambridge University Press

——— (1975b). The natural mortality of plaice. *J. Cons. Int. Explor. Mer. 36*: 144–9

——— (1977). The problems of stock and recruitment. In: J.A. Gulland (ed.), *Fish Population Dynamics*, John Wiley, London, pp. 116–33

——— (1978). The present state of acoustic survey. *J. Cons. Int. Explor. Mer. 38*: 28–32

——— (1980). The decline of the herring stocks and the gadoid outburst. *J. Cons. Int. Explor. Mer. 39*: 70–81

Cushing, D.H. and J.G.K. Harris (1973). Stock and recruitment and the problem of density dependence. *Rapp. Proc-Verb. Int. Comm. Explor. Mer. 164*: 142–55

Cushing, D.H. and J.W. Horwood (1977). Development of a model of stock and recruitment. In: J.H. Steele (ed.), *Fisheries Mathematics*, Academic Press, London, pp. 21–36

Daan, N. (1978). Changes in cod stocks and cod fisheries in the North Sea. In: G. Hempel (ed.), *North Sea Fish Stocks – Recent Changes and their Causes, Rapp. Proc-Verb. Int. Comm. Explor. Mer. 172*: 39–57

Dabrowski, K. (1975). Point of no return in the early life of fishes. An energetic attempt to define the food minimum. *Wiadomosci Ekologiczne 21*: 277–93

Dabrowski, K., S. Hassard, J. Quinn, T.J. Pitcher and A.M. Flinn (1980). Effect of *Geotrichum candidum* protein substitution in pelleted fish feed on the growth of rainbow trout (*Salmo gairdneri Rich*.) and on utilization of the diet. *Aquaculture 21*: 213–32

Delong, D.C., J.E. Halver and E.T. Mertz (1962). Nutrition of salmonid fishes. 10. Quantitative threonine requirements of chinook salmon at two temperatures. *J. Nutr. 76*: 174–8

Deriso, R.B. (1980). Harvesting strategies and parameter estimation for an age-structured model. *Can J. Fish. Aquat. Sci. 37*: 268–82

Diana, J.S. and W.C. Mackay (1979). Timing and magnitude of energy deposition and loss in the body, liver and gonads of the Northern pike, *Esox lucius*. *J. Fish. Res. Board Can. 36*: 481–7

Dickson, R.R. (1971). A recurrent and persistent pressure-anomaly pattern as the principal cause of the intermediate scale hydrographic variation in the European shelf seas. *Dtsch hydrogr. Z. 24*: 97–119

Dickson, R.R., J.G. Pope and M.J. Holden (1974). Environmental influences on the survival of North Sea cod. In: J.H.S. Blaxter (ed.), *The Early Life History of Fish*, Springer-Verlag, Berlin, pp. 69–80

Donaldson, L.R. and D. Menasveta (1961). Selective breeding of chinook salmon. *Trans. Amer. Fish. Soc. 90*: 160–4

Dzuik, L.J. and F.W. Plapp (1973). Insecticide resistance in mosquito fish from Texas. *Bull. Eur. Contam. Toxicol. 9*: 15–19

Eddie, G.O. (1977). The harvesting of krill. *FAO GLO/SO/77/2*: 1–76

Edwards, R.C.C., D.M. Finlayson and J.H. Steele (1972). An experimental study of the oxygen consumption, growth, and metabolism of the cod (*Gadus morhua L.*). *J. Exp. Mar. Biol. Ecol. 8*: 299–309

Edwardson, W. (1976). Energy demands of aquaculture. *Fish Farming International* (December 1976): 10–13

Elliott, J.M. (1976). The energetics of feeding, metabolism and growth of brown trout (*Salmo trutta L.*) in relation to body weight, water temperature and ration size. *J. Anim. Ecol. 45*: 923–48

——— (1979). Energetics of freshwater teleosts. In: P.J. Miller (ed.), *Fish Phenology: Anobolic Adaptiveness in Teleosts*, Symp. Zool. Soc. Lond. 44, pp. 29–61

Elliott, J.M. and L. Persson (1978). The estimation of daily rates of food consumption for fish. *J. Anim. Ecol. 47*: 977–99

Emlen, J.M. (1973). *Ecology: An Evolutionary Approach*, Addison-Wesley, Mass.

Erberhardt, L.L. (1977). Relationship between two stock-recruitment curves. *J. Fish. Res. Board Can. 34*: 425–42

Estes, J.A. (1979). Exploitation of marine mammals – *r selection of k strategists? J. Fish. Res. Board Can. 36*: 1009–17

Everhart, W.H., A.W. Eipper and W.D. Youngs (1975). *Principles of Fishery Science*, Comstock Press, New York, 288pp.

Everson, I. (1977). The living resources of the southern ocean. *FAO GLO/SO/77/1*: 1–156

Falconer, D.S. (1960). *Introduction to Quantitative Genetics*, Oliver and Boyd, Edinburgh

Fänge, R. and D. Grove (1979). Digestion. In: W.S. Hoar, D.J. Randall and J.R. Brett (eds), *Fish Physiology Volume 8: Bioenergetics and Growth*, Academic Press, New York, pp. 161–260

FAO (1978). Fishery commodity situation and outlook. FAO: Committee on Fisheries 12th Session, COFI/78/inf. 5, pp. 1–15

FAO (1979). *Yearbook of Fishery Statistics*, Volume 46, FAO, Rome

Farkas, D.F. (1978). Is freezing here to stay? A review of technological change in the quick freezing industry. Can we predict the future from the past? *Proc. 1st Inter. Frozen Food Industries Conference*: 1–13. (mimeo)

Favro, L.D., P.K. Kuo and J.F. Macdonald (1979). Population-genetic study of the effect of selective fishing on the growth rate of trout. *J. Fish. Res. Board Can. 36*: 552–61

Ferguson, A., K-J.M. Himberg and G. Svardson (1978). Systematics of the Irish pollan (*Coregonus pollan* Thompson): an electrophoretic comparison with other Holarctic Coregoninae. *J. Fish. Biol. 12*: 221–33

Firth, F.E. (1969). *The Encyclopedia of Marine Resources*, Van Nostrand Reinhold Co. New York

Fisher, R.A. (1958). *The Genetical Theory of Natural Selection*, 2nd edn, Dover, New York

Fisher, S.G. and G.E. Likens (1973). Energy flow in Bear Brook, New Hampshire: an integrative approach to stream ecosystem metabolism. *Ecol. Monogr. 43*: 421–39

Fletcher, R.I. (1978). Time-dependent solutions and efficient parameters for stock production models. *Fishery Bulletin, Seattle 76*: 377–88

Foerster, J.W. and S.P. Reagan (1977). Management of the Northern Chesapeake Bay American Shad Fishery. *Biological Conservation 12*: 179–201

Foerster, R.E. (1968). The sockeye salmon. *Bull. Fish. Res. Board Can. 162*: 1–422

Fox, P.J. (1978). Preliminary observations on different reproductive strategies in the bullhead (*Cottus gobio L.*) in northern and southern England. *J. Fish. Biol. 12*: 5–11

Fox, W.W. (1970). An exponential surplus yield model for optimizing by exploited fish populations. *Trans. Amer. Fish. Soc. 99*: 80–8

Frost, W.E. (1954). The food of pike, *Esox lucius L.*, in Windermere. *J. Anim. Ecol. 23*: 339–60

Frost, W.E. and C. Kipling (1967). A study of reproduction, early life, weight-length relationship and growth of the pike, *Esox lucius L.*, in Windermere. *J. Anim. Ecol. 36*: 651–93

Fry, F.E.J. (1947). Effects of the environment on animal activity. *Univ. Toronto Stud. Biol. Ser. 55*: 1–62

——— (1971). The effect of environmental factors on the physiology of fish. In: W.S. Hoar and D.J. Randall (eds.), *Fish Physiology Volume 6*, Academic Press, New York, pp. 1–98

Fryer, G. and T.D. Iles (1972). *The Cichlid Fishes of the Great Lakes of Africa*, Oliver and Boyd, Edinburgh

Gadgil, M. and W.H. Bossert (1970). Life historical consequences of natural selection. *Amer. Nat. 104*: 1–24

Gall, G.A.E. and S.J. Gross (1978). A genetic analysis of the performance of three rainbow trout broodstocks. *Aquaculture 15*: 113–27

Gallucci, V.F. and T.J. Quinn (1979). Reparameterising, fitting and testing a simple growth model. *Trans. Amer. Fish. Soc. 108*: 14–25

Garrod, D.J. (1973). Management of multiple resources. *J. Fish. Res. Board Can. 30*: 1977–85

——— (1977). The North Atlantic cod. In: J.A. Gulland (ed.), *Fish Population Dynamics*, Wiley, London, pp. 216–42

Garrod, D.J. and B.W. Jones (1974). Stock and recruitment relationship in the N.E. Atlantic cod stock and the implications for management of the stock. *J. Cons. Int. Explor. Mer. 36*: 35–41

Garrod, D.J. and J.M. Colebrook (1978). Biological effects of variability in the North Atlantic ocean. *Rapp. Proc-Verb. Reun. Int. Comm. Explor. Mer. 173*: 128–44

Garrod, D.J. and B.J. Knights (1979). Fish stocks: their life history characteristics and responses to exploitation. In: P.J. Miller (ed.), *Fish Phenology; Anabolic Adaptiveness in Teleosts*, Symp. Zool. Soc. Lond. 44, pp. 361–82

Gerking, S.D. (1955). Endogenous nitrogen excreton of bluegill sunfish. *Physiol. Zool. 28*: 283–9

Gibson, R.M. (1980). Optimal prey size selection by three-spined sticklebacks (*Gasterosteus aculeatus*): a test of the apparent size hypothesis. *Z. fur Tierpsychol. 52*: 291–307

Gjedrem, T. (1975). Possibilities for genetic gain in salmonids. *Aquaculture 6*: 23–9

Glen, D. (1974). Unilever salmon farm in a Scottish sea loch. *Fish Farming International 1*: 12–23

Gompertz, B. (1825). On the nature of the function expressive of the law of human mortality, and on a new mode of determining the value of life contingencies. *Phil. Trans. Roy. Soc. Lond. 115*: 515–85

Graham, H.W. and R.L. Edwards (1962). The world biomass of marine fishes. In: E. Heen and R. Kreuzer (eds.), *Fish in Nutrition*, Fishing

News Books, Farnham, pp. 3–8
Graham, M. (1935). Modern theory of exploiting a fishery and application to North Sea trawling. *J. Cons. Int. Explor. Mer. 10*: 264–74
——— (1956). *Sea Fisheries*, Edward Arnold, London
Greenwood, P.H. (1974). Cichlid fishes of Lake Victoria, West Africa: the biology and evolution of a species flock. *Bull. Brit. Mus. Nat. Hist. Suppl. 6*, London
——— (1975). *A History of Fishes* (J.R. Norman), 3rd edn, Ernest Benn, London
Greenwood, P.H. and J.M. Gee (1969). A revision of the Lake Victoria *Haplochromis* species (Pisces, Cichlidae), Part 8. *Bull. Br. Mus. Nat. Hist. (Zool.) 18*: 1–65
Gulland, J.A. (1961). Fishing and the stocks of fish at Iceland. *Fishery Invest. Lond. Ser. II, 23*: 1–52
——— (1965). Estimation of mortality rates. Annex to Artic fisheries working group report. *ICES CM No. 3*: 1
——— (ed.) (1970). *The Fish Resources of the Ocean*, Fishing News Books, Farnham
——— (1974). *The Management of Marine Fisheries*, Scientechnica, Bristol
——— (1977). Analysis of data and development of models. In: J.A. Gulland (ed.), *Fish Population Dynamics*, Wiley, London, pp. 67–95
——— (1978a). Fisheries management: new strategies for new conditions. *Trans. Amer. Fish. Soc. 107*: 1–11
——— (1978b). Review of the state of world fishery resources. *FAO Fisheries Circular No. 710. FIRM/C710*: 1–42
Gulland, J.A. and M.A. Robinson (1973). Economics of fishery management. *J. Fish. Res. Board Can. 30*: 2040–50
Halver, J.E. (1979). The nutritional requirements of cultivated warmwater and coldwater fish species. In: T.V.R. Pillay and W.A. Dill (eds.), *Advances in Aquaculture*, Fishing News Books, Farnham, pp. 574–80
Halver, J.E., D.C. DeLong and E.T. Mertz (1959). Methionine and cystine requirements of chinook salmon. *Fed. Proc. Fed. Amer. Soc. Exp. Biol. 18*: 2076
Harden-Jones, F.R. (1968). *Fish Migration*, Edward Arnold, London
Hardin, G. (1968). The tragedy of the commons. *Science 131*: 1243–7
Harding, D., J.H. Nichols and D.S. Tungate (1978). The spawning of plaice (*Pleuronectes platessa L.*) in the southern North Sea and English Channel. In: G. Hempel (ed.), *North Sea Fish Stocks –*

Recent Changes and their Causes. Rapp. Proc-Verb. Reun. Int. Comm. Explor. Mer. 172: 102–113

Hardman, J.M. and A.L. Turnbull (1974). The interaction of spatial heterogeneity, predator competition and the functional response to prey density in a laboratory system of wolf spiders and fruit flies. *J. Anim. Ecol. 43*: 155–71

Hardy, A.C. (1924). The herring in relation to its animate environment. Part 1: The food and feeding habits of the herring. *Fishery Invest. Lond. Ser. II*: 3

——— (1958). *The Open Sea. Volume 2: Fish and Fisheries*, Collins, London

Harris, G.P. (1980). Temporal and spatial scales in phytoplankton ecology. Mechanisms, methods, models, and management. *Can. J. Fish. Aquat. Sci. 37*: 877–900

Harris, J.G.K. (1975). The effect of density-dependent mortality on the shape of the stock and recruitment curve. *J. Cons. Int. Explor. Mer. 36*: 144–9

Hart, P.J.B. (1971). A comparative ecological study of five species of fish in the River Nene, Northamptonshire. PhD thesis, University of Liverpool

——— (1974). The distribution and long-term changes in abundance of larval *Ammodytes marinus* (Riatt) in the North Sea. In: J. Blaxter (ed.), *The Early Life History of Fish*, Springer-Verlag, Berlin, pp. 171–82

Hart, P.J.B. and T.J. Pitcher (1973). Population densities and growth of five species of fish in the River Nene, Northamptonshire. *Fisheries Management 4*: 69–86

Hart, P.J.B. and B. Connellan (1979). On pike predation in a fishery. *Proc. 1st Brit. Freshwater Fisheries Conf. 1*: 182–95

Hannesson, R. (1975). Fishery dynamics: a North Atlantic cod fishery. *Can. J. Econ. 8*: 151–73

Hilborn, R. (1979). Comparison of fisheries control systems that utilise catch and effort data. *J. Fish. Res. Board Can. 36*: 1477–89

Hill, H.W. and R.R. Dickson (1978). Long-term changes in North Sea hydrography. In: G. Hempel (ed.), *North Sea Fish Stocks – Recent Changes and their Causes. Rapp. Proc-Verb. Reun. Int. Comm. Explor. Mer. 172*: 310–34

Hiltz, D.F., D.H. North, B.S. Lall and R.A. Keith (1977). Storage life of refrozen silver hake (*Merluccius bilinearis*) processed as fillets and minced flesh from thawed, stored round-frozen fish. *J. Fish. Res. Board Can. 34*: 2369–73

Hjort, J. (1914). Fluctuations in the great fisheries of northern Europe viewed in the light of biological research. *Rapp. Proc-Verb. Reun. Int. Comm. Explor. Mer. 20*: 1–228
Hobson, E.S. (1979). Interactions between piscivorous fishes and their prey. In: R.H. Stroud and H. Clepper (eds), *Predator-Prey Systems in Fisheries Management*, Sport Fishing Institute, Washington, DC, pp. 231–42
Holden, M.J. (1978). Long-term changes in landings of fish from the North Sea. In: G. Hempel (ed.), *North Sea Fish Stocks – Recent Changes and their Causes. Rapp. Proc-Verb. Reun. Int. Comm. Explor. Mer. 172*: 11–26
Holling, C.S. (1965). The functional response of predators to prey density and its role in mimicry and population regulation. *Mem. Ent. Soc. Can. 45*: 5–60
Holt, S.J. and L.M. Talbot (1978). New principles for the conservation of wild living resources. *Wildlife Monographs 59*: 1–33
Hoogland, R., D. Morris and N. Tinbergen (1956). The spines of sticklebacks (*Gasterosteus* and *Pygosteus*) as means of defence against predators (*Perca* and *Esox*). *Behaviour 10*: 205–36
Horne, J. (1971). Some notes on fish handling and processing. *MAFF Torry Advisory Note no. 50*: 1–14
Houde, E.D. (1975). Effects of stocking density and food density on survival, growth and yield of laboratory-reared larvae of sea bream, *Archosargus rhomboidalis* (Sparidae). *J. Fish. Biol. 7*: 115–27
——— (1977a). Food concentration and stocking density effects on survival and growth of laboratory-reared larvae of bay anchovy *Anchoa mitchilli* and lined sole *Archirus lineatus*. *Mar. Biol. 43*: 333–41
——— (1977b). Abundance and potential yield of the Atlantic thread herring *Opisthonema oglinum*, and aspects of its early life history in the Eastern Gulf of Mexico. *Fishery Bulletin, Seattle 75*: 493–512
——— (1978). Critical food concentrations for larvae of three species of sub-tropical marine fishes. *Bull. Mar. Sci. 28*: 395–411
Huffaker, C.B. (1958). Experimental studies on predation: dispersion factors and predator-prey oscillations. *Hilgardia 27*: 343–83
Hunter, J.R. (1972). Swimming and feeding behaviour of larval anchovy, *Engraulis mordax*. *Fishery Bulletin, Seattle 70*: 821–38
Huntsman, G.R. (1979). Predation's role in structuring reef fish communities. In: R.H. Stroud and H. Clepper (eds), *Predator-Prey Systems in Fisheries Management*, Sport Fishing Institute,

Washington, DC, pp. 103–8
Huppert, D.D. (1979). Control of mixed stocks and multipurpose fleets. *J. Fish. Res. Board Can. 36*: 845–54
Hynes, H.B.N. (1970). *The Ecology of Running Waters*, Liverpool University Press, Liverpool
Idyll, C.P. (1973). The anchovy crisis. *Scientific American 228*: 22–9
Iles, T.D. (1974). The tactics and strategy of growth in fishes. In: F.R. Harden-Jones (ed.), *Sea Fisheries Research*, Elek Books, London, pp. 331–45
Jalabert, B. (1976). *In-vitro* oocyte maturation and ovulation in rainbow trout (*Salmo gairdneri*), northern pike (*Esox lucius*) and goldfish (*Carassius auratus*). *J. Fish. Res. Board Can. 33*: 974–88
—— (1977). Controle de la reproduction par les facteurs extremes chez les poissons. *Oceanis 2*: 141–50
Jeffers, J.N.R. (1978). *An Introduction to Systems Analysis with Ecological Applications*, Edward Arnold, London
Jensen, A.L. (1973). Relation between simple dynamic pool and surplus production models for yield from a fishery. *J. Fish. Res. Board Can. 30*: 998–1002
Jhingran, V.G. (1978). Systems of polyculture of fishes in the inland waters of India. *J. Fish. Res. Board Can. 33*: 905–10
Jolly, G.M. (1965). Explicit estimates from capture-recapture data with both death and immigration. Stochastic model. *Biometrika 52*: 225–47
Jones, R. (1976). Growth of fishes. In: D.H. Cushing and J.J.Walsh (eds), *The Ecology of the Seas*, Blackwell, Oxford, pp. 251–79
—— (1978). Competition and coexistence with particular reference to gadoid fish species. In: G. Hampel (ed.), *North Sea Fish Stocks – Recent Changes and their Causes. Rapp. Proc-Verb. Reun. Int. Comm. Explor. Mer. 172*: 292–300
Jones, R. and W.B. Hall (1973). A simulation model for studying the population dynamics of some fish species. In: M.S. Bartlett and R.W. Hiorns (eds), *The Mathematical Theory of the Dynamics of Biological Populations*, Academic Press, London, pp. 35–59
—— (1974). Some observations on the population dynamics of the larval stages of common gadoids. In: J.H.S. Blaxter (ed.), *The Early Life History of Fish*, Springer-Verlag, Berlin, pp. 87-102
Joseph, J. and J.W. Greenhough (1979). *International Management of Tuna, Porpoise and Billfish: Biological, Legal and Political Aspects*, University of Washington Press, Seattle, 253pp.
Joyner, T. (1975). Towards a planetary aquaculture – the seas as range and cropland. *Mar. Fish. Rev. 37*: 5–10

Kincaid, H.L., W.R. Bridges and B. von Limbach (1977). Three generations of selection for growth rate in fall-spawning rainbow trout. *Trans. Amer. Fish. Soc. 106*: 621–8

Kipling, C. and W.E. Frost (1969). Variations in the fecundity of pike, *Esox lucius L.*, in Windermere. *J. Fish. Biol. 1*: 221–37

Kolhonen, J. (1974). Fishmeal: international market situation and the future. *Mar. Fish. Rev. 36*: 36–40

Krebs, C.J. (1979). *Ecology: The Experimental Analysis of Distribution and Abundance*, 2nd edn, Harper and Row, New York

Krebs, J.R. (1978). Optimal foraging: decision rules for predators. In: J.R. Krebs and N.B. Davies (eds.), *Behavioural Ecology: An Evolutionary Approach*, Blackwell, Oxford, pp. 23-63

Kuo, C-M., C.E. Nash and Z.H. Shehadeh (1974). The effects of temperature and photoperiod on ovarian development in captive grey mullet (*Mugil cephalus L.*). *Aquaculture 3*: 25–43

Kuo, C-M. and C.E. Nash (1975). Recent progress on the control of ovarian development and induced spawning of the grey mullet (*Mugil cephalus L.*). *Aquaculture 5*: 19–29

Kutty, M.K. (1968). Estimation of the age of exploitation at a given fishing mortality. *J. Fish Res. Board Can. 25*: 1291–4

Laird, L.M. and B. Stott (1978). Marking and tagging. In: T. Bagenal (ed.), *Methods for Assessment of Fish Production in Fresh Waters*, IBP Handbook No. 3, 3rd edn, Blackwell, Oxford, pp. 84–100

Larkin, P.A. (1977). An epitaph for the concept of MSY. *Trans. Amer. Fish. Soc. 107*: 1–11

Larkin, P.A. and J.G. Macdonald (1968). Factors in the population biology of the sockeye salmon. *J. Anim. Ecol. 37*: 229–58

Larsen, K. (1966). Studies on the biology of Danish stream fishes. II. The food of pike (*Esox lucius*) in trout streams. *Medd. Dan. Fisk. Hav. NS4*: 271–326

Larsson, A., B-E. Bengtsson and O. Svanberg (1976). Some haematological and biochemical effects of cadmium on fish. In: A.P.M. Lockwood (ed.), *Effects of Pollutants on Aquatic Organisms*, SEB Seminar Series 2, Cambridge University Press, pp. 35–45

Lasker, R. (1978). The relation between oceanographic conditions and larval anchovy food in the Californian current: identification of factors responsible for recruitment failure. *Rapp. Proc-Verb. Reun. Int. Comm. Explor. Mer. 173*: 212–30

Lasker, R., H.M. Feder, G.H. Theilacker and R.C. May (1970). Feeding, growth and survival of *Engraulis mordax* larvae reared in the laboratory. *Mar. Biol. 5*: 345–53

Last, J.M. (1980). The food of 25 species of fish larvae in the W. Central North Sea. *MAFF Lowestoft, Fisheries Technical Report 60*: 1–44

Laughlin, R. (1965). Capacity for increase; a useful population statistic. *J. Anim. Ecol. 34*: 77–91

Laurence, G.C. (1974). Growth and survival of haddock. *Melanogrammus aeglefinus*, larvae in relation to planktonic prey concentration. *J. Fish. Res. Board Can. 31*: 1415–19

——— (1977). A bioenergetic model for the analysis of feeding and survival potential of winter flounder, *Psuedopleuronectes americanus*, larvae during the period from hatching to metamorphosis. *Fishery Bulletin, Seattle 75*: 529–46

Law, R. (1979a). Optimal life histories under age-specific predation. *Amer. Nat. 114*: 399–417

——— (1979b). Harvest optimisation in populations with age distributions. *Amer. Nat. 114*: 250–9

Leach, G. (1975). *Energy and Food Production*, Institute for Environment and Development, London

Lear, W.H. (1975). Evaluation of the transport of Pacific pink salmon (*Oncorhynchus gorbuscha*) from British Columbia to Newfoundland. *J. Fish. Res. Board Can. 32*: 2343–56

LeCren, E.D. (1951). The length-weight relationship and seasonal cycle in gonad weight and condition in the perch (*Perca fluviatilis*). *J. Anim. Ecol. 20*: 201–19

LeCren, E.D., C. Kipling and J.C. McCormack (1977). A study of the numbers, biomass and year-class strengths of perch (*Perca fluviatilis L.*) in Windermere 1941–66. *J. Anim. Ecol. 46*: 281–307

Lewontin, R.C. (1974). *The Genetic Basis of Evolutionary Change*, Columbia University Press, New York

Lindbergh, J.M. (1979). The development of a commercial Pacific salmon culture business. In: T.V.R. Pillay and W.A. Dill (eds), *Advances in Aquaculture*, Fishing News Books, Farnham, pp. 441–7

Lockwood, S.J. (1974). The use of the von Bertalanffy growth equation to describe the seasonal growth of fish. *J. Cons. Int. Explor. Mer. 35*: 175–9

Longhurst, A.R. (1969). Species assemblages in tropical demersal fisheries. In: *Proc. Symp. Oceanogr. Fisheries Trop. Ass.*, UNESCO, Paris, pp. 147–68

Love, M. (1970). *The Chemical Biology of Fishes*, Academic Press, London

Lowe-McConnell, R.H. (1962). The fishes of the British Guyana continental shelf. *J. Linn. Soc. (Zool.) 44*: 669–700

——— (1975). *Fish Communities in Tropical Freshwater*, Longman, London

——— (1977). *Ecology of Fishes in Tropical Waters*, Studies in Biology No. 76, Edward Arnold, London

Macan, T.T. (1966). The influence of predation on the fauna of a moorland fishpond. *Arch. Hydrobiol. 61*: 432–52

——— (1970). *Biological Studies of the English Lakes*, Longmans, London

MacArthur, R.H. and E.O. Wilson (1967). *The Theory of Island Biogeography*, Princeton University Press, New Jersey

Macdonald, P.D.M. and T.J. Pitcher (1979). Age groups from size-frequency data: a versatile and efficient method of analysing distribution mixtures. *J. Fish. Res. Board Can. 36*: 987–1001

Macdonald, R.D.S. (1979). Inshore fishing interests on the Atlantic coast: their response to extended jurisdiction by Canada. *Marine Policy* (July 1979): 171–89

Macer, C.T. (1974). Industrial fisheries. In: E.R. Harden-Jones (ed.), *Sea Fisheries Research*, Elek Books, London, pp. 193–222

McFarlane, G.A. and W.G. Franzin (1978). Elevated heavy metals: a stress on a population of white suckers, *Catostomus commersoni*, in Hamell Lake, Sasketchewan. *J. Fish. Res. Board Can. 35*: 963–970

McNeil, W.J. (1979). Review of transplantion and artificial recruitment of anadromous species. In: T.V.R. Pillay and W.A. Dill (eds), *Advances in Aquaculture*, Fishing News Books, Farnham, pp. 547–54

Mann, R.H.K. (1973). Observations on the age, growth, reproduction and food of the roach, *Rutilus rutilus*, in two rivers in southern England. *J. Fish. Biol. 5*: 707–36

——— (1974). Observations on the age, growth, reproduction and food of the dace, *Leuciscus leuciscus* (*L*.), in two rivers in southern England. *J. Fish. Biol. 6*: 237–53

——— (1976a). Observations on the age, growth, reproduction and food of the pike, *Esox lucius* (*L*.), in two rivers in southern England. *J. Fish. Biol. 8*: 179–97

——— (1976b). Observations on the age, growth, reproduction and food of the chub, *Squalius cephalus* (*L*.), in two rivers in southern England. *J. Fish. Biol. 8*: 265–88

Mann, R.H.K. and C.A. Mills (1979). Demographic aspects of fish fecundity. In: P.J. Miller (ed.), *Fish Phenology; Anabolic Adaptiveness in Teleosts*, Symp. Zool. Soc. Lond. 44, pp. 161–77

Marr, J.C. (1956). The critical period in the early life history of marine

fishes. *J. Cons. Int. Explor. Mer. 21*: 160–70
Marshall, N.B. (1971). *Explorations in the Life of Fishes*, Harvard University Press, Cambridge, Mass.
Marten, G.G. (1979a). Predator removal: effect on fisheries yields in Lake Victoria. *Science 203*: 646–7
——— (1979b). Impact of fishing on the inshore fishery of Lake Victoria. *J. Fish. Res. Board Can. 36*: 891–900
Matty, A.J. and P. Smith (1978). Evaluation of a yeast, a bacterium and an alga as a protein source for rainbow trout. I. Effect of protein level on growth, gross conversion efficiency and protein conversion efficiency. *Aquaculture 14*: 235–46
May, R.C. (1974). Larval mortality in marine fishes and the critical period concept. In: J.H.S. Blaxter (ed.), *The Early Life History of Fishes*, Springer-Verlag, Berlin, pp. 3–19
May, R.M. (1976). Estimating r: a pedagogical note. *Amer. Nat. 110*: 496–9
——— (1980). The economics and management of commercial fisheries. *Nature 287*: 675–6
May, R.M., J.R. Beddington, J.W. Horwood and J.G. Shepherd (1978). Exploiting natural populations in an uncertain world. *Math. Biosc. 42*: 219–52
May, R.M., J.R. Beddington, C.W. Clark, S.J. Holt and R.M. Laws (1979). Management of multispecies fisheries. *Science 205*: 267–77
Mayr, E. (1963). *Populations, Species and Evolution*, Belknap Press, Cambridge, Mass.
Medford, B.A. and W.C. MacKay (1978). Protein and lipid contents of gonads, liver, and muscle of northern pike (*Esox lucius*) in relation to gonad growth. *J. Fish. Res. Board Can. 35*: 213–19
Mitchell, C.L. (1979). Bioeconomics of commercial fisheries manangement. *J. Fish. Res. Board Can. 36*: 699–704
Moav, R. (1979). Genetic improvement in aquaculture industry. In: T.V.R. Pillay and W.A. Dill (eds), *Advances in Aquaculture*, Fishing News Books, Farnham, pp. 610–22
Moav, R., T. Brody and G. Hulatin (1978). Genetic improvement of wild fish populations. *Science 201*: 1090–4
Mohn, R.K. (1980). Bias and error propagation in logistic production models. *Can. J. Fish. Aquat. Sci. 37*: 1276–83
Morgan, N.C. and D.S. McLusky (1974). A summary of the Loch Leven IBP results in relation lake management and future research. *Proc. Roy. Soc. (Edinburgh) B: 74*: 407–16
Murphy, G.I. (1977). Clupeiods. In: J.A. Gulland (ed.), *Fish Population*

Dynamics, John Wiley, London, pp. 283–308
Needler, A.W.H. (1979). Evolution of Canadian fisheries management towards economic rationalisation. *J. Fish Res. Board Can. 36*: 716–24
Neilsen, L.A. and W.F. Schoch (1980). Errors in estimating mean weight and other statistics from mean length. *Trans. Amer. Fish. Soc. 109*: 319–22
Nelson, J. (1976). *Fishes of the World*, John Wiley, New York
Nikolaev, V.M. (1978). *Bulletin Statistique des Peches Maritimes*, Cons. Int. Explor. Mer., Charlottenlund Slot, Copenhagen
Noble, R. (1975). Growing fish in sewage. *New Scientist 67*: 259–61
Numann, W. (1972). The Bodensee: effects of overexploitation and eutrophication on the salmon community. *J. Fish. Res. Board Can. 29*: 833–47
O'Brien, W.J., N.A. Slade and G.L. Vinyard (1976). Apparent size as the determinant of prey selection by bluegill sunfish (*Lepomis machrochirus*). *Ecology 57*: 1304–30
O'Connell, C.P. (1976). Histological criteria for diagnosing the starving condition in early post-yolk-sac larvae of the northern anchovy, *Engraulis mordax*. *J. Exp. Mar. Biol. Ecol. 25*: 285–312
O'Connors, H.B., L.F. Wurster, C.D. Powers, D.C. Biggs and R.G. Rowland (1978). Polychlorinated biphenyls may alter marine trophic pathways by reducing phytoplankton size and production. *Science 201*: 737–9
OECD (1979). *Review of Fisheries in OECD Member Countries 1978*, Organisation for Economic Co-operation and Development, Paris
Orach-Meza, F.L. and S.B. Saila (1978). Application of a polynomial distributed lag model to the Maine Lobster fishery. *Trans. Amer. Fish. Soc. 107*: 402–11
Parker, R.R. and P.A. Larkin (1959). A concept of growth in fishes. *J. Fish. Res. Board Can. 16*: 721–45
Paulik, G.J. (1971). Anchovies, birds and fishermen in the Peru current. In: W.W. Murdoch (ed.), *Environment: Resources, Pollution and Society*, Sinauer, Stamford, Conn., pp. 156–85
——— (1973). Studies of the possible form of the stock-recruitment curve. *Rapp. Proc-Verb. Reun. Int. Comm. Explor. Mer. 164*: 302–15
Paulik, G.J. and L.E. Gales (1964). Allometric growth and the Beverton and Holt yield equation. *Trans. Amer. Fish. Soc. 93*: 369–87
Paulik, G.J., A.S. Hourston and P.A. Larkin (1967). Exploitation of multiple stocks by a common fishery. *J. Fish. Res. Board Can. 24*:

2527–37

Pawson, M.G. and S.J. Lockwood (1980). Mortality of mackerel following physical stress, and its probable cause. *ICES Symposium on The Biological Basis of Pelagic Fish Stock Assessment No. 27*: 1–5

Pella, J.J. and P.K. Tomlinson (1969). A generalised stock production model. *Bull. Inter-Amer. Trop. Tuna Comm. 13*: 421–96

Persson, P-E. (1979). The source of muddy odour in bream (*Abramis brama*) from the Porvoo sea (Gulf of Finland). *J. Fish. Res. Board Can. 36*: 883–90

Peterman, R.M. (1978). Testing for density-dependent marine survival in Pacific Salmonids. *J. Fish. Res. Board Can. 35*: 1434–50

——— (1980). Dynamics of native Indian food fisheries on salmon in British Columbia. *Can. J. Fish. Aquat. Sci. 37*: 561–6

Petersen, C.G.J. (1896). The yearly immigration of young plaice into the Limfjord from the German Sea. *Rep. Dan. Biol. Sta. 6*: 1–48

Pianka, E.R. (1970). On r and k selection. *Amer. Nat. 104*: 592–7

——— (1978). *Evolutionary Ecology*, 2nd edn, Harper and Row, New York

Pierce, R.J. and T.E. Wissing (1974). Energy cost of food utilisation in the bluegill (*Lepomis macrochirus*). *Trans. Amer. Fish. Soc. 103*: 38–45

Pillay, T.V.R. and W.A. Dill (eds) (1979). *Advances in Aquaculture*, Fishing News Books, Farnham

Pitcher, T.J. (1971). Population dynamics and schooling behaviour in the minnow, *Phoxinus phoxinus* (*L*.). DPhil thesis, University of Oxford

——— (1977). An energy budget for a rainbow trout farm. *Environmental Conservation 4*: 59–65

——— (1980). Some ecological consequencies of fish school volumes. *Freshwater Biology 10*: 539–44

Pitcher, T.J. and P.D.M. Macdonald (1973a). Two models for seasonal growth in fishes. *J. Appl. Ecol. 10*: 599–606

——— (1973b). An integration method for fish population fecundity. *J. Fish. Biol. 4*: 549–53

Pitcher, T.J. and G.J.A. Kennedy (1977). The longevity and quality of fin marks made with a jet inoculator. *Fisheries Management 8*: 16–18

Pitcher, T.J., B.L. Partridge and C.S. Wardle (1976). A blind fish can school. *Science 192*: 963–5

Pitcher, T.J., G.J.A. Kennedy and S. Wirjoatmodjo (1979). Links between the behaviour and ecology of fishes. *Proc. 1st Brit. Freshwater Fisheries Conf. 1*: 162–75

Pope, J.G. (1972). An investigation of the accuracy of virtual population analysis. *ICNAF Research Bulletin 9*: 65–74

——— (1977). Estimating fishing mortality: its precision and implication for fisheries. In: J.H. Steele (ed.), *Fisheries Mathematics*, Academic Press, New York, pp. 63–76

Popiel, J. and J. Josinski (1973). Industrial fisheries and their influence on catches for human consumption. *J. Fish. Res. Board Can. 30*: 2254–9

Purdom, C.E. (1974a). Breeding the domestic fish. In: *Fish Farming in Europe*, Oyez Int. Bus. Communications, London, pp. 61–8

——— (1974b). Variation in fish. In: F.R. Harden-Jones (ed.), *Sea Fisheries Research*, Elek Books, London, pp. 347–55

——— (1979). Genetics of growth and reproduction in teleosts. In: P.J. Miller (ed.), *Fish Phenology; Anabolic Adaptiveness in Teleosts*, Symp. Zool. Soc. Lond. 44, pp. 207–17

Pütter, A. (1920). WachstumsahnlichKerten. *Pfluegers. Arch. Gesamte Physiol. Meuschen Tiere 180*: 298–340

Pyke, G.H., H.R. Pulliam and E.L. Charnov (1977). Optimal foraging: a selective review of theory and tests. *Q. Rev. Biol. 52*: 137–54

Radovich, J. (1962). Effects of sardine spawning stock size and environment in year class production. *Calif. Fish and Game 48*: 123–40

Reay, P.J. (1979). *Aquaculture*, Studies in Biology No. 106, Edward Arnold, London

Reich, K. (1975). Multispecies fish culture (polyculture) in Israel. *Bamidgeh. Bull. Fish. Cult. Israel. 27*: 85–99

Ricker, W.E. (1946). Production and utilisation of fish populations. *Ecol. Monogr. 16*: 373–91

——— (1954). Stock and recruitment. *J. Fish. Res. Board Can. 11*: 559–623

——— (1973). Linear regressions in fishery research. *J. Fish. Res. Board Can. 30*: 409–34

——— (1975). Computation and interpretation of biological statistics of fish populations. *Bull. Fish. Res. Board Can. 191*: 1–382

——— (1979). Growth rates and models. In: W.S. Hoar, D.J. Randall and J.R. Brett (eds), *Fish Physiology Volume 8: Bioenergetics and Growth*, Academic Press, New York, pp. 677–743

Ricker, W.E. and H.D. Smith (1975). A revised interpretation of the history of the Skeena River sockeye. *J. Fish. Res. Board Can. 32*:

1369–81

Ricklefs, R.E. (1979). *Ecology*, 2nd edn, Thomas Nelson, Sunbury-on-Thames

Roberts, R.J. (1978). *Fish Pathology*, Balliere Tindall, London

Roedel, P.M. (1975). A summary and critique of the symposium on optimum yield. In: Optimum sustainable yield as a concept in fisheries management. *Amer. Fish. Soc*. Special Publication No. 9, pp. 78–89

Roff, D.A. and D.J. Fairbairn (1980). An evaluation of Gulland's method for fitting the Schaefer model. *Can. J. Fish. Aquat. Sci. 37*: 1229–35

Romer, A.S. (1962). *The Vertebrate Body*, 3rd edn, W.B. Saunders, Philadelphia

Rothschild, B.J. (1977). Fishing effort. In: J.A. Gulland (ed.), *Fish Population Dynamics*, Wiley, London, pp. 96–115

Rothschild, B.J. and A. Suda (1977). Population dynamics of tuna. In: J.A. Gulland (ed.), *Fish Population Dynamics*, Wiley, London, pp. 309–34

Rothschild, B.W. (1973). Questions of strategy in fishery management and development. *J. Fish. Res. Board Can. 30*: 2017–30

Rounsefell, G.A. (1958). Factors causing decline in sockeye salmon of Karluk River Alaska. *Fishery Bulletin, Seattle 58*: 83–169

Russell, E.S. (1931). Some theoretical consideration on the overfishing problem. *J. Cons. Int. Explor. Mer. 6*: 3–20

Ryther, J.H. (1969). Photosynthesis and fish production in the sea. *Science 166*: 72

Sahrhage, D. and G. Wagner (1978). On fluctuations in the haddock population of the North Sea. *Rapp. Proc-Verb. Reun. Int. Comm. Explor. Mer. 172*: 72–85

Sarig, S. (1979). Fish diseases and their control in aquaculture. In: T.V.R. Pillay and W.A. Dill (eds), *Advances in Aquaculture*, Fishing News Books, Farnham, pp. 190–7

Saunders, R.L. (1963). Respiration of the Atlantic cod. *J. Fish. Res. Board Can. 20*: 373–86

Schaefer, M.B. (1954). Some aspects of the dynamics of populations important to the management of commercial marine fisheries. *Bull. Inter-Amer. Trop. Tuna Commission 1*: 27–56

——— (1970). Men, birds, and anchovies in the Peru current – dynamic interactions. *Trans. Amer. Fish. Soc. 99*: 461–7

Schaffer, W.M. (1979). The theory of life-history evolution and its application to Atlantic salmon. In: P.J. Miller (ed.), *Fish*

Phenology; Anabolic Adaptiveness in Teleosts, Symp. Zool. Soc. London. 44, pp. 307–26

Schnute, J. (1977). Improved estimates from the Schaefer production model: theoretical considerations. *J. Fish. Res. Board Can. 34*: 583–603

Scott, A. (1979). Development of economic theory on fisheries regulation. *J. Fish. Res. Board Can. 36*: 725–41

Scott, D.B.C. (1979). Environmental timing and the control of reproduction in teleost fish. In: P.J. Miller (ed.), *Fish Phenology; Anabolic Adaptiveness in Teleosts*, Symp. Zool. Soc. Lond. 44, pp. 105–32

Seber, G.A.F. (1965). A note on the multiple-recapture census. *Biometrika 52*: 249–59

——— (1973). *The Estimation of Animal Abundance*, Griffen, London

Seneviratne, G. (1977). Sex hormones encourage a salmon run. *New Scientist 73*: 264–5

Shelbourne, J.E. (1957). The feeding and condition of plaice larvae in good and bad plankton patches. *J. Mar. Biol. Assoc. UK 36*: 539–52

Sikorski, Z.E. (1980). Structure of proteins of fish and shellfish. In: J.J. Connell (ed.), *Advances in Fish Science and Technology*, Fishing News Books, Farnham, pp. 78–90

Silliman, R.P. (1975). Selective and unselective exploitation of experimental populations of *Tilapia mossambica. US Wildlife Service, Fish. Bull. 73*: 495–507

Silvert, W. (1978). The price of knowledge: fisheries management as a research tool. *J. Fish. Res. Board Can. 35*: 208–12

Sinclair, W.F. (1978). Management alternatives and strategic planning for Canadian fisheries. *J. Fish. Res. Board Can. 35*: 1017–30

Sinha, V.R.P. (1979). New trends in fish farm management. In: T.V.R. Pillay and W.A. Dill (eds), *Advances in Aquaculture*, Fishing News Books, Farnham, pp. 123–6

Smith, V.L. (1969). On models of commercial fishing. *J. Polit. Econ. 77*: 181–98

Smith, P.E. and R. Lasker (1978). Position of larval fish in an ecosystem. *Rapp. Proc-Verb. Reun. Int. Comm. Explor. Mer. 173*: 77–84

Södergren, S. (1976). Ecological effects of heavy metal discharge in a salmon river. *Rep. Inst. Freshwater Res. Drottningholm 55*: 91–131

Soleim, P.A. (1942). Arsaker Til rike og Fattige arganger av sild.

Fiskeridir. Skr. Ser. Havunders 7: 1–39

Solomon, D.J. and A.E. Brafield (1972). The energetics of feeding, metabolism and growth of perch. (*Perca fluviatilis L.*). *J. Anim. Ecol. 41*: 699–718

Somerton, D.A. (1980). A computer technique for estimating the size of sexual maturity in crabs. *Can. J. Fish. Aquat. Sci. 37*: 1488–94

Southward, A.J. (1980). The Western English Channel – an inconstant ecosystem? *Nature 285*: 361–6

Southwood, T.R.E. (1977). Habitat, the templet for ecological strategies? *J. Anim. Ecol. 46*: 337–66

——— (1978). *Ecological Methods*, 2nd edn, Chapman and Hall, London

Southwood, T.R.E., R.M. May, M.P. Hassell and G.R. Conway (1974). Ecological strategies and population parameters. *Amer. Nat. 108*: 791–804

Spangler, G.R., N.R. Payne, J.E. Thorpe, J.M. Byrne, H.A. Regier and W.J. Christie (1977). Responses of percids to exploitation. *J. Fish. Res. Board Can. 34*: 1983–8

Stanley, J.G. (1974). Energy balance of white amur fed *Egeria. Hyacinth Control J. 12*: 62–6

Stearns, S.C. (1976). Life history tactics: a review of the ideas. *Q. Rev. Biol. 51*: 3–47

Steele, J.H. (1974). *The Structure of Marine Ecosystems*, Blackwell, Oxford

——— (1976). Patchiness. In: D.H. Cushing and J.J. Walsh (eds), *The Ecology of the Seas*, Blackwell, Oxford, pp. 98–115

Takafuji, A. (1977). Effect of rate of successful dispersal of a phytoserid mite on the persistence in the interactive system between predator and prey. *Res. Pop. Ecol. 18*: 210–22

Talbot, J.W. (1978). Changes in plaice larval dispersal in the last fifteen years. In: G. Hempel (ed.), *North Sea Fish Stocks – Recent Changes and their Causes. Rapp. Proc-Verb. Reun. Int. Comm. Explor. Mer. 172*: 114–23

Talling, J.F. (1965). Comparative problems of phytoplankton production and photosynthetic production in a tropical and a temperate lake. *Mem Inst. Ital. Hydrobiol. 18 (supp)*: 399–424

Tang, Y.A. (1979). Planning, design and construction of a coastal milkfish farm. In: T.V.R. Pillay and W.A. Dill (eds), *Advances in Aquaculture*, Fishing News Books, Farnham, pp. 104–17

Tarr, H. (1969). Nutritional value of fish muscle and problems associated with its preservation. *Can. J. Inst. Food Tech. 2*: 42–5

Tranter, D.J. (1976). Herbivore production. In: D.H. Cushing and J.J. Walsh (eds), *The Ecology of the Seas*, Blackwell, Oxford, pp. 182–224

Uhler, R.S. (1980). Least squares regression estimates of the Schaefer production model: some Monte Carlo results. *Can. J. Fish. Aquat. Sci. 37*: 1284–94

Valdivia, J.E.G. (1978). The anchoveta and El Niño. *Rapp. Proc-Verb. Reun. Int. Comm. Explor. Mer. 173*: 196–202

Vince, S., Valela, and N. Backus (1976). Predation by the salt marsh killifish in relation to prey size and habitat structure: consequencies for prey distribution. *J. Exp. Mar. Biol. Ecol. 23*: 255–66

von Bertalanffy, L. (1957). Quantitative laws in metabolism and growth. *Q. Rev. Biol. 32*: 217–31

Walter, G.G. (1973). Delay-differential equation models for fisheries. *J. Fish. Res. Board Can. 30*: 939–45

——— (1978). A surplus yield model incorporating recruitment and applied to a stock of Atlantic mackerel. *J. Fish. Res. Board Can. 35*: 229–34

Walters, C.J. (1969). A generalised simulation model for fish population studies. *Trans. Amer. Fish. Soc. 98*: 505–12

——— (1980). Comment on Deriso's delay-difference population model. *Can. J. Fish. Aquat. Sci. 37*: 2365

Ware, D.M. (1980). Bioenergetics of stock and recruitment. *Can. J. Fish. Aquat. Sci. 37*: 1012–24

Warner, R.R., D.R. Robertson and E.G. Leigh (1975). Sex change and sexual selection. *Science 190*: 633–8

Webb, D.C. (1975). Marine fish farming – environmental aspects of salmon farming. *Oceanology International 75*: 173–82

Webb, P.W. (1975). Hydrodynamics and energetics of fish propulsion. *Bull. Fish. Res. Board Can. 190*: 1–158

——— (1978). Partitioning of energy into metabolism and growth. In: S.D. Gerking (ed.), *Biological Basis of Freshwater Fish Production*, Blackwell, Oxford, pp. 184–214

Werner, E.E. (1972). On the breadth of diet in fishes. PhD thesis, Michigan State University

——— (1979). Niche partitioning by food size in fish communities. In: R.H. Stroud and H. Clepper (eds), *Predator Prey Systems in Fisheries Management*, Sport Fishing Institute, Washington, DC, pp. 311–22

——— (1980). Niche theory in fisheries ecology. *Trans. Amer. Fish. Soc. 109*: 257–60

Werner, R.G. and J.H.S. Blaxter (1980). Growth and survival of larval herring, *Clupea harengus*, in relation to prey density. *Can. J. Fish. Aquat. Sci. 37*: 1063–9

Wetzel, R.C. (1975). *Limnology*, W.B. Saunders, Philadelphia

Wiborg, K.F. (1976). Larval mortality in marine fishes and the critical period concept. *J. Cons. Int. Explor. Mer. 37*: 11

Wilen, J.E. (1979). Fishermen behaviour and the design of efficient fisheries regulation programmes. *J. Fish. Res. Board Can. 36*: 855–8

Williams, G.C. (1966). *Adaptation and Natural Selection*, Princeton University Press, New Jersey

Wilson, E.O. (1975). *Sociobiology: The New Synthesis*, Belknap Press, Cambridge, Mass.

Wilson, E.O. and W.H. Bossert (1971). *A Primer of Population Biology*, Sinauer, Sunderland, Mass.

Wilson, J.P.F. (1979). The biology and population ecology of the pollan, *Coregonus autumnalis* pollan Thompson, in Lough Neagh, Northern Ireland. DPhil thesis, New University of Ulster, Northern Ireland

Wilson, J.P.F. and T.J. Pitcher (in prep.). The management of the fishery for pollan in Lough Neagh, Northern Ireland

Winberg, G.G. (1960). *Rate of Metabolism and Food Requirements of Fish*. Beloruss. State Univ., Minsk (Fish. Res. Board Can. Transl. Series No. 194)

Wirjoatmodjo, S. (1980). Growth, food and movement of a flounder, *Platichthys flesus* (L.), population in an estuary. DPhil thesis, New University of Ulster, Northern Ireland

Woodhead, A.D. (1979). Senescence in fishes. In: P.J. Miller (ed.), *Fish Phenology; Anabolic Adaptiveness in Teleosts*, Symp. Zool. Soc. Lond. 44, pp. 179–205

Youngs, W.D. and D.S. Robson (1978). Estimation of population number and mortality rates. In: T.B. Bagenal (ed.), *Methods for the Assessment of Fish Production in Freshwaters*, IBP Handbook No. 3, 3rd edn, Blackwell, Oxford, pp. 137–64

Zaika, V.Y. and N.A. Ostrovskaya (1972). Indicators of the availability of food to fish larvae. I. The presence of food in the intestine as an indicator of feeding conditions. *J. Ichthyol. 12*: 94–103

INDEX OF FISH NAMES

GENERAL INDEX